SEMA SIMAI Springer Series

Volume 38

As of 2013, the SIMAI Springer Series opens to SEMA in order to publish a joint series aiming to publish advanced textbooks, research-level monographs and collected works that focus on applications of mathematics to social and industrial problems, including biology, medicine, engineering, environment and finance. Mathematical and numerical modeling is playing a crucial role in the solution of the complex and interrelated problems faced nowadays not only by researchers operating in the field of basic sciences, but also in more directly applied and industrial sectors. This series is meant to host selected contributions focusing on the relevance of mathematics in real life applications and to provide useful reference material to students, academic and industrial researchers at an international level. Interdisciplinary contributions, showing a fruitful collaboration of mathematicians with researchers of other fields to address complex applications, are welcomed in this series.

THE SERIES IS INDEXED IN SCOPUS

Alberto d'Onofrio • Antonio Fasano •
Federico Papa • Carmela Sinisgalli
Editors

Problems in Mathematical Biophysics

A Volume in Memory of Alberto Gandolfi

Forewords by Alessandro Bertuzzi, Alberto Pettorossi, and Riccardo Gandolfi

Springer

Editors
Alberto d'Onofrio
Department of Mathematics, Informatics and Geosciences
University of Trieste
Trieste, Italy

Antonio Fasano
FIAB SpA, Firenze, Italy and University of Florence
Firenze, Italy

Federico Papa
Institute for Systems Analysis and Computer Science "Antonio Ruberti"
CNR
Roma, Italy

Carmela Sinisgalli
Institute for Systems Analysis and Computer Science "Antonio Ruberti"
CNR
Roma, Italy

ISSN 2199-3041 ISSN 2199-305X (electronic)
SEMA SIMAI Springer Series
ISBN 978-3-031-60772-1 ISBN 978-3-031-60773-8 (eBook)
https://doi.org/10.1007/978-3-031-60773-8

This Springer imprint is published by the registered company Springer Nature Switzerland AG
The registered company address is: Gewerbestrasse 11, 6330 Cham, Switzerland

Alberto Gandolfi in 2018 during the summer holidays in Trentino Alto Adige

Foreword

This foreword contains four short tributes in memory of Alberto Gandolfi and a Eulogy. These writings were shared with the editors by scientists and close friends of Alberto upon learning about the creation of this book. The editors believe that, owing to their personal nature, these reminiscences serve as a fitting foreword and well complement this volume dedicated to a great scientist and man.

In Memory of Alberto Gandolfi

I met Alberto in the mid-seventies in what was then the Institute of Automatics of the Engineering Faculty in Rome. Initially, Alberto took an active part in the study of antigen–antibody binding and the modeling of immune response in cooperation with Carlo Bruni, Giorgio Koch, and Alfredo Germani. With the last of them, in the early eighties he dealt with topics in mathematics and systems theory.

In the same period of time, he began to address the modeling of cell cycle and cell proliferation in culture, as depicted by the DNA-fluorescence distributions measured by flow cytofluorimetry. In the nineties, he also addressed the analysis of biparametric DNA-BrdUrd distributions, when Carmela Sinisgalli began to cooperate with us. Among others, Giuseppe Starace provided data and knowledge about the experimental procedures.

A further step in tumor modeling occurred in the study, published in 2000, of cell kinetics in a tumor cord, a cylindrical arrangement of tumor cells growing around a central blood vessel. The evolution of the cord under treatment was characterized in articles co-authored by Antonio Fasano and Alberto d'Onofrio. In later years, the tumor spheroid model was also investigated, focusing attention on spheroid evolution during treatment and the formation of the necrotic core.

It is with great emotion that I mention here the model for the kinetic heterogeneity of a tumor cell population, developed few years later and co-authored by Ovide Arino, who could not see the paper completed and published. During a conference in Poland in September 2003, Ovide unexpectedly passed away and Alberto, who was also attending that conference, sent me a desperate text message to tell me the news. I still have a vivid memory of that moment.

Finally, in different research lines pursued in cooperation with Serenella Salinari and Geltrude Mingrone for the experimental and clinical studies, the pharmacokinetics of dicarboxylic acids and other compounds were explored.

Alberto had an excellent scientific talent and rare human qualities. He was an extremely generous person and he was always ready to help others. I have learned a lot working with him, not only in the specific topics we dealt with in our research, but also from the way he related to his colleagues and to people in general. Discussing arts and music with him, as well as science or politics, has always been stimulating and instructive. In this long period of time, his friendship has meant so much to me. It is a great privilege to have been his friend and colleague and to have known his beloved family.

Institute for Systems Analysis and Computer Science "Antonio Ruberti," CNR
Rome, Italy
November 2023

Alessandro Bertuzzi

Memories of a Good Friend

I shared the mature researchers' room on the third floor of the IASI Institute of the National Research Council with Alberto during the last years of his life. Also, Sandro Bertuzzi had a desk in our room. The three of us were "of a certain age."

We were bound by mutual esteem and sincere friendship. Many years ago, Sandro has been an assistant professor for the Automatic Control course I took in 1970 for my degree in Electronic Engineering. The teacher of the course was Professor Antonio Ruberti to whom the IASI Institute itself is entitled.

I remember Sandro's stimulating, beautiful tutorials. I met Alberto, Sandro's collaborator, at IASI some years later when he became a researcher there.

Alberto was rather quiet. He was very much focused in his work. He was always respectful and friendly. He chose, out of the three desks available in the room, the desk that was furthest away from the window light. What a beautiful example of self-denial and generosity!

Alberto knew of my interests in Formal Logic, Theoretical Computer Science, and Artificial Intelligence. I remember that a few times, after lunch, when Sandro wasn't in the room, he asked me some challenging questions that I could not escape. I couldn't say: *"I don't have time now"* or: *"Let's talk about it this evening."* It was necessary to try hard and to answer with prudence, balance, and maturity. I felt they were valuable questions, for him and for me, as well.

I remember, in particular, two questions:

Alberto, what do you think truth is?

(my name is Alberto too) and:

What is freedom? Is a human person free?

I don't know what drove Alberto to ask me such questions, but I felt I had to put all my effort into answering them in a "correct" manner. I felt I had to humbly accept that I did not have a definite answer. Yet, I wished to explore a way to an answer that was valid for both of us: "together toward truth." This is, after all, the essence of a life devoted to research.

These questions, I believe, stemmed from his being a seeker "all the way" and "at all times." "Always" in all areas. Sometimes, it seems to me, he wanted to explore the validity of some books he was reading at home or, perhaps, during the long hours of the dialysis in hospital (in fact, at least twice a week, before the kidney transplant, Alberto had to do dialysis). The problem came from the fact that on questions of that kind one cannot make a mathematical model and one can rarely rely on experiments that confirm the results of the deductions made.

Indeed, the two of us, in our respective scientific fields were used to make mathematical and/or logical models and then perform experiments to verify the conclusions.

Alberto and I were used to discuss the concept of scientific truth and the validity of physical laws. What is the epistemological status of the physical laws when they are in danger of being falsified by the results of new experiments? The "improvements" of physical laws are or are not directed toward a definitive, conclusive statement? I also talked to him about "truth in mathematics" and Gödel's incompleteness theorem in Peano's arithmetic. Alberto was well aware of these results.

The question of freedom was equally important for him and challenging for me.

I remember that we shared the belief that people are free, even though they may be conditioned by their upbringing, the cultural and social environment in which they live. They can truly choose according to values that they pursue, or wish to pursue, in their life. I remember one day telling Alberto what another Alberto, St. Albert the Great, Bavarian (Lauingen, 1193—Cologne, 15 November 1280), great philosopher, theologian, and scientist, had said about mankind:

> *"Der Mensch steht in der Mitte der Schöpfung -*
> *zwischen Stoff und Geist, zwischen Zeit und Ewigkeit"*
> *(Man stands in the middle of creation -*
> *between matter and spirit, between time and eternity).*

These few memories of Alberto are in my heart, together with many others, and when I look at his desk with his chair "now empty," I see him still sitting there, with me, keeping me company, stimulating me in my work.

At the end of the day, he would leave me and go out of the room saying: "OK, ... see you tomorrow!", as if to say: "I can't do more today than what I have done!"

Thank you, thank you so much, Alberto!

Please, continue to keep me company!

Department of Civil Engineering and Computer Science Engineering, University of Rome Tor Vergata
Rome, Italy

Institute for Systems Analysis and Computer Science "Antonio Ruberti," CNR
Rome, Italy

Alberto Pettorossi

November 2023

A Memory of Alberto Gandolfi, My Father

Looking through the bookshelves of my father's library, I am surprised still today by the great number of his interests and I pause to think what a wide universe of knowledge I have lost with him. Very many history volumes (especially history of the twentieth century), a really vast set of publications on ancient and modern philosophy, literature, politics, and religion. The latter subject looked accurately investigated: annotated editions of the Gospel, Church Fathers, reflections on free will and on the historical figure of Jesus of Nazareth. Books upon books, mixed with the ones—often with titles I couldn't understand—related to his work as a scientist.

Did my father believe in God? I can't answer that question, but I can report something he loved to say: "if a superior mind like Saint Augustine believed in God, there must be something I am not able to understand." The desire for knowledge was the center of my father's life, an irrepressible desire that was continuously self-nourishing. His books are nearly always underlined with his typical and gentle light pencil, a sign indicating that he rarely had the intention of just reading, while he was very often studying those writings intensively.

The last summer we spent in our holiday home in Puglia, absorbed in reading a book, he didn't realize that the sun had hit a part of his arm for a long time, causing a burn. Pointing to the burn he said jokingly "Look, do you see how I have reduced myself for the sake of science?".

Looking at his volumes today, I notice that one category is less represented than others: history of art. An apparent paradox, since the passion for art moved his life constantly and shaped our relationship. More than in books, he studied art with his eyes. Ever since I was a child, Saturdays and Sundays were exclusively dedicated to endless walks around Rome, to discover (and rediscover) monuments and museums. "Shall we go to see some churches today?", "There is an interesting exhibition I would like to see," "I wish to see the medieval mosaics of Santa Pudenziana again." Starting from these sentences, on Saturday morning we used to plan the weekend and I, as a child, followed him on these "pilgrimages" between baroque churches and collections of ancient paintings.

His passion lives on in me; today, I am an art historian working for the Italian Ministry of Culture, a different life choice was impossible to me. We grew up together on the topic; while studying I used what I learned from him, and the more I knew the subject, the more I fueled his desire to know with my observations. His passion became mine, and mine became his. We influenced each other in an omnivorous search that was reflected in his purchases of works of art, which later became our joint purchases, and are now only mine.

His small collection continues to be nurtured, without interruption and has never stopped growing in his memory. I have seen his taste (our taste) change countless times, to the point of creating small or very small independent and often antithetical nuclei. Seventeenth-century painting fascinated him and a canvas with a Saint Francis in meditation on the entrance wall of our house still welcomes those who enter today. He smiled amusedly when we commented on the saint's gaze turned upward and we said to ourselves: "he doesn't even look at us when we come home, he's busy with much more important reflections."

The auction houses were a fundamental stop on our walks and the foreign ones represented an excellent alternative to spend a few hours in front of the computer on rainy days. I remember his emails with a link to some work of art and a terse sentence "what do you think?"; I remember waiting for a painting that had to travel from Spain or England, the visits to the restorer ("a dirty painting is more fun, who knows what we'll find below!"), and the message exchanges to try to guess the author of a Madonna reading from the early eighteenth century (we succeeded!).

Without any fear of contradiction, he loved very much at the same time contemporary art, with an evolution of taste that took him from the neo-figurative painting of the nineties of the last century, to the "School of Piazza del Popolo" and to the abstract painting of the fifties. He loved Franco Angeli's works very much and before buying a beautiful blue and gold Half dollar he saw dozens of works to find one that was certainly authentic in terms of quality and collector history ("too many fakes, I don't trust them..."). Art graphics and small ink-on-paper works covered the walls of his bedroom, always with a common thread, their "political" weight. He loved the works by Emilio Vedova dedicated to the fight against Franco's regime in Spain, a splendid drawing by Salvatore Scarpitta was linked to the famous magazine Civiltà delle Macchine and he loved the reflection on totalitarianism by Franco Angeli.

My father was fascinated by ancient and modern engravings, in particular etchings and burin engravings; a sixteenth-century print by Luca di Leida coexisted absolutely well with works created twenty or thirty years ago.

In all this mare magnum there was also a place for Russian socialist realism. I convinced him to buy a huge portrait of Lenin for a few hundred euros, but after a few days he asked me to take it away because—always said with his self-deprecating smile—"his stare makes my socialist conscience feel uneasy."

In the last part of his life he became very keen on ancient books, particularly on the sixteenth-century editions and on the beautiful late eighteenth-century editions by Giambattista Bodoni. We leafed through the catalogs and every attempt I made to point out a book that was only beautiful because of its antique binding was rejected: "No! The content is boring...". He didn't even want to consider books about the

Italian royal family (the Savoys), while he loved those about Tuscany and everything that had a bizarre or curious topic.

In his youth he loved Caucasian carpets and Russian icons fascinated him; he had even covered the walls with the former, whereas over many years he was unable to find even one ancient icon worth purchasing. Lately he had been interested in Italian painting of the 1930s and African sculpture, but unfortunately there wasn't enough time to see the realization of these new aesthetic and intellectual curiosities.

Many projects, many satisfactions, many journeys of thought, mind, and eyes, always experienced under the watchful gaze of my mother Angela, the only true backbone of our family, practical and rational, who with the love of a mother and a wife has allowed my father, me, and my brother Enrico to live our passions.

Alberto Gandolfi passed away suddenly, while he was spending his holidays in his beloved mountains of Alto Adige (in the north of Italy, on the border with Austria), two days later I was supposed to join him to look for some ancient Tyrolean objects and to visit together the churches of the Alps so familiar to us. Shortly before his death he was going through a book on quantum physics in relation to the concept of eternity.

Roma, Italy
December 2023

Riccardo Gandolfi

All Started with a Floppy Disk

On the 20th of July 2019, just 20 days before his unexpected passing, Alberto Gandolfi wrote his final email to me. It was a remarkably kind email in which he expressed his delight upon receiving some excellent news I shared with him about my family. The closing lines of his email read as follows: "Should you happen to visit Rome, both my wife and I would be delighted to meet you and your family. I will be in Rome from August 1 to 11. In any case, I hope you enjoy your holidays. *Un grande abbraccio*, Alberto."

We didn't have the opportunity to meet because I ultimately couldn't make it to Rome. When I received the heart-wrenching email from Carmela Sinisgalli, informing me of Alberto's sudden and tragic passing, I broke down in tears, much like a three-year-old who has lost their father. Alberto, indeed, was my scientific father. The vast majority of my scientific knowledge and identity is owed to him.

Our friend Urszula Ledzewicz provided me with the most insightful definition of who Alberto was: "Alberto is not an Italian: he is a British Gentlemen." I think that she was perfectly right (this definition also aptly applies to Sandro Bertuzzi, who was Alberto's close colleague and collaborator.).

Alberto was undoubtedly a true Roman at heart, but his family's roots in Tuscany endowed him with a distinct accent from his family's hometown: the charming city of Lucca. In 1973, he earned a five-year degree in Electronic Engineering, specializing in Control Theory. In terms of today's educational structure, this would be equivalent to holding a Master of Science (MSc) in Applied Mathematics and Physics, reflecting the outstanding study framework of Italian universities in the 1970s.

It may sound unusual, but during that era, Italy didn't offer Ph.D. programs, so exceptionally bright students like Alberto were placed as research fellows for several years. As a "research fellow," Alberto began his work on applying systems theory to biology at a research center in Automatic Control and Computer Science, which had been established by one of the prominent figures in Control Theory in Rome, Antonio Ruberti. Many years later, this center evolved into the "Istituto di Analisi dei Sistemi ed Informatica" under the Italian National Research Council.

During that period, there was a significant focus within the Italian control community on modeling physiological feedback mechanisms. This particular focus aligned seamlessly with Alberto's deep personal interest in biomedicine. It marked the inception of what would become a long and outstanding career in the field of science for him.

Alberto was a truly multifaceted man with a profound passion for the humanities, particularly history and art. He had a small yet remarkable collection of paintings, with a special affection for the seventeenth century, but he also appreciated contemporary art. During our travels to international conferences, we often took the opportunity to explore local museums. I have a vivid memory of our visit to the "Old Masters Gallery" Museum in Dresden, where Alberto felt entirely at home—he knew that museum like the back of his hand. On another occasion, he beamed with paternal pride as he shared that his son, Riccardo, had recently rescued an old painting from a trash can in the heart of Rome, close to a nobleman's palace. And there was a time when he expressed deep concern about one of his paintings entrusted to an art conservator for a delicate and somewhat risky restoration. Alberto's love for art and the stories he shared about it added a rich and colorful dimension to his already impressive personality.

Alberto harbored another passionate interest in politics, a facet of his character that few of his international colleagues might have suspected. He held a strong left-leaning political inclination and was deeply engaged in the political scene in Rome. Evidently, he successfully passed on this passion to his son, Enrico. Alberto took immense pride in Enrico's active involvement in politics and his election to the local council, even if it was for a different and somewhat less left-leaning party, which he found somewhat amusing.

Alberto had a series of historical coworkers such as Sandro Bertuzzi (with whom he shared an office for decades, even after their retirement), Mimmo Iannelli, Andrea Pugliese, Paolo Ubezio, Alfredo Germani, besides of course Antonio Fasano, Carmela Sinisgalli, Federico Papa (his last disciple), myself, and many others.

Alberto had an unbounded love for biology, medicine, and biomathematics, and, more broadly, for applied mathematics. His extensive list of published subjects is truly impressive: Cell Cycle and Proliferation, Tumor Cords Dynamics, Mathematical Physiology, Immunology, Stochastic Calculus (in the Control Theory flavor), Statistics: Theoretical and Applied (he was a true wizard of data fitting and of educated guesswork), Applied Statistical Physics, Systems Biology, Tumor Angiogenesis, Mathematical Epidemiology and Immuno-epidemiology, "Epidemiology," Radiotherapy.

His home was the IASI CNR. I distinctly recall the heart-wrenching phone call he made to inform me that he had been abruptly compelled to retire. His sadness was profound. He was well aware that he could still have a desk at IASI CNR to continue his work, but he carried the tragic realization that he would transition from being a host to becoming a guest. Despite being dearly beloved by all his colleagues, the distinction between a host and a guest was not lost on him.

As you can imagine, I hold deep admiration and gratitude for Alberto, and our friendship was profound. However, as strange as it may seem, during my initial two years at IASI CNR, first as a predoc and then in my first year of my PhD, I harbored a certain fear of both Alberto and Sandro. I had read some of their many papers and greatly desired to engage in discussions with them, but I lacked the courage to approach them. My interactions were limited to a simple "good morning" when we crossed paths at the Institute. It's quite likely that I would have had no contact with them if it weren't for a fortuitous computer incident.

Indeed, one afternoon in 1997, like many other PhD students, I found myself in the spacious computer room on the third floor of IASI CNR. Most researchers only entered this room when their computers were "out of service," and this was also the case for Alberto, as he later explained. He logged into a computer and began his work. Then, suddenly, I noticed him in a bit of distress, apparently dealing with a computer problem. Seizing the opportunity, I approached him and said, "Professor Gandolfi, can I help you?" He informed me that he had mistakenly inserted his floppy disk the wrong way, and to my amazement, I was able to recover the data from it. He expressed his gratitude, and from there, we engaged in conversation. Eventually, we moved to his office, where Sandro was also present, and Sandro began asking me questions, exchanging ideas, and discussing potential projects. This chance encounter marked the beginning of a 21-year-long friendship and a happy and vast scientific collaboration.

Alberto imparted several of his key skills to me, primarily through the mechanism of teaching by example. First and foremost, he instilled in me the importance of being active in a multitude of fields. He found it both amusing and intellectually stimulating, and I couldn't fathom Alberto as a scientist confining himself to a single micro-area of study for his entire career. It seemed impossible! Secondly, he taught me the challenging art of "seeing" connections and common patterns among seemingly disparate phenomena and mathematical subjects. These were links that often remained invisible to others but were distinctly visible to both of us. Thirdly, he emphasized the duty of assisting young researchers to the best of one's ability and treating them as equals. Most importantly, Alberto impressed upon me that a genuine love for biology and medicine is the essential foundation for making meaningful contributions to mathematical biophysics. I will forever be immensely grateful for these remarkable gifts that Alberto bestowed upon me.

I finish my short and incoherent notes by simply saying "*Ciao, Alberto the great! Au revoir la-haut!*".

Department of Mathematics, Informatics and Geosciences Alberto d'Onofrio
University of Trieste
Trieste, Italy
November 2023

Eulogy

Alberto Gandolfi was born in Lucca (Tuscany) in 1948. He had a leading position at the Istituto di Analisi dei Sistemi ed Informatica "Antonio Ruberti" (IASI) of the Italian National Research Council (CNR) in Rome.

He was a bright scientist, author of many important papers, mainly in the field of biomathematics. He is also remembered as a man of great education, of natural kindness, and very sensitive to the persons surrounding him. He died suddenly and untimely in August 2019. Many of the authors of the papers here collected share for sure this picture of that rare man. The editors of this book were particularly affectionate to him. Working with Alberto was a great pleasure and a continuous source of inspiration, and talking with him was a great human experience because of his thoughtfulness and his deep knowledge in various fields of humanities. A great and extraordinary person that all the contributors to this book want to celebrate with an affectionate smile.

For a tragic coincidence, while this book was in its final stage of preparation, his son Enrico died unexpectedly. He was not yet 40 years old. He had a position in the Rome City Council, where he had started a promising political career. Due to such a sad, incredible circumstance, here we like to remember him too and express our deep condolence to the beloved Alberto's wife Angela and son Riccardo, who suffered such great losses in a short time span.

Trieste, Italy — Alberto d'Onofrio
Firenze, Italy — Antonio Fasano
Roma, Italy — Federico Papa
Roma, Italy — Carmela Sinisgalli
November 2023

Preface

This book project stems from our desire to pay a tribute to the memory of our friend and colleague, Alberto Gandolfi. We pursued this goal by assembling a comprehensive compilation of scientific papers in the field of Mathematical Modeling of Biophysical phenomena, which was the research area he nurtured, with persevering and careful enthusiasm, since 1975.

Alberto's untimely passing away in August 2019 left a void still deeply felt by all who had the privilege of collaborating with him. The esteemed authors who kindly agreed to contribute to this volume held him in high regard for his unwavering scientific rigor, a quality he complemented with an immense passion for biology, medicine, and the natural sciences as a whole. His memory lives on in the work and dedication of those who continue to be inspired by his legacy.

The remarkable scientific standing of the contributing authors is, we believe, a good evidence to the quality of this book. Encompassing both theoretical and practical aspects of the Mathematical Biophysics, the covered topics span a spectrum of differential problems, including stochastic, nonlinear, quasilinear, delayed, and related solution analyses, with application in physics, biology, and medicine, including population dynamics, tumor growth and control, immunology, epidemiology, ecology, and others. As a result, we believe the book offers a comprehensive and current overview of compelling subjects and challenges within the realm of mathematical biophysics.

During his intense 44-year long career, Alberto played a pivotal role in catalyzing a profound shift in the attitudes toward mathematical biophysics. He was also a witness of the many "revolutions" in our field, transformations which gained momentum especially in the present century.

There have been, notably, at least two significant scientific transitions in Mathematical Biophysics. The initial transition marked a gradual shift from a phase in which mathematical biophysics held primary appeal within the most practically oriented sectors of the Science Technology Engineering Math (STEM) community. This transition led to a subsequent phase, perhaps even more crucial from a "sociological" perspective. In this latter phase, an encouraging number of biomedical scientists became actively engaged in quantitative research.

Furthermore, the noteworthy development of research groups (frequently headed by biomedical scientists who have transitioned into computational sciences) is becoming integral part of medical and biological research centers.

The second transition is an internal transformation within the STEM realm. From the perspective of applied mathematical physicists, there is a growing realization that the utilization of classical mechanics and nonlinear analysis in the examination of biological, medical, and social phenomena presents a spectrum of novel and formidable challenges situated at the forefront of contemporary mathematics.

This trend is reflected in the rising quantity of scientific articles in the field of mathematical biophysics not only in computational biology or bioinformatics scientific journals but also in mathematical, computer science, and theoretical physics journals.

More notable revolutions have taken place. First, the advancement of nonlinear analysis in conjunction with the proliferation and increasing efficiency at low costs of computers has led to the dissemination of "interpretable" nonlinear models. Furthermore, thanks to the further progress in both scientific calculus and hardware, we witnessed the emergence of multiscale modeling, which allows for the explicit representation and embedding of microstructures, including individual cells. Lastly, Alberto could also witness the nascent stages of the Machine Learning Revolution that we are presently experiencing.

This book comprises contributions that delve into the personal research achievements of eminent scientists in the field, all of whom were deeply respected by Alberto Gandolfi. In particular, the objective of this publication is to offer carefully chosen assortments of cutting-edge findings in specific key areas.

Venturing into the field of mathematical biophysics demands the development of a mindset that extends far beyond the typical applications of mathematics and physics. In this field, the learning process transcends basic and refined STEM concepts and methods. Indeed, it often necessitates a deep understanding of the interconnectedness of seemingly diverse biological phenomena, all intricately tied to biological, clinical, or environmental problems.

In the chapters of this book, we are confident that the authors have effectively conveyed not only their research findings but, more significantly, their peculiar perspective and approach to problem-solving.

The great majority of contributions in this book, if not all of them, transcend traditional categorizations within narrow domains. This holds true from both biological and biophysical perspectives, as well as from a mathematical standpoint. They deal with a diverse range of subjects, including oncology, epidemiology, neurosciences, and biochemistry. The mathematical aspect is equally varied, with the use of a wide array of tools such as continuous Markov chains, partial differential equations, kinetic theory, applied statistical mechanics, noise-induced transitions, and numerous others. For this reason, we have chosen not to categorize this book into sections based on specific topics but instead to arrange the contributions alphabetically following the family name of the first authors.

We want to finally stress that numerous challenges lie ahead for the field of Mathematical Biophysics, beginning from the judicious and productive utilization of and, to some extent, integration with Machine Learning. The future of our scientific community hinges on the imperative need for ever closer collaboration between biological, medical, and environmental scientists and their STEM counterparts.

Trieste, Italy — Alberto d'Onofrio
Firenze, Italy — Antonio Fasano
Roma, Italy — Federico Papa
Roma, Italy — Carmela Sinisgalli
November 2023

Contents

Controlling the Spread of an Epidemic in Agriculture—the Case of *Xylella Fastidiosa*: A Research Project

Sebastian Aniţa, Vincenzo Capasso, Matteo Montagna, and Simone Scacchi

Abstract In this paper, an extension is reported of the spatially structured mathematical model for a *Xylella fastidiosa* epidemic, presented in previous papers by the same authors, concerning the dynamics of a predator, i.e. *Zelus renardii*, which has been recently suggested as a possible biocontrol agent of the *X. fastidiosa* insect vector *Philaenus spumarius*. By taking into account previous investigations, the choice of the functional response to predation is indeed crucial; this significantly depends upon the role of the predator with respect to the insect vector of the epidemic. Possible choices adopted in literature are the Holling type II for a specialist predator, and the Holling type III for a generalist predator. The analysis reported here shows that a specialist predator, whenever identified, would indeed lead to the eventual eradication of a *Xylella* epidemic. Unfortunately our investigations have shown that the adoption of the generalist predator *Z. renardii* is not an efficient control strategy to eradicate a *Xylella* epidemic. For the time being, weed cut and/or selected cultivar have been confirmed as effective control measures. We sketch future research directions, mainly concerning the role of seasonality with respect to possible control interventions.

S. Aniţa
Faculty of Mathematics, "Alexandru Ioan Cuza" University of Iaşi, Iaşi, Romania

"Octav Mayer" Institute of Mathematics of the Romanian Academy, Iaşi, Romania
e-mail: sanita@uaic.ro

V. Capasso (✉)
ADAMSS (Centre for Advanced Applied Mathematical and Statistical Sciences), Università degli Studi di Milano "*La Statale*", Milano, Italy
e-mail: vincenzo.capasso@unimi.it

M. Montagna
Department of Agricultural Sciences, and BAT Center (Interuniversity Center for Studies on Bioinspired Agro-Environmental Technology), Università degli Studi di Napoli "*Federico II*", Portici, Italy
e-mail: matteo.montagna@unina.it

S. Scacchi
Department of Mathematics, Università degli Studi di Milano "*La Statale*", Milano, Italy
e-mail: simone.scacchi@unimi.it

A. d'Onofrio et al. (eds.), *Problems in Mathematical Biophysics*, SEMA SIMAI Springer Series 38, https://doi.org/10.1007/978-3-031-60773-8_1

Highlights

- Prediction and control of the socio-economic important Olive Quick Decline Syndrome (OQDS).
- Regional control.
- A mathematical model based on reaction-diffusion systems.
- The role of possible insect predators in controlling a *Xylella* epidemic.
- The role of seasonality with respect to possible control interventions.

1 Introduction

This project is part of a long term research concerning "regional control". In real conditions, where populations of the organisms that have to be controlled are structured, it is practically impossible, due to the limited resources, to control the relevant system by global interventions in the whole habitat. In geographical epidemiology (for human populations, or in agriculture) this might also be due to unavoidable logistic restrictions. This has motivated, about 20 years ago, the authors SA and VC to propose regional control methods, along the principle of "Think Globally—Act Locally". They have shown that, under suitable assumptions, it is possible to eradicate an epidemic in the whole relevant habitat by interventions only in a suitably chosen bounded subset of it (see [2–4, 14]).

With the project presented here we wish to propose extensions of our own recent literature concerning the control of the current devastating olive tree disease known as olive quick decline syndrome (OQDS) caused by the bacterium *Xylella fastidiosa*. The death of an infected plant is due to the occlusion of xylem vessels induced by the formation of a bacterial micro film that inhibits the flux of water.

In 2013, *X. fastidiosa* subsp. *pauca* (hereafter only *Xylella fastidiosa*) has resulted infesting a wide area of olive tree plantations in Apulia (Italy) [27]. In this Region the main vector of *X. fastidiosa* has been identified in the meadow spittlebug *Philaenus spumarius* (Hemiptera, Aphrophoridae), a xylem sap-feeding specialist [15, 27]. *P. spumarius* nymphs develop on herbaceous vegetation, confined in an own produced protective foam, while adults move to the tree canopies in late spring/early summer. Based on field surveys performed in the Apulia region, adults acquire the *X. fastidiosa* inoculum from infected olive trees. The possibility that weeds represent a reservoir of the bacterium cannot be excluded [8, 18]. *X. fastidiosa* possesses a wide range of host plants [21] and, to our knowledge, no studies were conducted for assessing the presence of this bacterium on weeds in Southern Italy.

The insect vector acquire *X. fastidiosa* together with the xylem meal from infected plants, then bacteria are inoculated into a new host during the insect meals and undergo multiplication after inoculation.

Typical symptoms are leaf scorch, dieback of twigs, branches and possibly of the whole plant [16].

The current practice (Regione Puglia: Delibera della Giunta Regionale del 24 Ottobre 2018 n. 1890—http://ulivi.sit.puglia.it/portal/portale gestione agricoltura/Documenti) consists of uprooting any infected tree and all other trees in its nontrivial neighborhood, either infected or uninfected, which causes serious social and economic drawbacks; hence alternative management strategies are urgently needed.

Our investigations [6, 9, 11], have shown that the most promising target for an effective and cost-efficient control of the *X. fastidiosa* epidemic is represented by agricultural management practices consisting of the regular removal of weeds within and in the surroundings of olive orchards. A further interesting strategy, as expected, is represented by the use of more resistant cultivar; besides the economic effort required to replace olive trees, this strategy has a severe impact on the farm productivity and economy due to the low production of new plantations during the first years after the transplanting.

Motivated by recent papers [19, 23, 25], where *Zelus renardii* (Hemiptera, Reduviidae) has been proposed as a possible control agent of *P. spumarius*, in [7] an extension of our previous spatially structured mathematical model has been proposed, to include the dynamical behavior of the predator. The fact that *Z. renardii* has been reported to be a generalist predator implies the choice of an Holling type III functional response of predation in the mathematical model. As a consequence, it has been shown that the introduction of *Z. renardii* is not an efficient control strategy to eradicate a *Xylella* epidemic. Instead, a specialist predator or a parasitoid, whenever identified, would lead to the eventual eradication of a *Xylella* epidemic; as a matter of fact, in this case the appropriate choice for the predation functional response would be an Holling type II. Similar problems have been analyzed in [5].

In either cases it has been confirmed that a significant reduction of the weed biomass can lead to the eradication of the vector population, independently of the presence of predators.

A warning has emerged concerning the introduction of a non native predator; it may have a catastrophic impact on the relevant ecosystem (see e.g. https://www.cno.it/zelus-renardii-il-suo-impiego-preoccupa-gli-scienziati).

Up to now in our models we had not yet taken into account the role of seasonality in the parameters of the relevant agro-ecosystem, because of possible consequent mathematical complexities.

In the first part, Sects. 2, 3, 4, and 5, we present a review of the most relevant achievements concerning our *Xylella* project. In the last section we provide a view on our future investigations concerning the analysis of the impact of seasonality of parameters on the qualitative behavior of the integrated mathematical model olive trees—weeds—vector insects—predators. On the other hand we wish to analyze the impact of a periodic inflow of predators in an olive orchard on the reduction, and possible eradication of the epidemic.

In Sect. 2 we shall present the structure of our model.

In Sect. 3 we shall present possible control strategies.

In Sects. 4 and 5 we shall present the action of possible predators on the control of a *Xylella* epidemic.

In Sect. 6 we shall present a possible extension of the model including seasonality of the relevant parameters.

2 Our Model

Since the feeding behavior and metabolic processes are qualitatively similar for both nymphs and adults [20], only one stage of active vectors of the infection has been considered.

In the spatially structured model the population of vector insects and the population of infected olive trees have been included; weeds will be taken into account as a given environmental parameter.

For the habitat of the relevant agro-ecosystem, we take a bounded domain $\Omega \subset \mathbb{R}^2$, with a sufficiently smooth boundary $\partial\Omega$.

Let $C_I(x,t)$ denote the spatial density of the total population of insects, $s_1(x,t)$ the spatial density of susceptible insects, and $i_1(x,t)$ the spatial density of infected insects, so that

$$C_I(x,t) = s_1(x,t) + i_1(x,t).$$

Let $C_T(x,t)$ denote the spatial density of the total biomass of olive trees, $s_2(x,t)$ the spatial density of the biomass of healthy trees, and $i_2(x,t)$ the spatial density of the biomass of infected trees, so that

$$C_T(x,t) = s_2(x,t) + i_2(x,t).$$

2.1 *Vector: Philaenus Spumarius*

Due to the fact that bacteria reside only in the foregut of an adult insect, the latter generate only healthy offspring (see e.g. [1, 26], and references therein). The parameter r denotes the birth rate of new insects.

As discussed in [6, 9, 11], reproduction of insects in the relevant ecosystem may occur only if bushes and other plants are present, explaining the presence of a "carrying capacity" $M(x)$ in the logistic term; it describes the biomass of all such plants, that we will simply call *weeds*, either healthy or infected.

Both healthy and infected insects experience natural mortality at a rate n; χ is a scale parameter.

Finally we assume that insects may diffuse in the habitat (with constant diffusion coefficient to avoid purely technical complications) [22].

As far as the "local incidence" for insects, at point $x \in \Omega$, and time $t > 0$, as in previous models [3, 12], we assume that it is due to bites to infected olive trees at any point $x' \in \Omega$ of the habitat, within a spatial neighborhood of x represented by a suitable probability kernel $k(x, x')$, depending on the specific structure of the local ecosystem. Hence the "local incidence rate (i.r.)" for insects, at point $x \in \Omega$, and time $t \geq 0$, is taken as (see e.g. [13])

$$(i.r.)_I(x, t) = s_1(x, t) \int_\Omega k(x, x') i_2(x', t) dx' .$$

Hence the spatial dynamics of susceptible insects is expressed by the following equation

$$\begin{aligned}\frac{\partial s_1}{\partial t}(x, t) = d\Delta s_1(x, t) + rC_I(x, t)[M(x) - \chi s_1(x, t)] - ns_1(x, t) \\ -\beta s_1(x, t) \int_\Omega k(x, x') i_2(x', t) dx',\end{aligned} \tag{1}$$

while the spatial dynamics of infective insects is expressed by the following equation

$$\begin{aligned}\frac{\partial i_1}{\partial t}(x, t) = d\Delta i_1(x, t) - ni_1(x, t) - r\chi i_1(x, t) C_I(x, t) \\ +\beta s_1(x, t) \int_\Omega k(x, x') i_2(x', t) dx'.\end{aligned} \tag{2}$$

Both Eqs. (1) and (2) act in a spatial domain $\Omega \subset \mathbb{R}^2$, at times $t \in (0, +\infty)$. They are subject to suitable initial conditions. We assume homogeneous Neumann boundary conditions.

By summing Eqs. (1) and (2) we obtain an equation for the total insect population $C_I(x, t)$

$$\frac{\partial C_I}{\partial t}(x, t) = d\Delta C_I(x, t) + rC_I(x, t)[M(x) - \chi C_I(x, t)] - nC_I(x, t). \tag{3}$$

2.2 *Olive Trees*

For the olive trees we refer to their canopies, so that we may consider pruning and regrowth. Healthy trees (canopy) are produced at constant net regrowth rate q, get infected by contact with infected insects at rate ζ, or by human activities, such as pruning, grafting, logging, at rate b.

For trees, in view of their long survival, we may neglect natural mortality, but not mortality due to human activities, at a rate ℓ. Infected trees experience an additional disease-related mortality rate μ.

Hence the spatial dynamics of trees is expressed by the following equations

$$\begin{aligned}\frac{\partial s_2}{\partial t}(x,t) &= (q-\ell)s_2(x,t) - s_2(x,t)\frac{C_T(x,t)}{C^0} \\ &\quad -(\zeta i_1(x,t) + b\ell i_2(x,t))s_2(x,t) + \alpha i_2(x,t); \end{aligned} \tag{4}$$

$$\begin{aligned}\frac{\partial i_2}{\partial t}(x,t) &= -\mu i_2(x,t) - \ell i_2(x,t) - i_2(x,t)\frac{C_T(x,t)}{C^0} \\ &\quad +(\zeta i_1(x,t) + b\ell i_2(x,t))s_2(x,t) - \alpha i_2(x,t). \end{aligned} \tag{5}$$

Both Eqs. (4) and (5) act in the spatial domain $\Omega \subset \mathbb{R}^2$, at times $t \in (0, +\infty)$. They are subject to suitable initial conditions.

2.3 *Predators*

The evolutionary dynamics of the population of predators, denoted by $Z(x,t)$, in absence of preys may be described as follows

$$\frac{\partial Z}{\partial t}(x,t) = d_2 \Delta Z(x,t) + r_2 Z(x,t)\left[1 - \frac{Z(x,t)}{K(x) + cM(x)}\right], \tag{6}$$

for $x \in \Omega, t > 0$, subject to suitable boundary and initial conditions.

The parameter r_2 denotes the natural birth rate of new predators.

We have assumed that the carrying capacity of the predators, in addition to a general parameter $K(x)$, depends upon the additional environmental parameter $M(x), x \in \Omega$, that has been already introduced above.

In order to describe mathematically the contribution of *P. spumarius* to the dynamics of the predator population, we have to take into account the functional response $g(C_I)$ of the predator population to a prey population.

In [29, p.83] (see also [5], [24, p.38]) it was evidenced that, in case the relevant predator acts as a specialist one for the prey, the functional response of predation may be taken of the Holling type II form (see [29]) as follows

$$g_{II}(C_I(x,t))Z(x,t) = \frac{\gamma_1 C_I(x,t)}{1 + \gamma_2 C_I(x,t)} Z(x,t), \tag{7}$$

with $\gamma_1, \gamma_2 > 0$.

In case the relevant predator acts as a generalist one for the prey, the functional response of predation follows an S-shaped curve (Holling type III), according to which the predator dedicates its predation only to large quantities of a specific prey:

$$g_{III}(C_I(x,t))Z(x,t) = \frac{\gamma_1(C_I(x,t))^2}{1+\gamma_2(C_I(x,t))^2}Z(x,t), \tag{8}$$

with $\gamma_1, \gamma_2 > 0$.

According to what reported in [23], *Z. renardii* is a generalist predator, and its efficiency in the predation *P. spumarius* is rather low. Unfortunately, to our knowledge specialist predators for the *P. spumarius* have not been identified yet. Anyhow, for a matter of completeness, both cases have been analyzed [7].

If we include the impact of predation in Eq. (3) we have

$$\begin{aligned}\frac{\partial C_I}{\partial t}(x,t) = d\Delta C_I(x,t) + rC_I(x,t)[M(x) - \chi C_I(x,t)] - nC_I(x,t) \\ -pg(C_I(x,t))Z(x,t),\end{aligned} \tag{9}$$

where g is either g_{II} or g_{III} introduced in (8) and (7).

As a consequence Eqs. (1), (2) can be rewritten as follows

$$\begin{aligned}\frac{\partial s_1}{\partial t}(x,t) = d\Delta s_1(x,t) + rC_I(x,t)[M(x) - \chi s_1(x,t)] - ns_1(x,t) \\ -pg(C_I(x,t))Z(x,t)\frac{s_1(x,t)}{C_I(x,t)} - \beta s_1(x,t)\int_\Omega k(x,x')i_2(x',t)dx'\end{aligned} \tag{10}$$

$$\begin{aligned}\frac{\partial i_1}{\partial t}(x,t) = d\Delta i_1(x,t) - ni_1(x,t) - pg(C_I(x,t))Z(x,t)\frac{i_1(x,t)}{C_I(x,t)} \\ -r\chi i_1(x,t)C_I(x,t) + \beta s_1(x,t)\int_\Omega k(x,x')i_2(x',t)dx'.\end{aligned} \tag{11}$$

Assumptions

- $\Omega \subset \mathbb{R}^2$ is a bounded domain with a smooth boundary;
- $M \in L^\infty(\Omega)$; $M(x) \geq 0$ a.e. $x \in \Omega$;
- $K \in L^\infty(\Omega)$, $K(x) \geq k_0$ a.e. in Ω (here k_0 is a positive constant);
- $k \in L^\infty(\Omega \times \Omega)$, $k(x,x') \geq 0$ a.e. in $\Omega \times \Omega$; normalized to β;
- $d,\ r,\ r_2,\ \chi, n,\ q,\ \ell,\ C^0,\ c, \mu,\ \zeta,\ b,\ \alpha,\ \beta$ are positive constants (Table 1).

3 Control, Eradicability

Among possible controls the following ones might be considered: (a) increasing the insect vector death rate n, by insecticides and/or treated nets; (b) reduce the contact rate β, of trees towards insects by installing treated nets; (c) decrease infection rate from tools to trees b, by disinfection; (d) decrease carrying capacity of insects M, by eliminating weeds in the olive orchards; (e) decrease the infection rate of trees ζ, acting on the choice of more resistant cultivar.

Table 1 Model parameters

Symbol	Description
r	*P. spumarius* birth rate
r_2	*Z. renardii* birth rate
χ	*P. spumarius* intraspecific competition rate
n	*P. spumarius* natural mortality rate
q	Healthy trees (canopy) regrowth rate
C^0	Trees carrying capacity parameter
$K(x)$	Baseline *Z. renardii* carrying capacity parameter
$M(x)$	Spatial distribution of the weed biomass
c	Scale parameter of the impact of the weed biomass on the carrying capacity of *Z. renardii*
ℓ	Elimination rate of trees by pruning or logging
b	Infection rate of trees by infected tools
μ	Infected trees mortality rate
α	Infected trees recovery rate
β	*P. spumarius* infection rate by infected trees
ζ	Trees infection rate by infected *P. spumarius*

In [9, 11] and [6] only (d) and (e) have been considered as feasible control practices:

$C1$: Weed cut : $M \to M(1-\gamma_{21})$; decrease carrying capacity of insects, by eliminating weeds in the olive orchards.

$C2$: Choice of the cultivar: $\zeta \to \zeta(1-\gamma_{22})$; decrease infection rate of trees, acting on the cultivar.

Later in [7] the possible introduction of a predator of the insect vector *P. spumarius* in olive orchards was analyzed as an additional control strategy for the eradication of a *Xylella* epidemic.

As anticipated in the Introduction, for various reasons, including cost reduction, one may consider the implementation of the proposed control measures only on a suitable subregion $\omega \subset \Omega$ of the whole habitat, including areas already affected by the epidemic, augmented by a preventive confinement area; hence, in the evolution equation for predators, we have introduced an additional term $u(x) \geq 0$, acting only in the $\omega \subset \Omega$. This control term may be understood as a continuous infusion of predators at position $x \in \omega \subset \Omega$ (we take $u \in L^\infty(\omega)$, $u(x) \geq 0$ a.e. $x \in \omega$).

Actually the control might also be time dependent, but for the time being it has been taken constant in time.

Hence the dynamics of the predator population would be described by

$$\frac{\partial Z}{\partial t}(x,t) = d_2 \Delta Z(x,t) + r_2 Z(x,t)\left[1 - \frac{Z(x,t)}{K(x)+cM(x)}\right] + g(C_I(x,t))Z(x,t) + \mathbb{I}_\omega(x)u(x). \tag{12}$$

Denote $Q = \Omega \times (0, +\infty)$ and $\Sigma = \partial\Omega \times (0, +\infty)$.

Let us consider only the populations of insects, C_I for the *P. spumarius*, and Z for its predator, for a general functional response of predation:

$$\begin{cases} \dfrac{\partial C_I}{\partial t} = d\Delta C_I + rC_I(M - \chi C_I) - nC_I - pg(C_I(x,t))Z(x,t), & (x,t) \in Q \\ \dfrac{\partial Z}{\partial t} = d_2\Delta Z + r_2 Z\left[1 - \dfrac{Z}{K + cM}\right] + g(C_I(x,t))Z(x,t) + \mathbb{I}_\omega u, & (x,t) \in Q \\ \dfrac{\partial C_I}{\partial n}(x,t) = \dfrac{\partial Z}{\partial n}(x,t) = 0, & (x,t) \in \Sigma \\ C_I(x,0) = C_{I0}(x),\ Z(x,0) = Z_0(x), & x \in \Omega. \end{cases} \tag{13}$$

Via standard methods of analysis for reaction-diffusion systems, we may state the following theorem:

Theorem 1 *For any $C_{I0}, Z_0 \in L^\infty(\Omega)$, $C_{I0}(x), Z_0(x) \geq 0$ a.e. $x \in \Omega$ and for any $u \in L^\infty(\omega)$, $u(x) \geq 0$ a.e. $x \in \omega$, we get that (13) has a unique solution.*

By Rayleigh's principle [10, p. 516], we may claim that the following eigenvalue problem

$$\begin{cases} -d\Delta\varphi - rM\varphi + n\varphi = \lambda\varphi, & x \in \Omega \\ \dfrac{\partial\varphi}{\partial n}(x) = 0, & x \in \partial\Omega. \end{cases} \tag{14}$$

admits

$$\tilde{\lambda}_1 = \inf_{\varphi \in H^1(\Omega), \|\varphi\|_{L^2(\Omega)} \neq 0} \frac{\int_\Omega (d|\nabla\varphi|^2 - rM|\varphi|^2 + n|\varphi|^2)dx}{\int_\Omega |\varphi|^2 dx} \tag{15}$$

as its principal eigenvalue, to which we may associate a positive eigenfunction $\tilde{\varphi}_1$.

It is not difficult to show that, if $\tilde{\alpha}$ is a positive real number, the function

$$C_2(x,t) = \tilde{\alpha}\tilde{\varphi}_1(x)\exp(-\tilde{\lambda}_1 t)$$

is the unique solution to the following linear parabolic problem:

$$\begin{cases} \dfrac{\partial C}{\partial t} = d\Delta C + rMC - nC, & (x,t) \in Q \\ \dfrac{\partial C}{\partial n}(x,t) = 0, & (x,t) \in \Sigma \\ C(x,0) = \tilde{\alpha}\tilde{\varphi}_1(x), & x \in \Omega \end{cases} \tag{16}$$

It is also clear that, if $\tilde{\lambda}_1 > 0$, then $C_2(t) \longrightarrow 0$ in $L^\infty(\Omega)$, as $t \to +\infty$.

Now, if we take $\tilde{\alpha}$ a sufficiently large constant, by comparison results for parabolic equations, we may state that

$$0 \leq C_I(x,t) \leq C_2(x,t) \quad \text{a.e. } x \in \Omega,\ \forall t \geq 0.$$

As a consequence, we get that $C_I(t) \longrightarrow 0$ in $L^\infty(\Omega)$, as $t \to +\infty$.

On the other hand, if $\tilde{\lambda}_1 = 0$, then there exists two positive constants $\tilde{\alpha}, \overline{\alpha}$ such that

$$0 \leq C_I(x,t) \leq \frac{\tilde{\alpha}}{t+\overline{\alpha}}\tilde{\varphi}_1(x) \quad \text{a.e. } x \in \Omega,\ \forall t \geq 0.$$

As a consequence, we get that $C_I(t) \longrightarrow 0$ in $L^\infty(\Omega)$, as $t \to +\infty$.

We may then state the following theorem.

Theorem 2 *If $\tilde{\lambda}_1 \geq 0$, then for any $u \in L^\infty(\omega)$, $u(x) \geq 0$, we get that*

$$C_I(t) \longrightarrow 0 \quad \text{in } L^\infty(\Omega),\ \text{as } t \to +\infty.$$

This result expresses, in mathematical terms, the fact that for $\tilde{\lambda}_1 \geq 0$ the total population of vector insects goes extinct, independently of the presence of predators in the ecosystem. We may notice that, according to (15), we may make $\tilde{\lambda}_1 \geq 0$ by reducing M, the weed biomass, which confirms the results of our previous papers [6, 9, 11].

It is clear that an additional presence of predators, makes the decay of $C_I(t)$ to 0 even faster.

Let now $\tilde{\lambda}_1 < 0$. In this case, in absence of predators we cannot claim that in general the vector population will eventually tend to zero. Additional investigations have been required. In particular the qualitative behavior of System (13) differs with respect to the two cases of specialist and generalist predators.

4 Specialist Predators

For a specialist predator System (13) becomes

$$\begin{cases} \dfrac{\partial C_I}{\partial t} = d\Delta C_I + rC_I(M - \chi C_I) - nC_I - p\dfrac{\gamma_1 C_I}{1+\gamma_2 C_I}Z, & (x,t) \in Q \\ \dfrac{\partial Z}{\partial t} = d_2\Delta Z + r_2 Z\left[1 - \dfrac{Z}{K+cM}\right] + \dfrac{\gamma_1 C_I}{1+\gamma_2 C_I}Z + \mathbb{I}_\omega u, & (x,t) \in Q \\ \dfrac{\partial C_I}{\partial n}(x,t) = \dfrac{\partial Z}{\partial n}(x,t) = 0, & (x,t) \in \Sigma \\ C_I(x,0) = C_{I0}(x),\ Z(x,0) = Z_0(x), & x \in \Omega. \end{cases} \tag{17}$$

Consider the following nonlinear parabolic problem

$$\begin{cases} \dfrac{\partial Z_1}{\partial t} = d_2\Delta Z_1 + r_2 Z_1\left[1 - \dfrac{Z_1}{K+cM}\right] + \mathbb{I}_\omega u, & (x,t) \in Q \\ \dfrac{\partial Z_1}{\partial n}(x,t) = 0, & (x,t) \in \Sigma \\ Z_1(x,0) = 0, & x \in \Omega. \end{cases} \tag{18}$$

Since $u(x) > 0$ a.e. $x \in \omega$, we get that for any $t_0 > 0$, there exists $m_{t_0} > 0$ such that

$$Z_1(x,t_0) \geq m_{t_0}, \quad \text{a.e. } x \in \Omega.$$

It can be shown (see [7]) that the parabolic problem (18) admits a unique positive steady state $Z^* \in L^2(\Omega)$, i.e. a unique positive solution of the following semilinear elliptic problem

$$\begin{cases} d_2\Delta Z^* + r_2 Z^*(1 - \dfrac{Z^*}{K+cM}) + \mathbb{I}_\omega u = 0, & x \in \Omega \\ \dfrac{\partial Z^*}{\partial n}(x) = 0, & x \in \partial\Omega. \end{cases} \tag{19}$$

A regularity result for the solutions to the elliptic equations implies that $Z^* \in H^2(\Omega) \subset C(\overline{\Omega})$. Moreover

$$Z_1(t) \longrightarrow Z^* \quad \text{in } L^\infty(\Omega),$$

as $t \to +\infty$.

The principal eigenvalue $\lambda_1(\varepsilon)$ to

$$\begin{cases} -d\Delta\varphi - rM\varphi + n\varphi + \dfrac{p\gamma_1(Z^* - \varepsilon)}{1 + \gamma_2(C^* + \varepsilon)}\varphi = \lambda\varphi, \ x \in \Omega \\ \dfrac{\partial\varphi}{\partial n}(x) = 0, \qquad\qquad\qquad\qquad\qquad\qquad x \in \partial\Omega, \end{cases} \tag{20}$$

can be expressed as follows

$$\lambda_1(\varepsilon) = \inf_{\varphi \in H^1(\Omega), \|\varphi\|_{L^2(\Omega)} \neq 0} \times \frac{\int_\Omega \left(d|\nabla\varphi|^2 - rM|\varphi|^2 + n|\varphi|^2 + \dfrac{p\gamma_1(Z^* - \varepsilon)}{1 + \gamma_2(C^* + \varepsilon)}|\varphi|^2 \right) dx}{\int_\Omega |\varphi|^2 dx}. \tag{21}$$

If $\lambda_1(\varepsilon)$ is positive, then

$$C_\varepsilon(t) \longrightarrow 0 \quad \text{in } L^\infty(\Omega),$$

as $t \to +\infty$ and so

$$C_I(t) \longrightarrow 0 \quad \text{in } L^\infty(\Omega),$$

as $t \to +\infty$.

On the other hand, by Rayleigh's principle we get that

$$\lim_{\varepsilon \to 0+} \lambda_1(\varepsilon) = \lambda_1(0).$$

To summarize, we get the following result

Theorem 3 *If* $\tilde{\lambda}_1 < 0$ *and if* $\lambda_1(0) > 0$, *then*

$$C_I(t) \longrightarrow 0 \quad in\ L^\infty(\Omega),\ as\ t \to +\infty.$$

Remark 1 According to Theorem 2, if $\tilde{\lambda}_1 \geq 0$ (which may occur for a small value of M, the weed biomass) the vector population vanishes for large times, independently of the presence of predators.

While, according to Theorem 3, for $\tilde{\lambda}_1 < 0$ eradication of the vector population is possible upon the additional information that $\lambda_1(0) > 0$.

The form of $\lambda_1(0)$, given by Rayleigh's formula (21), shows that if the inflow of specialist predators in the relevant habitat is sufficiently large (or we extend the region of intervention ω), eradication of the total vector population is possible.

5 Generalist Predators

In this case System (13) has to be changed into the following one, in which an Holling type III response to predation occurs. In this case (C_I, Z) is the solution to

$$
\begin{cases}
\dfrac{\partial C_I}{\partial t} = d\Delta C_I + rC_I(M - \chi C_I) - nC_I - p\dfrac{\gamma_1 C_I^2}{1+\gamma_2 C_I^2}Z, & (x,t) \in Q \\[2ex]
\dfrac{\partial Z}{\partial t} = d_2\Delta Z + r_2 Z\left[1 - \dfrac{Z}{K + cM}\right] + \dfrac{\gamma_1 C_I^2}{1+\gamma_2 C_I^2}Z + \mathbb{I}_\omega u, & (x,t) \in Q \\[2ex]
\dfrac{\partial C_I}{\partial n}(x,t) = \dfrac{\partial Z}{\partial n}(x,t) = 0, & (x,t) \in \Sigma \\[2ex]
C_I(x,0) = C_{I0}(x),\ Z(x,0) = Z_0(x), & x \in \Omega
\end{cases}
\tag{22}
$$

We are considering the case in which $\tilde{\lambda}_1 < 0$, as in Theorem 3 concerning specialist predators. In that case we have been able to show that, under the additional assumption of a large inflow of predators in the relevant habitat, the vector population will eventually go extinct. For generalist predators the result is negative, as we will show in the following theorem.

Theorem 4 *If $\tilde{\lambda}_1 < 0$, then there is no $u \in L^\infty(\omega)$, $u(x) \geq 0$ a.e. $x \in \omega$ such that*

$$C_I(t) \longrightarrow 0 \quad \text{in } L^\infty(\Omega), \text{ as } t \to +\infty.$$

The above results, in particular Theorem 2 has confirmed the peculiar role of the parameter M, representing the weed biomass existing in the relevant habitat: independently of the introduction of predators, cutting weeds has been confirmed as a cost effective practice for controlling a *Xylella* epidemic in olive orchards (see the outcomes of [6, 9, 11]).

It would be of great interest, from an agronomical point of view, to relate the mathematical threshold parameters, $\tilde{\lambda}_1$ and $\lambda_1(0)$, to some relevant parameters of intervention, such as $M(x)$ (the spatial distribution of the weed biomass) and $u(x)$ (the spatial distribution of the predators inflow). Unfortunately this is a nontrivial mathematical problem, from a purely analytical point of view. A practical response to this issue can be obtained via adequate computational experiments, as it has been reported in [7].

On the other hand, the values of $\tilde{\lambda}_1$ and $\lambda_1(0)$ depend, in addition to the possible parameters of intervention $M(x)$ and $u(x)$, upon intrinsic parameters of the relevant agro-ecosystem, which are not yet known numerically. As mentioned in the concluding remarks, it is hopeful that our results may stimulate additional field experiments aimed at the identification of real values of all relevant parameters.

6 Future Directions

An important extension of the above model concerns the natural seasonality of the dynamics of a *Xylella* epidemic, the ecology and the hemimetabolous post-embrionic development of insect vector being the main issue [8, 17, 30].

It is clear indeed that both the environmental parameters related to the carrying capacity of the relevant agro-ecosystem, i.e. M and K should be taken as time dependent, with a suitable periodicity, related to annual seasonality—say high values in Spring and Fall, low values in Winter and Summer.

Considering their role in the reproduction rates of both vector and predator insects, this may drive agronomic practices tending to keep both quantities at sufficiently low values all over the year; actually the practice of weed cut, i.e. reduction of M, has been since long implemented in olive orchards in large areas of the Apulia region, but unfortunately not everywhere.

It is clear that all vitality parameters of both insects and trees depend on time in a seasonal way.

The above discussion leads to the following updated mathematical assumptions on the parameters of an agro-ecosystem.

Assumptions
Assume a period $T = 1$ year.

- $M \in L^\infty(\Omega \times [0, +\infty))$; $M(x, t + T) = M(x, t)$ and $M(x, t) \geq M_0 > 0$, for $x \in \Omega$, $t \geq 0$.
- $K \in L^\infty(\Omega \times [0, +\infty))$; $K(x, t + T) = K(x, t)$ and $K(x, t) \geq K_0 > 0$, for $x \in \Omega$, $t \geq 0$.
- $r(t)$, $r_2(t)$, $n(t)$, $q(t)$, $\ell(t)$, are periodic with period 1 year. All these parameters are taken nonnegative and belonging to $C([0, +\infty)$. Moreover we may take $r(t) \geq r_0 > 0$; $\widetilde{r}_2(t) \geq \widetilde{r}_0 > 0$; for $t \geq 0$.

Of course, in the numerical simulations, the graph of the above functions during each period of 1 year should take into account possible seasonal variations.

With this in mind, System (13), without inflow, can be rewritten in the following form, for a general functional response of predation:

$$\begin{cases} \dfrac{\partial C_I}{\partial t}(x,t) = d\Delta C_I(x,t) + r(x,t)C_I(x,t)(M(x,t) - \chi C_I(x,t)) \\ \qquad -n(x,t)C_I(x,t) - pg(C_I(x,t))Z(x,t), \quad (x,t) \in Q \\ \dfrac{\partial Z}{\partial t}(x,t) = d_2\Delta Z(x,t) + r_2(x,t)Z(x,t)\left[1 - \dfrac{Z(x,t)}{K(x,t) + cM(x,t)}\right] \\ \qquad +g(C_I(x,t))Z(x,t), \quad (x,t) \in Q \\ \dfrac{\partial C_I}{\partial t}(x,t)(x,t) = \dfrac{\partial Z}{\partial t}(x,t)(x,t) = 0, \quad (x,t) \in \Sigma \\ C_I(x,0) = C_{I0}(x), \ Z(x,0) = Z_0(x), \quad x \in \Omega. \end{cases}$$

By splitting the vitality parameter $r_2(x,t)$ into a birth rate $\widetilde{r}_2$ and a death rate $\widetilde{n}_2$, a more appropriate evolution equation for $Z(x,t)$ would be

$$\begin{aligned} \frac{\partial Z}{\partial t}(x,t) &= d_2\Delta Z(x,t) + \widetilde{r}_2(x,t)Z(x,t)(1 - \frac{Z(x,t)}{K(x,t) + cM(x,t)}) \\ &\quad -\widetilde{n}_2(x,t)Z(x,t) + g(C_I(x,t))Z(x,t), \quad (x,t) \in Q, \end{aligned} \tag{23}$$

with $r(t), n(t), \widetilde{r}_2(t), \widetilde{n}_2(t),\ q(t)$, periodic with period 1 year.

According to [18] (see also [8, 28]), female insects lay eggs in autumn; juvenile stages occur from February to early May, while they have been on weeds in the olive orchard; adults appear in May-June and move to the canopy of olive trees; after August they move back to the weeds, for mating and oviposition. Adults live until autumn. The phenology of *P. spumarius* may vary, depending on the latitude and the seasonal variations in terms of the climatic conditions. By taking all these facts into account, a possible conjecture would be that, by eliminating weeds in February-March-April (so to eliminate nymphs) and in August-September (so to reduce the suitable oviposition site), the insect vector population may be eradicated from the relevant orchard.

Future investigations will include the solution of suitable optimal control problems, with respect to both the parameters of the dynamical system and the subregion of intervention $\omega \subset \Omega$. As mentioned in the above conjecture, seasonal pulse interventions in olive orchards should be expected to be a good control strategy in terms of cost/effectiveness.

It is clear that, for a realistic optimal control problem, relevant participating costs need to be expressed: e.g. production, losses, and management costs.

Once again, validation of the models proposed here represents a key issue: although we have tried to render explicit the assumptions underlying our models, the latter have not yet been validated, by comparison with empirical data. Therefore we caution that it is far from being conclusive for *X. fastidiosa* subsp. *pauca*—*P. spumarius* - olive tree epidemics. However it is desirable that, with additional features that make it more realistic, our models might provide the foundations for designing optimal control strategies by public decision makers.

It is hopeful that the proposed mathematical models themselves may stimulate additional experiments, aimed at identifying real values for all relevant parameters.

References

1. Almeida, R.P.P., Blua, M.J., Lopes, J.R.S., Purcell, A.H.: Vector transmission of Xylella fastidiosa: applying fundamental knowledge to generate disease management strategies. Ann. Entomol. Soc. Am. **98**, 775–786 (2005)
2. Aniţa, S., Capasso, V.: A stabilizability problem for a reaction-diffusion system modelling a class of spatially structured epidemic systems. Nonlin. Anal. Real World Appl. **3**, 453–464 (2002)
3. Aniţa, S., Capasso, V.: A stabilization strategy for a reaction-diffusion system modelling a class of spatially structured epidemic systems (*think globally, act locally*). Nonlin. Anal. Real World Appl. **10**, 2026–2035 (2009)
4. Aniţa, S., Capasso, V.: Stabilization of a reaction-diffusion system modelling a class of spatially structured epidemic systems via feedback control. Nonlin. Anal. Real World Appl. **13**, 725–735 (2012)
5. Aniţa, S., Casas, J., Suppo, C.: Impulsive spatial control of invading pests by generalist predators. Math. Med. Biol. **31**, 284–301 (2014)
6. Aniţa, S., Capasso, V., Scacchi, S.: Controlling the spatial spread of a Xylella epidemic. Bull. Math. Biol. **83**, 32 (2021)
7. Aniţa, S., Capasso, V., Montagna, M., Scacchi, S.: Predators as a possible strategy for controlling a Xylella epidemic? Math. Model. Nat. Phenom. **7**, 279–305 (2022)
8. Ben Moussa, I.E., et al.: Seasonal fluctuations of sap-feeding species infected by Xylella fastidiosa in Apulian olive groves of Southern Italy. J. Econ. Entomol. **109**, 1512–1518 (2016)
9. Beretta, E., Capasso, V., Scacchi, S., Brunetti, M., Montagna, M.: Prevention and control of OQDS (olive quick decline syndrome) outbreaks caused by Xylella fastidiosa. J. Theor. Biol. **542**, 111118 (2022)
10. Brezis, H.: Sobolev Spaces and Partial Differential Equations. Springer, New York (2011)
11. Brunetti, M., Capasso, V., Montagna, M., Venturino, E.: A mathematical model for Xylella fastidiosa epidemics in the Mediterranean regions. Promoting good agronomic practices for their effective control. Ecol. Modell. **432**, 109–204 (2020). https://doi.org/10.1016/j.ecolmodel.2020.109204
12. Capasso, V.: Asymptotic stability for an integro-differential reaction-diffusion system. J. Math. Anal. Appl. **103**, 575–588 (1984)
13. Capasso, V.: Mathematical Structures of Epidemic Systems, 2nd corrected printing. Lecture Notes Biomathematics, vol. 97. Springer-Verlag, Heidelberg (2008)
14. Capasso, V., Aniţa, S. The interplay between models and public health policies: regional control for a class of spatially structured epidemics (Think globally, act locally). Math. Biosci. Eng. **15**, 1–20 (2018)
15. Cariddi, C., Saponari, M., Boscia, D., De Stradis, A., Loconsole, G., Nigro, F., Porcelli, F., Potere, O., Martelli, G.P.: Isolation of a Xylella fastidiosa strain infecting olive and oleander in Apulia, Italy. J. Plant Pathol., **96**, 425–442 (2014)
16. Carlucci, A., Lops, F., Marchi, G., Mugnai, L., Surico, G.: Has Xylella fastidiosa "chosen" olive trees to establish in the Mediterranean basin? Phytopathol. Mediterr. **52**, 541–544 (2013)
17. Cornara, D., Saponari, M., Zeilinger, A.R., de Stradis, A., Boscia, D., Loconsole, G., Bosco, D., Martelli, G.P., Almeida, R.P.P., Porcelli, F.: Spittlebugs as vectors of Xylella fastidiosa in olive orchards in Italy. J. Pest. Sci. **90**, 521–530 (2017)
18. Cornara, D., Cavalieri, V., Dongiovanni, C., Altamura, G., Palmisano, F., Bosco, D., Porcelli, F., Almeida, R.P.P., Saponari, M.: Transmission of Xylella fastidiosa by naturally infected Philaenus spumarius (Hemiptera, Aphrophoridae) to different host plants. J. Appl. Entomol. **141**, 80–87 (2017)

19. Davranoglou, L.R.: Zelus renardii (Kolenati, 1856), a New World reduviid discovered in Europe (Hemiptera: Reduviidae: Harpactorinae). Entomol. Mon. Mag. **147**, 157–162 (2011)
20. Janse, J.D., Obradovic, A.: Xylella fastidiosa: its biology, diagnosis, control and risks (Minireview). J. Plant Pathol. **92**, S1.35–S1.48 (2010)
21. Jeger, M. et al.: EFSA PLH Panel, Updated pest categorisation of Xylella fastidiosa. EFSA J. **16**(7), e05357 (2018). https://doi.org/10.2903/j.efsa.2018.5357
22. Lago, C., Garzo, E., Moreno, A., Barrios, L., Martí-Campoy, A., Rodríguez-Ballester, F., Fereres, A.: Flight performance and the factors affecting the flight behaviour of Philaenus spumarius the main vector of Xylella fastidiosa in Europe. Sci. Rep. **11**, 17608 (2021). https://doi.org/10.1038/s41598-021-96904-5
23. Lahbib, N., Picciotti, U., Sefa, V., Boukhris-Bouhachem, S., Porcelli, F., Garganese, F.: Zelus renardii roaming in Southern Italy. Insects **13** (2022) 158.
24. Lewis, M.A., Petrovskii, S.V., Potts, J.R.: The Mathematics Behind Biological Invasions. Springer, Heidelberg (2016)
25. Liccardi, A., Fierro, A., Garganese, F., Picciotti, U., Porcelli, F.: A biological control model to manage the vector and the infection of Xylella fastidiosa on olive trees. PLoS ONE **15**, e0232363 (2020)
26. Redak, R.A., Purcell, A.H., Lopes, J.R.S., Blua, M.J., Mizell, R.F. III, Andersen, P.C.: The biology of xylem fluid-feeding insect vectors of Xylella fastidiosa and their relation to disease epidemiology, applying fundamental knowledge to generate disease management. Ann. Rev. Entomol. **49**, 243–270 (2004)
27. Saponari, M., Loconsole, G., Cornara, D., Yokomi, R.K., De Stradis, A., Boscia, D., Bosco, D., Martelli, G.P., Krugner, R., Rocelli, F.: Infectivity and transmission of Xylella fastidiosa by Philaenus spumarius (Hemiptera: Aphrophoridae) in Apulia, Italy. J. Econ. Entomol. **107**, 1316–1319 (2014)
28. Trkulja, V., Tomić, A., Iličić, A., Nožinić, M., Popović Milovanović, T.: Xylella fastidiosa in Europe: from the introduction to the current status. Plant Pathol. J. **38**, 551–571 (2022)
29. Turchin, P.: Complex Population Dynamics: A Theoretical/Empirical Synthesis, Monographs in Population Biology. Princeton University Press, Princeton, (2003)
30. Weaver, C.R., King, D.R.: Meadow Spittlebug. Research Bulletin, vol. 741. Ohio Agriculture Experiment Station, Wooster (1954)

Predicting and Explaining with Models: A Few Remarks on Mathematical Immunology

Clemente F. Arias, Federica Bertocchini, Francisco J. Acosta, Gerardo Oleaga, Cristina Fernández-Arias, and Miguel A. Herrero

Abstract In this work, we discuss the limitations of mathematical models to provide explanations and make useful predictions in Biology. As a benchmark to assess the utility of models, we first create two toy systems that mimic the individual behavior of T cells during immune responses. We use these microscopic toy systems to simulate the process that is usually followed in mathematical immunology. We formulate macroscopic models that simulate the dynamics of the toy systems and analyze their explicative power. The rationale for our approach is that: (i) the microscopic causes of the observed behavior are (by definition) perfectly known, which allows evaluating the performance of the macroscopic models; and (ii) the predictions made by these models can easily be tested by comparing them with numerical simulations of the toy systems they intend to simulate. By using this strategy, we show that models can reproduce an observed behavior while being incompatible with the biological rules that produce that behavior. Therefore, the ability of a mathematical model to describe a given dynamics does not imply that it can explain the causes underlying it. We also show that a mechanistic approach is unnecessary to provide valuable insight into immunological problems. Macroscopic models of an utterly phenomenological nature can be used to predict important aspects of immune responses using virtually no information about the microscopic details of the interactions between effector immune cells and pathogens.

C. F. Arias (✉) · F. Bertocchini
Centro de Investigaciones Biológicas Margarita Salas, Madrid, Spain
e-mail: tifar@ucm.es

F. J. Acosta
Departamento de Ecología, Universidad Complutense de Madrid, Madrid, Spain

G. Oleaga · C. Fernández-Arias
Departamento de Análisis Matemático y Matemática Aplicada, Universidad Complutense de Madrid, Madrid, Spain

M. A. Herrero
Departamento de Matemática Aplicada, Universidad Complutense de Madrid, Madrid, Spain

A. d'Onofrio et al. (eds.), *Problems in Mathematical Biophysics*, SEMA SIMAI Springer Series 38, https://doi.org/10.1007/978-3-031-60773-8_2

1 Introduction

The use of mathematical models to address immunological problems dates back at least to the 60s of the previous century [1]. Since then, a variety of modeling techniques have been applied to study different aspects of the immune system [1–3]. The strategy usually followed to model a particular aspect of the immune response can be summarized as follows. The first step consists in reducing the complexity of the problem of interest, selecting only a few variables assumed to account for available real-world observations [2]. This choice of variables is obviously constrained by previous knowledge about the system under scrutiny. In the particular case of the immune system, the response to infections involves the coordinated action of many different entities (from molecular signals to cells and tissues), and although many details of this response have been described in the literature, others remain largely unknown. Therefore, in the choice of variables the assumption is implicit that the elements responsible for the phenomenon of interest have been identified. This choice also defines the scope of the model and the scale (in space and time) at which the problem is approached.

The selection of variables is critical to achieve an operational balance between the complexity of the original biological system and that of the mathematical model intended to simulate it, which may dramatically escalate as the degree of detail increases [4]. Once the relevant variables have been selected, they are usually integrated into a conceptual model, often expressed in qualitative and verbal terms, that make explicit assumptions about their mutual interactions.

We remark that up to this point, the approach is shared by experimental and mathematically-oriented immunology studies alike. In the former, there is also a limit to the number of immune factors that can be jointly taken into account in a given experiment, which reduces the number of variables that can be conjointly analyzed. Therefore, the complexity of the studied system must be carefully dissected to select the aspects to be studied in each experimental setting. The goal of experimental studies is not merely to collect data about particular immunological processes but to formulate conceptual models in which those data can be interpreted to shed light on the mechanisms that cause the observed behaviors. These models can then be used to formulate new hypotheses as well as a guide to decide the direction of subsequent research.

The next step in the construction of mathematical models in immunology is the formalization of the conceptual models into mathematical equations that could be analytically or numerically solved. Such formalization provides a more powerful tool to understand the process under study than their informal counterparts [5]. This is especially so when dealing with highly dynamic problems such as those arising during immune responses, whose functional features can be hard to grasp without appropriate mathematical tools. For instance, mathematical modeling allows confirming whether the postulated conceptual models produce the expected behavior and are coherent with previous knowledge, something that is difficult to assess when models are stated in verbal terms only.

In fact, one of the most frequent reasons to resort to mathematical modeling in immunology is to discriminate between alternative explanations for the process under investigation [6, 7]. The rationale for this approach is that a valid mechanistic explanation of this process must be able to reproduce its observed behavior. Reciprocally, it is widely assumed that a model that succeeds in reproducing a given behavior has to be mechanistic, i.e. it should *explain* the underlying causes of that behavior [2, 8–11]. From this assumption, it follows that for mathematical models to be useful, they must first be validated, i.e. tested against real-world data [7].

The validation of a mathematical model, and hence the evaluation of its explanatory power, usually consists in finding the parameter values that best fit the solutions of the model to a given data set. A mathematical model that cannot be successfully parameterized is discarded as unable to provide useful information about reality. Conversely, a good fit is normally interpreted as the validation of the underlying immune hypotheses of a mathematical model [6, 7]. Thus, measures of goodness-of-fit to data have been traditionally considered a quantitative objective score to compare among alternative, "rival" candidates [2, 12–18]. According to this view, models that successfully describe available data can then be used to make predictions about previously unseen scenarios [18, 19] and the eventual validation of these predictions is considered as further proof of the validity of the model and, consequently of its biological assumptions [6, 20].

This approach relies on several questionable assumptions. The first, and most basic one, is that experimental data are a faithful representation of reality. However, the interpretation of the data is not always trivial and can be biassed by the assumptions of models in which this interpretation relies on [21, 22]. For instance, failure to experimentally measure causally important variables may result in the detection of spurious trends in the data, a phenomenon known as Simpson's paradox [23]. Furthermore, the parameterization of mathematical models of immunological problems is often based on data available in the literature. However, using published data to parameterize mathematical models is not straightforward, as the experimental setups used to produce those data are highly heterogeneous [2]. A second critical assumption generally taken for granted in mathematical immunology is that a good fit to data necessarily results from model features that are directly related to the biological process of interest. However, this is not always the case, since artificially increasing the complexity of the model can easily improve the statistical fit of a model [12, 18, 24]. Finally, predictive models do not need to be mechanistic, as shown by computational algorithms such as AlphaFold, which can predict the three-dimensional structure of proteins from their sequence without even attempting to explain how proteins actually fold [25]. In fact, this approach outperforms more mechanistic algorithms based on thermodynamic or kinetic simulation of protein physics [25].

The previous observations raise significant concerns about current approaches to mathematical modeling in Biology. Our aim in this work is to contribute to a better understanding of these issues by analyzing in detail a case study in Immunology. More precisely, in the following sections we focus on the T cell response to acute infection, and propose two microscopic toy systems that mimic the behavior of individual T cells during immune response. We then formulate mathematical models that simulate the dynamics of the toy systems. The rationale for this approach is that the mechanisms responsible for the behavior of the toy systems are fully known *a priori*, which allows comparing the explanations and the predictions provided by the models with the reality they intend to simulate.

By using this approach, we show that it is possible to formulate macroscopic models that reproduce an observed behavior and yet fail to account for the biological rules that produce that behavior, which suggests that using the goodness-of-fit as a criterion to evaluate the explicative power of a model may be misleading. We also show that a macroscopic model does not need to be explicative in order to make useful predictions about relevant aspects of the immune system. In particular an alternative, phenomenological model yields a remarkable result, namely the possibility for slowly proliferating pathogens to evade immune responses.

2 A Case Study: Modeling the Use of Mathematical Models in Immunology

The detection of a pathogen triggers a series of immune mechanisms in the organism intended to rapidly neutralize the infection. The responses to intracellular pathogens (such as viruses and bacteria) are mediated, among other immune mechanisms, by T cells. In the case of an infection, naive T cells are activated by antigen presenting cells (APCs) that have migrated from the infected tissue into nearby lymph nodes [26]. The architecture of lymph nodes facilitates close cell-to-cell interactions between APCs and naive T cells. In the course of these interactions, the T-cell receptor (TCR) of the naive T cell gets in contact with the antigens carried by APCs in their major histocompatibility complexes. Those T cells equipped with TCRs capable of recognizing the pathogen's antigens engage in long-term physical contacts with APCs known as immune synapses [26].

As a consequence of the immune synapse, the T cell activates and becomes an effector T cell. Effector T cells then proliferate massively, a process known as clonal expansion, and migrate to the infected tissue, where they recognize and selectively kill the host cells showing signs of infection. Once the infection is controlled, effector T cells undergo apoptosis (clonal contraction) and disappear, restoring the system to its initial equilibrium state (Fig. 1) [27]. Intriguingly, the contraction of T cell populations is not triggered by the disappearance of the pathogen but is delayed with respect to the end of the infection [28]. After clonal contraction, only around 5–10% of effector T cells survive in the form of memory T cells, providing an effective defense mechanism in case of future infections by the same pathogen. In this work, we will not consider the formation of memory T cells and we will focus instead on the dynamics of effector T cells.

The immune response of T cells is a paramount example of distributed system in which the behavior of the population (in terms of population dynamics) emerges

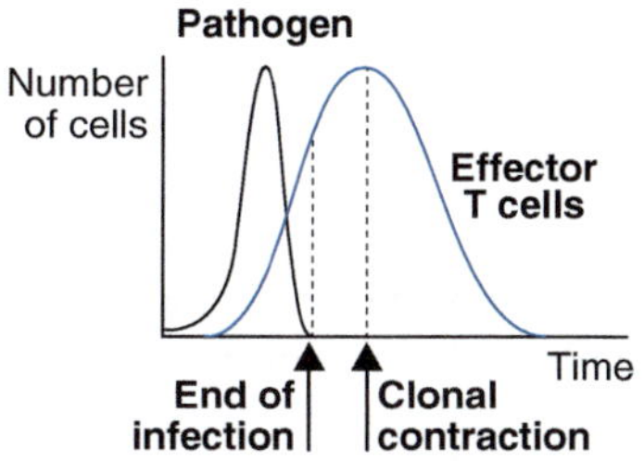

Fig. 1 Dynamics of effector T cell immune responses against acute infections. The presence of a pathogen triggers the activation and proliferation of T cells that migrate to the infected tissue. After the infection has been neutralized, T cells undergo massive apoptosis and disappear. Remarkably, T cells continue to proliferate after the pathogen has been cleared, so the contraction of the population is delayed with respect to the end of the infection

from decisions taken by individual cells of whether to divide or to undergo apoptosis. In the absence of a central organ of control, each cell must continuously evaluate the progression of the infection based on local information and react accordingly to ensure that the collective response is effective. We will next formulate two computational toy systems that reproduce this cellular logic. In later sections we will explore the ability of mathematical models to explain the behavior of both toy systems.

2.1 *Toy T-Cell Systems*

To formulate these systems we make the following basic assumptions. In both systems, toy T cells (hereafter referred to as T_t cells to differentiate them from real-world T cells) are instantaneously activated in the case of an infection. After activation, their effect on pathogen clearance follows the law of mass action:

$$P'(t) = \alpha P(t) - \beta P(t) n_T(t) \quad (1)$$

where $P(t)$ and $n_T(t)$ are the number of infectious agents and T_t cells respectively, and α and β are positive parameters.

The information conveyed by their interactions with the infectious agent is used by T_t to decide at any time whether they have to divide or die. The fate of T_t cells is controlled by two alternative proteins, c and a, that inhibit the machinery of cell division and apoptosis respectively. The effect of c is inspired in that of the Retinoblastoma protein, which is known to block the expression of genes needed for the progression of the cell cycle. In turn, the effect of a is similar to that of Bcl-2, a protein family that controls the decision of the cell to enter the program of apoptosis. Denoting by $c(t)$ and $a(t)$ the concentration of proteins in a given T_t at time t, the cell decisions to divide or to die are determined by the conditions:

$$c(t) = 0 \text{ and } a(t) = 0$$

respectively.

We further assume that the dynamics of c and a in each T_t are determined by proliferation and death membrane receptors (r_p and r_d respectively), according to the following equations:

$$\begin{cases} c'(t) = -\mu_{pc} r_p(t) \\ a'(t) = -\mu_{da} r_d(t), \end{cases} \tag{2}$$

where μ_{pc} and μ_{da} are positive parameters.

Finally, the number of receptors in the membrane of each T_t cell at any time is given by:

$$\begin{cases} r_p'(t) = \varphi_p(t) - \lambda_{pp} r_p(t) \\ r_d'(t) = \lambda_{pd} r_p(t) - \lambda_{dd} r_d(t), \end{cases} \tag{3}$$

where λ_{pp}, λ_{pd}, and λ_{dd} are positive parameters and $\varphi_p(t)$ represents the production of r_p receptors at time t.

Equations (2) and (3) define the "molecular algorithm" that determines the fate of T_t cells during infections (see Fig. 2). Based on this logic, we will now define two alternative toy systems that differ in the nature of the mechanisms that cause the expression of r_p receptors in T_t cells, i.e. in the particular form adopted by $\varphi_p(t)$. In the first system (toy System A), r_p receptors are produced as a consequence of the interaction between the T_t cell and the pathogen. according to this equation:

$$\varphi_p(t) = \lambda_{Pp} \rho P(t), \tag{4}$$

where λ_{Pp} and ρ are positive parameters and $\rho P(t)$ represents the number of interactions taking place at time t between the T_t cell and the infectious agent.

For the second toy system (labeled as toy System B), we consider a more complex scenario. In this case, another cell type exists that contributes the response and that we will label as "helper" cells (labeled as H_t). The contact with the pathogen has two alternative effects on H_t cells. First, it induces the expression of proliferation receptors as shown in Fig. 2. In addition, it activates the production of an intracellular molecule s according to the following dynamics:

$$s'(t) = \lambda_{Ps} \rho P(t) - \lambda_{ss} s(t), \tag{5}$$

where λ_{Ps}, ρ, and λ_{ss} are positive parameters and $\rho P(t)$ is the number of interactions with the pathogen perceived at time t by the H_t cell.

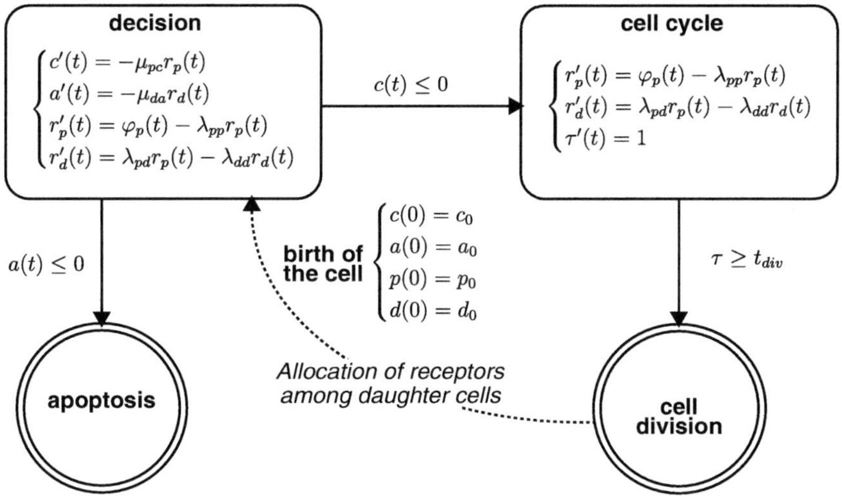

Fig. 2 **Intracellular mechanisms that determine the fate of T_t cells.** Each cell is formed with an initial amount of inhibitory molecules c and a and remains in the first stage of the cell cycle (labeled as **decision**) until one of these inhibitors disappears. The disappearance of the apoptosis inhibitor a triggers the death of the cell (final state **apoptosis**) whereas that of the cycle inhibitor c causes the entry of the cell into the next phase of the cell cycle (labelled as **cell cycle**). After this phase is completed (its duration is given by parameter t_{div}), the T_t cell divides (final state **cell division**), and its membrane receptors are randomly distributed among its two daughter cells

The signal is released from the cytoplasm of H_t cells to the extracellular space, where it fosters the proliferation of T_t cells. Denoting by $S(t)$ the amount of extracellular signal at time t, the dynamics of the second toy model respond to this model:

$$\begin{cases} s_i'(t) = \lambda_{Ps} \rho_i P(t) - \lambda_{ss} s_i(t) \text{ for } i = 1, \cdots, n_H(t) \\ S'(t) = \lambda_{HS} \sum_{i=1}^{n_H(t)} s_i(t) - \mu_{TS} S(t) n_T(t) \\ P'(t) = \alpha P(t) - \beta P(t) n_T(t) \end{cases} \tag{6}$$

where λ_{HS} and μ_{TS} are positive parameters, and $n_H(t)$ and $n_T(t)$ are the number of H_t and T_t cells respectively.

In turn, the presence of signal S in the extracellular space fosters the expression of r_p receptors in T_t cells:

$$\varphi_p(t) = \lambda_{Sp} S(t), \tag{7}$$

for a positive parameter λ_{Sp}.

The logic of the two toy systems formulated above is summarized in Fig. 3. The dynamics of these systems is implemented within the framework of hybrid dynamic systems, which are characterized by the combination of differential equations and

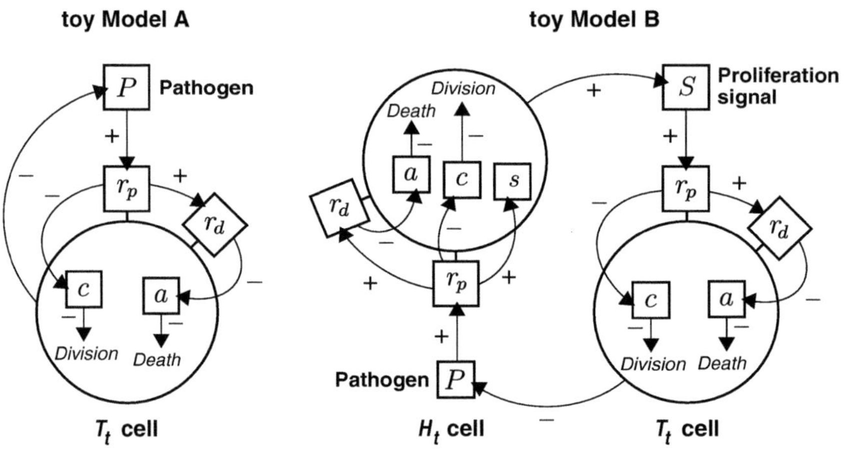

Fig. 3 Molecular mechanisms of cell fate determination in the two toy systems. See explanations in the text

discrete changes in the state of the system as illustrated in Fig. 2 (see [29] for further details). In the case of the previous toy models, the dynamics of intracellular molecules and receptors, as well as that of the pathogen and the extracellular signal S (in toy Model 2) are governed by differential equations. Simultaneously, the decision of a cell to divide or die entail an instantaneous transition in the system that translates into a change in the set of differential equations that model its behavior. The implementation of the toy Systems can be outlined as follows:

- The system is initialized with three T_t cells and, in the case of toy System B, with three additional H_t cells. For each of them, the initial value of inhibitors c and a is given by $c_{ini} = 100{,}000$ and $a_{ini} = 200{,}000$ respectively. The number of receptors p and d is set to zero. The initial number of antigens is 10,000.
- At each iteration, a system of equations is created that contains the equations shown in Fig. 2 corresponding to each cell of the population and Eq. (1) (toy System A) or (6) (toy System B) are numerically solved. The integration stops if $c_i(t) = 0$ or $a_i(t) = 0$ for some cell i. In the first case (cell division), cell i is replaced by two new cells, whose initial configuration is $c = c_{ini}$ and $a = a_{ini}$. The receptors of cell i are distributed randomly among the daughter cells. For each receptor, a random number R is calculated between 0.3 and 0.7. If r is the number of receptors in cell i, the number of receptors in both daughter cells is given by Rr and $(1 - R)r$. In the case $a = 0$ (apoptosis) the cell i is removed from the population.
- The iterations continue while the number of T_t cells is greater than zero.

Figure 4 shows the results of numerical simulations of the toy systems during acute infections.

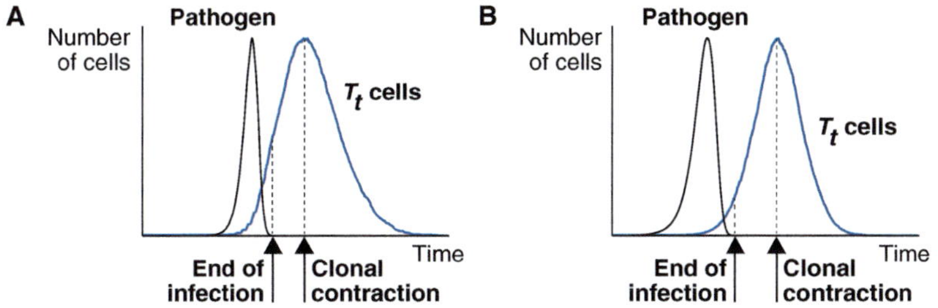

Fig. 4 Numerical simulations of the dynamics of the two toy systems. Both toy System A (left) and toy System B (right) display similar behaviors to that of T cell populations during acute infections. The values of the parameters used in these simulations are $t_{div} = 5$, $\lambda_{ap} = 0.0001$, $\lambda_{pp} = 0.0045$, $\lambda_{ps} = 0.0001$, $\lambda_{ss} = 0.002$, $\lambda_{pd} = 0.005$, $\lambda_{dd} = -0.005$, $\mu_{pc} = -0.001$, $\lambda_{ap2} = 0.05$, $\mu_{da} = -0.004$, $\alpha = 0.07$ (A) and 0.02 (B), $\beta = 0.001$, $\alpha_2 = 0.3$, $\beta_2 = 1000$, $\lambda_{ap} = 0.001$

2.2 *Mathematical Modelling of Toy System A*

In this section, we will simulate the process of formulating mathematical models in immunology by modeling the dynamics of toy System A. To that end, we will follow the standard procedure in the field as described above. First, we will define the appropriate scale of study and the variables to be used in the model, and formulate a conceptual model of how these variables interact with each other based on available information about the behavior of toy System A. Then, we will formalize this conceptual model in mathematical terms and test the resulting mathematical models against a data set. Finally, we will analyze the explanatory power of this approach.

To determine the scale of our mathematical model, we remark that immune cells are not spatially arranged forming tissues but circulate in the blood as independent individuals. In spite of their individuality, their function in the organism is only evident when considered at a population level. Therefore, although the ultimate mechanisms underlying the collective behavior of immune cells lie inside individual cells, it is their population-scale dynamics that holds the key to understanding their role in the organism. For this reason, immunological models often take a macroscopic approach in which immune cells are grouped into population-level compartments [13, 19]. Prototypically, the dynamics of these populations take the form of a balance equation [2, 5, 17, 30]:

$$\textit{Rate of change} = \textit{Proliferation} - \textit{Death} \pm \textit{Differentiation} \pm \textit{Migration}$$

To give explicit mathematical expressions to these processes, the immune response is often assumed to resemble ecological predator-prey models, with the immune cells preying on the pathogen [21, 31–34]. This analogy has lead to the widespread use of Lotka-Volterra-like ordinary differential equations to model the dynamics of immune responses [11, 35, 36].

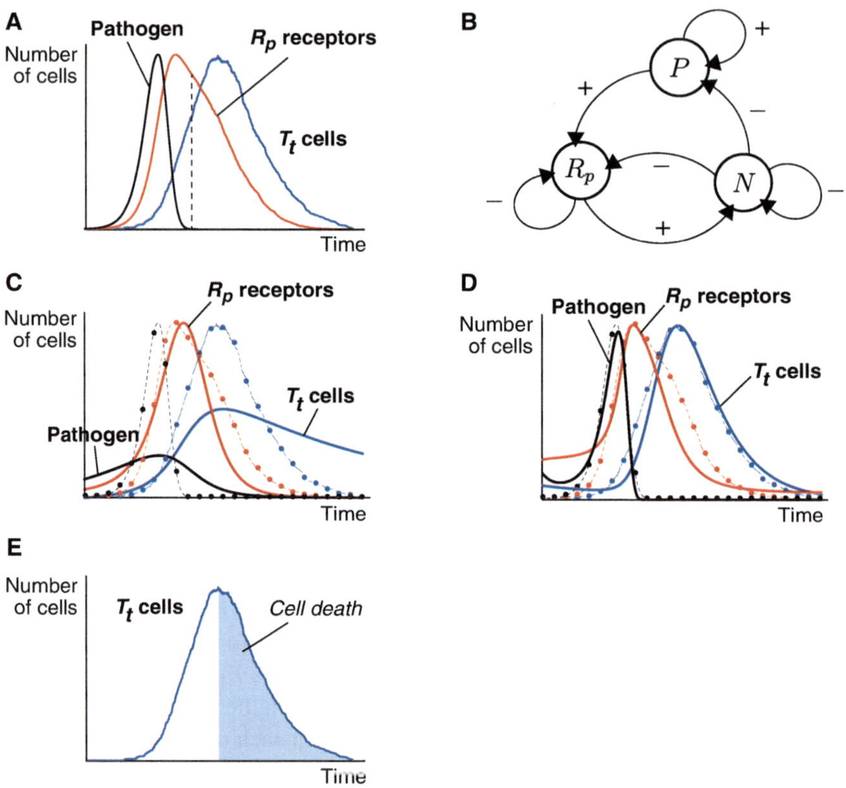

Fig. 5 Macroscopic Lotka-Volterra-type model of the dynamics of toy System A. (A) Dynamics of the pathogen, T_t cells, and R_p receptors during acute infections produced by toy System A. (B) Conceptual model of the functional relationships between the macroscopic variables that are used to simulate the behavior of toy System A. (C) (D) Fit of the solutions of macroscopic models (8) and (9) to "real" data (parameterization was performed using the DiffEqFlux.jl Julia library [37]). The "real" data obtained from simulations of toy System A are shown in dashed lines for reference (see (A)). The fit of Model (9) (D) is patently better than that of Model (8) (D). (E) In toy System A, the death of T_t cells only occurs in the last stage of the response (light blue area)

Once the modeling framework is established, the next step consists in selecting the variables to be included in the model. As discussed above, this choice is constrained by the previous knowledge about the studied system. Let us assume that in the case of toy System A, this knowledge stems exclusively from the experimental data shown in Figs. 4A and 5A. Specifically, Fig. 4A shows that the infection triggers the proliferation of T_t cells. In turn, T_t cells destroy the pathogen and lead to its eventual disappearance. Remarkably, T_t cells continue to proliferate after the pathogen has been effectively controlled. Additional data in Fig. 5 show the dynamics of expression of R_p receptors in T_t cells during an immune response (Fig. 5A). We assume that these receptors are known to foster proliferation in T_t cells but the rest of the information about the intracellular mechanisms outlined in Figs. 2 and 3 is not available. This scenario allows to explore the utility of mathematical modeling in cases where only partial information about the studied system is available, as it is usually the case in immunological studies.

The data in Figs. 4A and 5A seem to support a conceptual model in which the pathogen induces the expression of receptors R_p in T_t cells, thus fostering their proliferation. Additional assumptions are needed to account for the observed dynamics of receptors and T_t cells. First, the number of R_p receptors starts to decline before the pathogen disappears, which implies the existence of a mechanism that counteracts the effect of the pathogen. Since this decline coincides with the proliferation of T_t cells, it can be hypothesized to result from a negative feedback mechanism that controls the expression of receptors. Finally, the contraction of the T_t cell population can be modeled as an exponential decay in the absence of pathogens (Fig. 5B).

Can this conceptual model explain the observed dynamics of T_t cells, R_p receptors, and pathogens? As discussed above, this question would normally be addressed by translating the model into mathematical terms. To do that, let us denote by $R_p(t)$, $N(t)$, and $P(t)$ the number of receptors, T_t cells, and pathogens at time t respectively. The conceptual model shown in Fig. 5B can be written as

$$\begin{cases} R_p'(t) = \lambda_R P(t) R_p(t) - \delta_R R_p(t) N(t) \\ N'(t) = \lambda_N R_p(t) - \delta_N N(t) \\ P'(t) = \alpha P(t) - \beta P(t) N(t). \end{cases} \tag{8}$$

for positive parameters λ_R, δ_R, λ_N, δ_N, α, and β.

Alternatively, the same model could be modeled by the following equations:

$$\begin{cases} R_p'(t) = \lambda_R P(t) R_p(t) - \delta_R R_p(t) N(t) \\ N'(t) = \lambda_N R_p(t) N(t) - \delta_N N(t) \\ P'(t) = \alpha P(t) - \beta P(t) N(t) \end{cases} \tag{9}$$

We remark that both models have the same number of parameters, and that they only differ in their expression of T_t cells proliferation (highlighted in red in the previous equations). According to Model (8), receptors have a direct, linear effect on the growth of the T_t cell population. Model (9), on the other hand, assumes that it is the proliferation rate of the population ($N'(t)/N(t)$) that increases linearly with the number of receptors.

This difference is crucial for the performance of the models, defined as their ability to fit a data set (Figs. 5C and D). Whereas Model (9) can be parameterized in a way that closely describes the dynamics of toy System A, Model (8) fails to capture the main feature of those dynamics, particularly the delay in the contraction of the T_t cell population relative to the end of the infection. These results point out some key issues that emerge in the use of mathematical models in immunology. The most obvious one is that the mathematical formalization of a conceptual model does not need to be unique, which implies that the same model can be discarded or accepted as a satisfactory mechanistic explanation depending on its specific mathematical formulation.

As shown by Eqs. (8) and (9), the formulation of a mathematical model normally requires additional assumptions that are not explicit in the original conceptual model. Importantly, these assumptions may not be motivated by mechanistic arguments but can be imposed by the chosen modeling framework. For instance, in toy System A the effect of r_p receptors in the proliferation of T_t cells is mediated by the cell cycle inhibitor c, a factor that is not considered in Models (8) and (9). This effect is modeled instead as an abstract, population-level interaction between the macroscopic variables that represent the number of receptors in the population (R_p) and the number of T_t cells ($N(t)$). The particular mathematical expression of these interactions is typical of classical population dynamics models [21, 31–34]. In these models, such interactions are taken to affect the first derivative of the population sizes and to follow the law of mass action. Thus, they are normally modeled as linear functions (e.g. in migration or death) or as the product of variables (e.g. in predation).

The mechanistic nature of this approach is not evident. For instance, within the framework of classical population dynamics, proliferation and death rates are often assumed to be constant parameters [9, 17, 30, 38–42]. Hence the term $-\delta_N N(t)$, used to simulate the death of T_t cells in models (8) and (9). Numerical experiments on toy System A, however, show that the disappearance of T_t cells does not follow this logic. Instead, the rate of cell death is initially zero and becomes abruptly positive at a given moment after the clearance of the pathogen (Fig. 5E), a behavior that cannot possibly be simulated by assuming a constant death rate throughout the T_t cell response.

The rate of T_t cell is the population-level manifestation of the interactions between receptors (r_p and r_d) and inhibitory molecules (c and a) that take place at the scale of individual cells. The fact that this rate is variable implies that the outcome of these interactions changes as the infection progresses. A mechanistic model should explain the source of these changes, i.e. why the rate of cell death is different in earlier and later stages of the infection. In other terms, in a model intended to be explicative, this rate should be an output and not a parameter.

In summary, in this section we have simulated the process of creation of mathematical models in immunology. We have shown that describing a behavior using mathematical tools is not necessarily a synonym of explaining. Therefore, using the goodness-of-fit as an indicator of the mechanistic nature of a model may be misleading.

2.3 An Alternative Modeling Approach

In this section, we take a different approach to modeling the dynamic of toy Systems A and B. We begin by remarking that, despite the differences in the microscopic details between both toy systems, they exhibit similar population-level dynamics (Fig. 4). In both cases the population of T_t cells expands in response to the presence of the pathogen and contracts after the infection has been effectively neutralized. Moreover, both of them show inertia, i.e. contraction is delayed with respect to the disappearance of the pathogen. At a mathematical level, this feature can be naturally described by means of second order differential equations. In particular, and following [43], let us consider the following model:

$$\begin{cases} N''(t) = -kN(t) + \lambda P(t) \\ P'(t) = \alpha P(t) - \beta P(t)N(t) \end{cases} \quad \text{while } N(t) \geq 0 \text{ and } P(t) \geq P_{min} \tag{10}$$

where k, λ, α, and β are positive parameters. Let us assume that the infection is controlled when $P(t) \leq P_{min}$. In this case, the previous equations becomes:

$$\begin{cases} N''(t) = -kN(t) \\ P(t) = 0 \end{cases} \tag{11}$$

Figure 6.A shows that this model captures the basic qualitative features of the dynamics of toy systems A and B, included, as should be expected, the inertial behavior of T_t cell populations. We remark that this is a descriptive model and that it does not intend to account for the mechanisms responsible for those dynamics. In fact, the delayed contraction is the only information about T_t cells that is taken into account in the formulation of this model.

The question arises if such a descriptive approach is capable of making valuable predictions about the dynamics of toy systems A and B. Numerical simulations of the this model for different values of parameters α and β (which represent pathogen proliferation and clearance rates) show that it can give rise to two qualitatively different results (Fig. 6B). Other than the successful neutralization of the infection (see 1, 2, and 3 in Fig. 6B), Eqs. (10) and (11) and produce an alternative outcome in which the pathogen escapes the action of T_t cells (4, 5, and 6 in Fig. 6B).

Considering that Eqs. (10) and (11) are macroscopic models intended to simulate the dynamics of toy systems A and B, these results can be interpreted as explicit predictions about their behavior. As a matter of fact, these predictions are confirmed by numerical experiments of the toy Systems (Figs. 6C and D). Pathogens with low proliferation rates induce oscillations in the number of T_t cells or weak responses that end before the neutralization of the infection is complete.

The previous results show that descriptive models can provide valuable insight into unexpected aspects of the systems they intend to simulate and raise the question of whether these conclusions can be extrapolated to real-world T cell responses. In this regard, we remark that the clonal contraction of T cells is similar to that described by toy systems A and B (see Figs. 1 and 4). Particularly, the initiation of clonal contraction is also delayed with respect to the end of the infection [28]. Since this is the main assumption in which Eqs. (10) and (11) are based, we should expect for this behavior to occur also during T cell immune responses. The relevance of

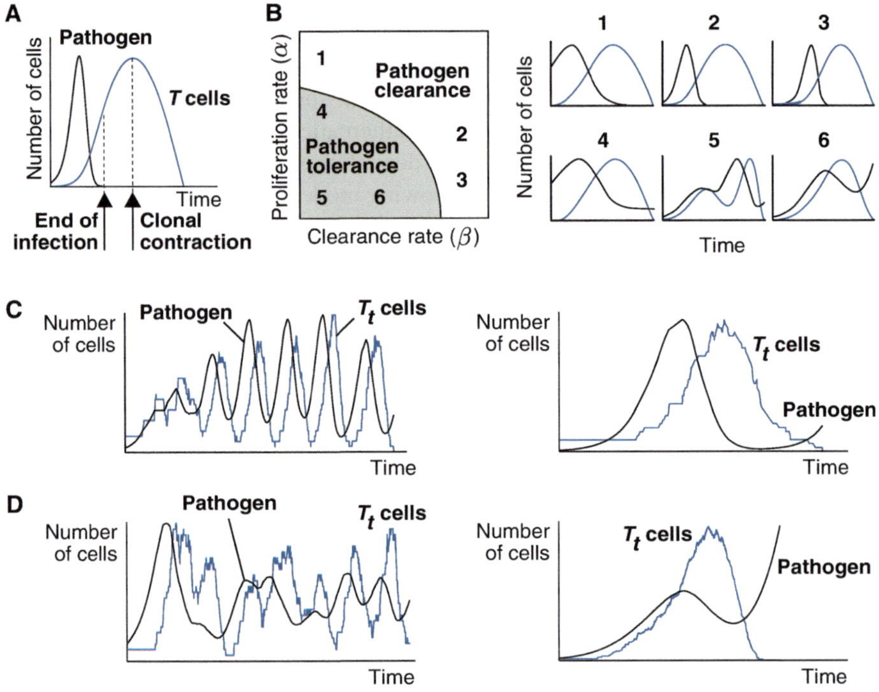

Fig. 6 Second order model of the dynamics of toy Systems A and B. (A) Equations (10) and (11) capture the inertial behavior of T_t cells in toy Systems. (B) This model suggests that this behavior only occurs for high values of parameters α and β. According to this model, a region exists where the pathogen can escape the action of the T_t cells (gray area). (C) (D) This prediction is confirmed for both toy Model A (C) and B (D)

this prediction in the context of T cells is not merely theoretical, since this would provide pathogens with a potential mechanism of immune evasion [44].

This point stresses the somewhat unexpected utility of toy systems as compared to their counterpart real-world systems. Testing the hypothesis of immune evasion by slow-growing pathogens using real-world data is not straightforward. This task is much simpler if toy systems are used instead (as shown in Fig. 6). These systems suggest that the microscopic details of the mechanisms involved in the interactions between T cells and pathogens may not be relevant for a crucial aspect of their macroscopic dynamics, such as delayed clonal contraction. Our results show that these macroscopic behaviors can be well described by simple models that do not intend to explain, only to describe. This approach provides an alternative framework to model the T cell immune response to acute infections [43], shedding light on different aspects of the immune response, from the existence of unexpected mechanisms of T-cell immune tolerance to new strategies for the rational design of T-cell based vaccines [45, 46].

3 Discussion

In any scientific field, making accurate predictions is much more pressing than understanding the underlying reasons why such predictions actually work. At a theoretical level, it is often assumed, however, that one needs to know the nuts and bolts of any system (physical or biological) in order to formulate successful predictions, that arguably would make use of such details. According to such view, explanations should precede predictions, and accordingly priority had to be given to digging deep for more and more details of any given problem, before daring to forecast how the corresponding dynamics will unfold in different practical situations.

While this point of view has considerable merit in itself, our results point out to the fact that (fortunately) such strategy needs not be mandatory. In particular, and taking as a showcase the issue of immune response to acute infections, we have seen that macroscopic mathematical models could provide useful predictions, even if they fail to explain the ultimate reasons for the behaviors they predict.

In fact, macroscopic models, often involving differential equations, are strategies of choice for practitioners in the field due to good reasons: mathematicians are usually familiar with them, there is already a wealth of theoretical and numerical results on differential equations ready for use and, most importantly, macroscopic variables being descriptors for averaged variables, one can dispense with the amount of microscopic detail required for model formulation at a smaller space scale. However, in most cases, the choice of a macroscopic model, even if based on a well-established reference fact, leaves plenty of room for arbitrary choices. Indeed, we have shown here that such a model can adequately describe observed behaviors even if it is not compatible with the molecular mechanisms that (by assumption) cause that behavior.

Interestingly, if macroscopic models are no longer required to fit data, they could still provide useful insight into the problem under consideration. In particular, we have shown that Eqs. (10) and (11) yield a remarkable result, namely the possibility for slowly proliferating pathogens to evade immune response.

References

1. Louzoun, Y.: The evolution of mathematical immunology. Immunol. Rev. **216**(1), 9–20 (2007)
2. Andrew, S.M., Baker, C.T., Bocharov, G.A.: Rival approaches to mathematical modelling in Immunology. J. Comput. Appl. Math. **205**(2), 669–686, (2007)
3. Ewald, J., Sieber, P., Garde, R., Lang, S.N., Schuster, S., Ibrahim, B.: Trends in mathematical modeling of host–pathogen interactions. Cell. Mol. Life Sci. **77**, 467–480 (2020)
4. Bonner, J.: First Signals: The Evolution of Multicellular Development. Princeton University Press, Princeton (2000)
5. Bocharov, G., Volpert, V., Ludewig, B., Meyerhans, A.: Basic principles of building a mathematical model of immune response. Math. Immunol. Virus Infections 15–34 (2018)

6. Antia, R., Ganusov, V.V., Ahmed, R.: The role of models in understanding CD8+ T-cell memory. Nat. Rev. Immunol. **5**(2), 101–111 (2005)
7. Eftimie, R., Gillard, J.J., Cantrell, D.A.: Mathematical models for Immunology: current state of the art and future research directions. Bull. Math. Biol. **78**, 2091–2134 (2016)
8. Xiaodan, M., LaMonte, M., Hovey, K., Freudenheim, J., Christopher, A., Genco, R. et al.: Physiology & behavior. HHS Public Acc. 139–48 (2017)
9. Bocharov, G., Quiel, J., Luzyanina, T., Alon, H., Chiglintsev, E., Chereshnev, V., Meier-Schellersheim, M., Paul, W.E., Grossman, Z.: Feedback regulation of proliferation vs. differentiation rates explains the dependence of CD4 T-cell expansion on precursor number. Proc. Natl. Acad. Sci. **108**(8), 3318–3323 (2011)
10. Armitage, P., Berry, G., Matthews, J.N.S.: Statistical Methods in Medical Research. John Wiley & Sons, Hoboken (2008)
11. Kareva, I., Luddy, K.A., O'Farrelly, C., Gatenby, R.A., Brown, J.S.: Predator-prey in tumor-immune interactions: a wrong model or just an incomplete one? Front Immunol **12**, 668221 (2021)
12. Pitt, M.A., Myung, I.J.: When a good fit can be bad. Trends Cogn. Sci. **6**(10), 421–425 (2002)
13. Flossdorf, M., Höfer, T.: Modeling and inference of cell population dynamics. In: Systems Immunology, pp. 251–264. CRC Press, Boca Raton (2018)
14. Baker, C.T., Bocharov, G., Ford, J.M., Lumb, P.M., Norton, S.J., Paul, C., Junt, T., Krebs, P., Ludewig, B.: Computational approaches to parameter estimation and model selection in Immunology. J. Comput. Appl. Math. **184**(1), 50–76 (2005)
15. Thomas-Vaslin, V., Altes, H.K., de Boer, R.J., Klatzmann, D.: Comprehensive assessment and mathematical modeling of T cell population dynamics and homeostasis. J. Immunol. **180**(4), 2240–2250 (2008)
16. Berger, C., Sommermeyer, D., Hudecek, M., Berger, M., Balakrishnan, A., Paszkiewicz, P.J., Kosasih, P.L., Rader, C, Riddell, S.R.: Safety of targeting ROR1 in primates with chimeric antigen receptor-modified T cells. Cancer Immunol. Res. **3**(2), 206–216 (2015)
17. Althaus, C.L., Ganusov, V.V., De Boer, R.J.: Dynamics of CD8+ T cell responses during acute and chronic lymphocytic choriomeningitis virus infection. J. Immunol. **179**(5), 2944–2951 (2007)
18. Myers, C.R.: Zen and the art of parameter estimation in systems biology. In: Systems Immunology, pp. 123–138. CRC Press, Boca Raton (2018)
19. Lythe, G., Molina-París, C.: Some deterministic and stochastic mathematical models of naïve T-cell homeostasis. Immunol. Rev. **285**(1), 206–217 (2018)
20. Asquith, B., Bangham, C.R.: The dynamics of T-cell fratricide: application of a robust approach to mathematical modelling in Immunology. J. Theor. Biol. **222**(1), 53–69 (2003)
21. De Boer, R.J., Perelson, A.S.: Quantifying T lymphocyte turnover. J. Theor. Biol. **327**, 45–87 (2013)
22. Gadhamsetty, S., Beltman, J.B., De Boer, R.J.: What do mathematical models tell us about killing rates during HIV-1 infection? Immunol. Lett. **168**(1), 1–6 (2015)
23. Carey, L.A.: Ccr 20th anniversary commentary: Simpson's paradox and neoadjuvant trials. Clin. Cancer Res. **21**(18), 4027–4029 (2015)
24. Ginzburg, L., Jensen, C.: Rules of thumb for judging ecological theories. Trends Ecol. Evol. **19**, 121–126 (2004)
25. Pahari, S., Sun, L., Alexov, E.: PKAD: a database of experimentally measured pka values of ionizable groups in proteins. Database **2019** (2019)
26. Murphy, K., Weaver, C.: Janeway's Immunobiology. Garland Science. New York (2016)
27. Haring, J.S., Badovinac, V.P., Harty, J.T.: Inflaming the CD8+ T cell response. Immunity **25**(1), 19–29 (2006)
28. Badovinac, V.P., Porter, B.B., Harty, J.T.: Programmed contraction of CD8+ T cells after infection. Nat. Immunol. **3**(7), 619–626 (2002)
29. Arias, C.F., Herrero, M.A., Acosta, F.J., Fernandez-Arias, C.: A mathematical model for a T cell fate decision algorithm during immune response. J. Theor. Biol. **349**, 109–120 (2014)

30. Lewkiewicz, S., Chuang, Y.-L., Chou, T.: A mathematical model of the effects of aging on naive T cell populations and diversity. Bull. Math. Biol. **81**, 2783–2817 (2019)
31. Wang, H., Peskin, C.S., Elston, T.C.: A robust numerical algorithm for studying biomolecular transport processes. J. Theor. Biol. **221**(4), 491–511 (2003)
32. Nowak, M.A., Bangham, C.R.: Population dynamics of immune responses to persistent viruses. Science **272**(5258), 74–79 (1996)
33. De Boer, R.J., Perelson, A.S.: Towards a general function describing T cell proliferation. J. Theor. Biol. **175**(4), 567–576 (1995)
34. Schleicher, J., Conrad, T., Gustafsson, M., Cedersund, G., Guthke, R., Linde, J.: Facing the challenges of multiscale modelling of bacterial and fungal pathogen–host interactions. Brief. Funct. Genom. **16**(2), 57–69 (2017)
35. Colyvan, M., Ginzburg, L.: Analogical thinking in Ecology: looking beyond disciplinary boundaries. Q. Rev. Biol. **85**(2), 171–182 (2010)
36. Bocharov, G.A., Grebennikov, D.S., Savinkov, R.S.: Mathematical Immunology: from phenomenological to multiphysics modelling. Russ. J. Numer. Anal. Math. Model. **35**(4), 203–213 (2020)
37. Rackauckas, C., Innes, M., Ma, Y., Bettencourt, J., White, L., Dixit, V.: Diffeqflux.jl - A Julia library for neural differential equations (2019). CoRR abs/1902.02376
38. Gerritsen, B., Pandit, A.: The memory of a killer T cell: models of CD8+ T cell differentiation. Immunol. Cell Biol. **94**(3), 236–241 (2016)
39. McLean, A.R.: Modelling T cell memory. J. Theor. Biol. **170**(1), 63–74 (1994)
40. Lorenzi, T., Chisholm, R.H., Melensi, M., Lorz, A., Delitala, M.: Mathematical model reveals how regulating the three phases of T-cell response could counteract immune evasion. Immunology **146**(2), 271–280 (2015)
41. Reynolds, J., Amado, I.F., Freitas, A.A., Lythe, G., Molina-París, C.: A mathematical perspective on CD4+ T cell quorum-sensing. J. Theor. Biol. **347**, 160–175 (2014)
42. León, K., Lage, A., Carneiro, J.: Tolerance and immunity in a mathematical model of T-cell mediated suppression. J. Theor. Biol. **225**(1), 107–126 (2003)
43. Arias, C.F., Herrero, M.A., Cuesta, J.A., Acosta, F.J., Fernández-Arias, C.: The growth threshold conjecture: a theoretical framework for understanding T-cell tolerance. R. Soc. Open Sci. **2**(7), 150016 (2015)
44. Vossen, M.T., Westerhout, E.M., Söderberg-Nauclér, C., Wiertz, E.J.: Viral immune evasion: a masterpiece of evolution. Immunogenetics **54**, 527–542 (2002)
45. Fernandez-Arias, C., Arias, C.F., Zhang, M., Herrero, M.A., Acosta, F.J., Tsuji, M.: Modeling the effect of boost timing in murine irradiated sporozoite prime-boost vaccines. PLoS ONE **13**(1), e0190940 (2018)
46. Arias, C.F., Herrero, M.A., Bertocchini, F., Acosta, F.J., Fernandez-Arias, C.: Modeling the dependence of immunodominance on T cell dynamics in prime-boost vaccines. Mathematics **9**(1), 28 (2020)

A Scenario-Based Study on Hybrid PDE-ODE Model for Cancer-on-Chip Experiment

Gabriella Bretti, Elio Campanile, Marta Menci, and Roberto Natalini

Abstract Recently, a growing interest in reproducing biological phenomena by in silico models has been registered. In this framework, the present work is inspired by new advancements in Organs-on-chip technology and, in particular, in Cancer-on-chip experiment, where tumor cells are treated with chemotherapy drugs and secrete chemical signals in the environment thus stimulating immune response.

Here we present a scenario-based study realized by a simulation algorithm based on a hybrid model consisting of a coupled PDE–ODE system describing the concentration of chemicals as well as immune cells migration in the Cancer-on-chip environment. The main goal of this work is to investigate numerically the qualitative effects on the overall dynamics of immune cells given by coefficients regulating cell alignment, growth rate and inflow of chemicals. To this aim, some simulated scenarios obtained by varying the values of corresponding model parameters are presented.

G. Bretti · R. Natalini (✉)
Istituto per le Applicazioni del Calcolo "Mauro Picone", Consiglio Nazionale delle Ricerche, Rome, Italy
e-mail: gabriella.bretti@cnr.it; roberto.natalini@cnr.it

E. Campanile
Fondazione The Microsoft Research, University of Trento Centre for Computational and Systems Biology (COSBI), Rovereto, Italy

Department of Mathematics, University of Trento, Trento, Italy
e-mail: elio.campanile@unitn.it

M. Menci
Università Campus Bio-Medico di Roma, Rome, Italy
e-mail: m.menci@unicampus.it

A. d'Onofrio et al. (eds.), *Problems in Mathematical Biophysics*, SEMA SIMAI Springer Series 38, https://doi.org/10.1007/978-3-031-60773-8_3

1 Introduction

In the last decade new in-vitro models based on Organs-on-chip (OOC) technology have been developed, consisting of microfluidic devices engineered to reproduce mechanics and surrounding physiochemical environment of entire living organisms with modular degree of complexity, see [15, 20, 24, 35] and references therein.

In this framework, we explore the possibility to bridge the biological in-vitro models and the mathematical-based in-silico models in order to gain further insights into the observed phenomena. Indeed, using data extracted from the OOC biological experiments, in-silico models can be developed and then be applied as predictive models of the observed phenomena. Such models, suitably calibrated against biological data, can be employed both for the optimal design of laboratory experiments and for the drug-testing procedures to see the effects of therapeutic treatments. Due to the non-trivial dynamics and evolution of the chemical field across time and space in the microfluidic environment, the distribution of chemical gradients is not easily measurable experimentally, thus it represents an unknown of this phenomenon.

Here we deal with Cancer-on-chip (COC) microfluidic co-culture systems, designed to better understand and interpret the interactions between tumor and immune systems, in the case of cancer cells exposed to a chemotherapy drug. In this experiment, immune cells (ICs) migration and interaction with cancer cells are activated by the chemical gradient secreted by Tumor Cells (TCs). This work focuses on numerical investigations of the qualitative effects produced on the overall dynamics of ICs by coefficients regulating cell alignment, growth rate and inflow of chemical substances. To this aim we show some scenarios obtained by the simulation algorithm by changing the values of corresponding model parameters.

Collective migration has been widely studied and described by mathematical models at different scales: macroscopic, with partial differential equations (PDEs) describing cell density or chemical gradient variations, see [16, 19]; microscopic, with agent-based or ordinary differential equations (ODEs) reproducing the movement of single cells. The evolution over time of chemical gradients in the environment can also be formulated with ODEs or PDEs, or by cellular automata incorporating local reaction-diffusion dynamics. A plethora of mathematical models for biological phenomena can be found in [1, 8, 18, 21, 23, 30, 34]. In this context, hybrid multiscale models are obtained combining the *macro* approach expressed by PDEs with the *micro* approach expressed by ODEs [2, 29]. See [28] for the literature reviews on hybrid models.

In recent publications of some of the authors, macroscopic modeling of COC experiment and related estimation techniques for model parameters have been proposed, see [5, 6]. A rather general model for the chip environment dynamics based on cellular automata approach was proposed in [7]. A discrete-in-continuous hybrid approach has also been formulated as a PDE reaction-diffusion partial model for the evolution of the chemicals, coupled with a ODE particle model for cell motion [4]. Analytical investigations concerning well-posedness and asymptotic behavior of solutions for this class of hybrid coupled system have been proposed in [14, 25]. Recently, the mean field limit of a general class of deterministic hybrid macro-micro models towards a non-local Vlasov equation and the formal derivation of a new non-local hydrodynamic macroscale model was proposed in [27].

1.1 Original Contribution of the Present Paper

Here, starting from the discrete-in-continuous model presented in [4], we enrich it with further features and we propose a scenario-based study showing how model parameters may affect both ICs dynamics and interactions with TCs.

As a whole, the model presented here is more general and is able to reproduce the observed biological phenomena with a higher level of realism with respect to the previous one. In greater detail, concerning cell interactions and organization, the model now takes into account an alignment mechanism among cells with an additional term in the equation of ICs movement. Moreover, differently from our previous work [4], in order to be more adherent to the experimentally observed dynamics, we take into account a slight Brownian motion for TCs.

It is worth noting that, in the proposed model, the chemotaxis equation describing the production of chemical signals is only influenced by the change in TCs positions but not by ICs dynamics. Then, the coupling between the chemical (continuum) scale and ICs (discrete) scale is only mono-directional, i.e. ICs dynamics is regulated by the production of thechemoattractant but not vice versa. In the future, considering a more detailed description of the different cell mechanisms, such as chemoattractant consumption caused by ICs and, most importantly, TCs death, eventually induced by ICs as well as caused by the effects of chemotherapy drugs, the bidirectional coupling between discrete and continuous scales will be accomplished.

1.2 Laboratory Experiment

For the development of our modeling we refer to in vitro microfluidic experiments described in [12, 32], where all the details about the laboratory settings can be found. The immune-oncology chip designed for the experiment consists of a complex geometry composed by cylindrical wells containing, separately, cell cultures of TCs and ICs, and allowing circulation of cells and chemical signals in all the chip interconnected areas. A picture of the whole microfluidic chip is reported in Fig. 1. In particular, wells are in contact with 3D rectangular compartments connected to each other. The monitored area of the microfluidic chip consists of two main culture chambers (see Panel (b) of Fig. 1), the one on the left used for plating adherent TCs and the one on the right hosting floating ICs, connected by a bridge of microchannels allowing chemical and physical contacts.

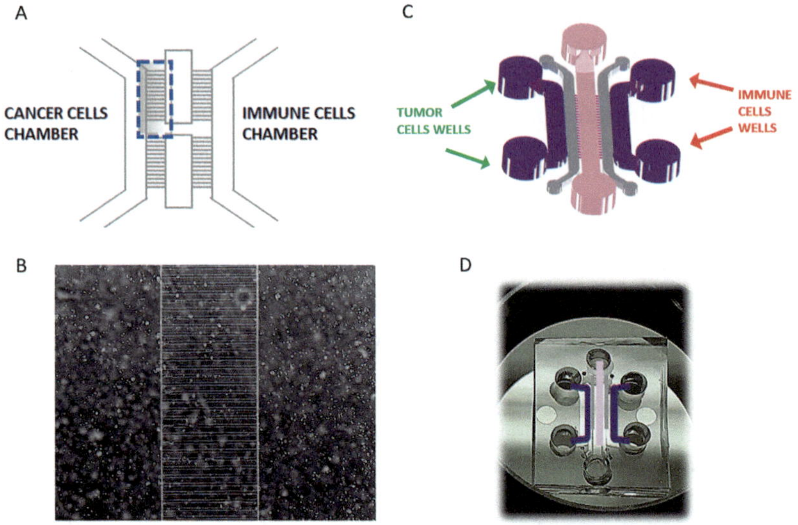

Fig. 1 Microfluidic chip environment. Panel (**a**): schematics of the 2D layout with three main culture chambers connected by arrays of microchannels, with the monitored area indicated by blue dotted rectangle. Panel (**b**): Chip area detected in video recordings. Panel (**c**): schematics of the 3D configuration of the whole chip. Panel (**d**): actual photograph of the microfluidic device. Credits: Re-editing of pictures in [4, 9] and [32]

Since in the video footage the experiment is recorded at a fixed height, the third spatial dimension in our framework is neglected. Moreover, TCs are cultured in adhesion and thus we are allowed to consider 2D models. ICs, biased by the detection of chemical signal produced by TCs, start migrating and, after crossing microchannels, enter in contact with dying cancer cells, and establish interactions leading to TCs killing events. The culture medium in the microfluidic environment is neutral, i.e. no exogenous substance is introduced. Timelapse microscope images are taken every 2 min over a period of 72 h of migration.

2 The Model

We consider the hybrid discrete-in-continuous model presented in [4], here enhanced with more features. This model gathers the advantages of both microscopic and macroscopic descriptions: cells are treated as discrete entities, whereas the concentration of the chemicals is modelled as a continuum. From the mathematical point of view, the dynamics of each cell is described by means of second order ordinary differential equations, and the evolution in time of the chemical concentration through reaction-diffusion partial differential equations. In particular, we consider a parabolic diffusion equation with terms of source and degradation. At cellular level, the microscopic approach allows to describe in fine detail the mechanical interactions influencing ICs dynamics, e.g. attraction and repulsion forces, and pattern formation resulting from alignment effects. The coupling between microscopic and macroscopic scale is realized through the gradient of chemical concentration: ICs move towards higher concentration of the chemical signal, which is released by TCs. The hybrid model reads as:

$$\dot{\mathbf{Y}}_j = \sigma_W \boldsymbol{\psi} \tag{1}$$

$$\partial_t \varphi = D\Delta\varphi + \xi F_4(\mathbf{Y}) - \eta\varphi \tag{2}$$

$$\ddot{\mathbf{X}}_i = \gamma F_1(\chi(\varphi)\nabla\varphi) + F_2(\mathbf{X}, \mathbf{Y}) + F_3(\mathbf{X}) + \beta F_5(\dot{\mathbf{X}}, \mathbf{X}) - \mu\dot{\mathbf{X}}_i \tag{3}$$

where $D, \xi, \eta, \gamma, \beta, \mu$ are positive constants and $F_n(\cdot), n = 1, 2, 3, 4, 5$ are suitable functions as reported hereafter. The position of the $i-$th immune cell at time t, denoted by $\mathbf{X}_i(t)$, $i = 1, \ldots, N_I$, is computed in (3), while $\mathbf{Y}_j(t)$ denoting the position of the $j-$th cancer cell at time t is computed in (1), for any $j = 1, \ldots, N_T$. At each time, the positions of the ICs are stored in $\mathbf{X} = [\mathbf{X}_1, \ldots, \mathbf{X}_{N_I}]$ and the ones of TCs in $\mathbf{Y} = [\mathbf{Y}_1, \ldots, \mathbf{Y}_{N_T}]$. The evolution in time of TCs positions is ruled by a Gaussian white noise $\boldsymbol{\psi}$ with uncorrelated components, and σ_W denotes the standard deviation of TCs positions. The presence of noise determines small changes in TCs coordinates over time, since they oscillate around their initial position. Moreover, this implies that the source term in the Eq. (2) for chemical signal φ, changes according to the positions assumed by TCs across time.

Initial conditions are given by the position and velocity of immune cells at the initial time $t = 0$, together with the initial concentration of the chemoattractant φ:

$$\mathbf{X}(0) = \mathbf{X}_0\,, \ \dot{\mathbf{X}}(0) = \mathbf{V}_0\,, \ \mathbf{Y}(0) = \mathbf{Y}_0\,, \ \varphi(\boldsymbol{x}, 0) = F_4(\mathbf{Y}(0)). \tag{4}$$

The system is solved on a domain $\Omega = [x_L, x_R] \times [y_B, y_T]$, which represents a subarea of the left chip compartment, see Figs. 1 and 2. This is motivated by the following reasons:

- to allow a detailed analysis of short range interactions and pattern formation in the chip environment;
- to reduce the computational cost of numerical simulations.

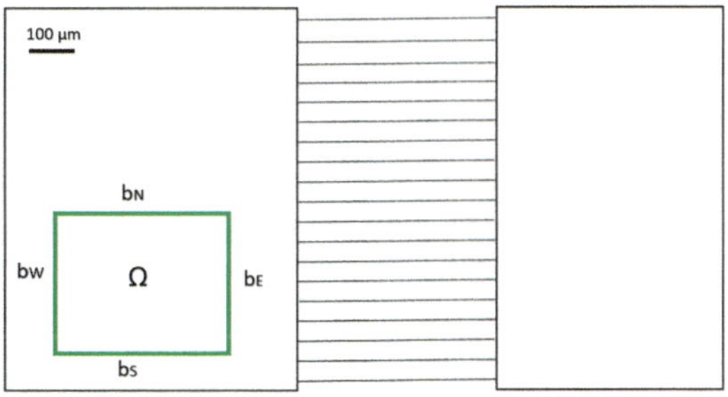

Fig. 2 Schematic representation of the experimentally monitored area shown in Fig. 1. The computational square domain Ω is depicted in green

We require the non-homogeneous Robin boundary condition:

$$D\frac{\partial \varphi}{\partial \mathbf{n}} + a\varphi = b, \text{ on } \partial\Omega, \tag{5}$$

where $a = (a_S, a_E, a_N, a_W)$ is the rate of exchange of chemoattractant with the external environment and $b = (b_S, b_E, b_N, b_W)$ is the parameter regulating the inflow/outflow of chemoattractant. Note that labels $S(South)$, $E(East)$, $N(North)$, $W(West)$ denote the four sides of domain Ω according to counterclockwise orientation: (a_S, b_S) are assumed on $y = y_B$, (a_E, b_E) on $x = x_R$, (a_N, b_N) on $y = y_T$ and (a_W, b_W) on $x = x_L$. Note that in the following, we will keep $a_S = a_E = a_N = a_W$ fixed, as discussed in Remark 1.

Function F_1 it is the chemotactic term related to the detection of a chemical signal by $i-$th cell in its neighborhood and it is computed as a weighted average over a ball of radius R_0 and centered in $\mathbf{X}_i$:

$$F_1(\chi(\varphi)\nabla\varphi) = \frac{1}{\mathcal{W}}\int_{B(\mathbf{X}_i,R_0)} \chi(\varphi)\nabla\varphi(\boldsymbol{x},t)w_i(\boldsymbol{x})d\mathbf{x}, \tag{6}$$

where $B(\mathbf{X}_i, R_0) := \{\boldsymbol{x} : \|\boldsymbol{x} - \mathbf{X}_i\| \leq R_0\}$, $\|\cdot\|$ being the Euclidean norm,

$$w_i(\boldsymbol{x}) = \begin{cases} 2exp\Big(-(\log 2)\|\boldsymbol{x} - \mathbf{X}_i\|^2/R_0^2\Big), & \text{if } \|\boldsymbol{x} - \mathbf{X}_i\| \leq R_0 \\ 0, & \text{otherwise} \end{cases} \tag{7}$$

is a truncated Gaussian weight function, and, independently of i:

$$\mathcal{W} := \int_{B(\mathbf{X}_i,R_0)} w_i(\boldsymbol{x})d\mathbf{x}. \tag{8}$$

The function $\chi(\varphi)$, representing the chemotactic sensitivity of ICs, is here assumed as in [22], i.e. $\chi(\varphi) = k_1/(k_2 + \varphi)^2$, where k_1 represents the cellular drift velocity, while k_2 is the receptor dissociation constant, indicating how many molecules are necessary to bind the receptors. Also known as receptor saturation function, the function mimics the effect of reducing cell migration in areas of high chemoattractant concentrations. In the following, R_0 is assumed to be larger than the IC radius R_I and a weighted average is introduced, in order to model the fact that the chemical signal is sensed more in the center of the cell and less at the edge of the cell extensions.

Function F_2 it includes a repulsion effect among TCs and ICs. In particular, repulsion occurs at a distance between the centers of two cells less than R_1, where R_1 is defined as the sum of the two radii of the immune and tumor cells, R_I and R_T respectively. With this formulation, repulsion occurs when an IC and a TC start being effectively overlapped. We assume

$$F_2(\mathbf{X}, \mathbf{Y}) = \sum_{j:\mathbf{Y}_j \in B(\mathbf{X}_i, R_1)\setminus\{\mathbf{X}_i\}} K(\mathbf{Y}_j - \mathbf{X}_i), \tag{9}$$

where the function K depends on the relative positions $\mathbf{Y}_j - \mathbf{X}_i$, namely:

$$K(\mathbf{Y}_j - \mathbf{X}_i) = -\omega_{repTI}\left(\frac{1}{\|\mathbf{Y}_j - \mathbf{X}_i\|} - \frac{1}{R_1}\right)\frac{\mathbf{Y}_j - \mathbf{X}_i}{\|\mathbf{Y}_j - \mathbf{X}_i\|}, \quad \text{if } \|\mathbf{Y}_j - \mathbf{X}_i\| \leq R_1, \tag{10}$$

with $\omega_{repTI} > 0$ is the parameter tuning the effect.

Function F_3 it includes adhesion-repulsion effects between ICs. In particular repulsion occurs if the distance between the centers of two ICs is less than R_3 and takes into account the effects of a possible cell deformation. Conversely, adhesion occurs at a distance greater than R_3 and less than $R_4 > R_3$, and it is due to a mechanical interaction between cells via filopodia. We assume:

$$F_3(\mathbf{X}) = \sum_{j:\mathbf{X}_j \in B(\mathbf{X}_i, R_4)\setminus\{\mathbf{X}_i\}} K(\mathbf{X}_j - \mathbf{X}_i), \tag{11}$$

where the function K depends on the relative positions $\mathbf{X}_j - \mathbf{X}_i$, namely

$$K(\mathbf{X}_j - \mathbf{X}_i) = \begin{cases} -\omega_{rep}\left(\dfrac{1}{\|\mathbf{X}_j - \mathbf{X}_i\|} - \dfrac{1}{R_3}\right)\dfrac{\mathbf{X}_j - \mathbf{X}_i}{\|\mathbf{X}_j - \mathbf{X}_i\|}, & \text{if } \|\mathbf{X}_j - \mathbf{X}_i\| \leq R_3 \\ \omega_{adh}\left(\|\mathbf{X}_j - \mathbf{X}_i\| - R_3\right)\dfrac{\mathbf{X}_j - \mathbf{X}_i}{\|\mathbf{X}_j - \mathbf{X}_i\|}, & \text{if } R_3 < \|\mathbf{X}_j - \mathbf{X}_i\| \leq R_4, \end{cases}$$

where $\omega_{rep}, \omega_{adh}$ are constants and $R_3 = 2R_I$ are chosen so that the repulsion occurs when two cells effectively overlap. Note that the first case gives a repulsion which goes as $1/r$, with r is the distance among the centers of two cells, while the second case is the Hooke's law of elasticity.

Friction The term $\mu\dot{\mathbf{X}}_i$ in the equation of the motion is due to cell adhesion to the substrate, with damping coefficient equal μ.

Function F_4 it models the production of chemoattractant by TCs and it is written as:

$$F_4(\mathbf{Y}) = \sum_{j=1}^{N_T} \chi_{B(\mathbf{Y}_j, R_T)}, \quad \chi_{B(\mathbf{Y}_j, R_T)}(\mathbf{x}) = \begin{cases} 1, & \text{if } \mathbf{x} \in B(\mathbf{Y}_j, R_T) \\ 0, & \text{otherwise,} \end{cases} \tag{12}$$

where N_T is the total number of TCs. In the previous formula R_T is the radius of a cancer cell, considering that the source of chemoattractant is defined by the dimension of a single cell.

Function F_5 it represents cell alignment. We assume a Cucker and Smale-like flocking term [10]:

$$F_5(\mathbf{X}, \dot{\mathbf{X}}) = \frac{1}{N_i} \sum_{j:\mathbf{X}_j \in B(\mathbf{X}_i, R_3)\setminus\{\mathbf{X}_i\}} \frac{1}{\left(1 + \frac{\|\mathbf{X}_j - \mathbf{X}_i\|^2}{R_3^2}\right)^{\sigma}} (\dot{\mathbf{X}}_j - \dot{\mathbf{X}}_i), \tag{13}$$

where $R_3 = 2R_I$ has been chosen as a suitable radius of influence based on [13], $\sigma > 0$ and $N_i := \text{card}\{j \neq i : \mathbf{X}_j \in B(\mathbf{X}_i, R_3)\}$ is the number of ICs inside the ball.

The system of Eqs. (1)–(2)–(3) can now be explicitly rewritten as:

$$\begin{cases} \dot{\mathbf{Y}}_j = \sigma_W \boldsymbol{\psi} \\ \partial_t \varphi = D\Delta\varphi + \xi \sum_{j=1}^{N_T} \chi_{B(\mathbf{Y}_j, R_T)} - \eta\varphi \\ \ddot{\mathbf{X}}_i = \frac{\gamma}{\mathcal{W}} \int_{B(\mathbf{X}_i, R_0)} \chi(\varphi(\mathbf{x}, t)) \nabla\varphi(\mathbf{x}, t) w_i(\mathbf{x}) d\mathbf{x} + \\ + \sum_{j:\mathbf{Y}_j \in B(\mathbf{X}_i, R_1)\setminus\{\mathbf{X}_i\}} K(\mathbf{Y}_j - \mathbf{X}_i) + \\ + \sum_{j:\mathbf{X}_j \in B(\mathbf{X}_i, R_4)\setminus\{\mathbf{X}_i\}} K(\mathbf{X}_j - \mathbf{X}_i) + \\ + \frac{\beta}{N_i} \sum_{j:\mathbf{X}_j \in B(\mathbf{X}_i, R_3)\setminus\{\mathbf{X}_i\}} \frac{1}{\left(1 + \frac{\|\mathbf{X}_j - \mathbf{X}_i\|^2}{R_3^2}\right)^{\sigma}} (\dot{\mathbf{X}}_j - \dot{\mathbf{X}}_i) - \mu \dot{\mathbf{X}}_i. \end{cases} \tag{14}$$

3 Numerical Approximation

In this Section we present numerical simulations based on finite difference schemes employed for Eqs. (2) coupled with (3), with the aim of showing the effect of variation of some key parameters of the model. We consider a square computational domain $\Omega = [x_L, x_R] \times [y_B, y_T]$ of monitored area with $x_L = y_B = 100\,\mu\text{m}$ and $x_R = y_T = 500\,\mu\text{m}$. Then the considered domain has horizontal and vertical size $L_x = L_y = 400\,\mu\text{m}$ and is depicted in Fig. 2 as a green square located at the bottom of left chamber. We introduce a discretization on L_x in $N_x - 1$ subintervals of length $\Delta x = \dfrac{L_x}{N_x - 1}$ and a discretization on L_y in $M_y - 1$ subintervals of length

$\Delta y = \dfrac{L_y}{M_y - 1}$. In all of performed tests, the time and spatial steps are chosen as $\Delta t = 10s$, i.e. representing 1/6 of the video footage timeframe (of 1 min) and $\Delta x = \Delta y = 2\,\mu$m, i.e. the half of ICs radius.

Then we introduce a Cartesian grid Ω_Δ consisting of grid points (x_n, y_m), where $x_n = x_L + n\Delta x$, for $n = 0, ..., N_x - 1$ and $y_m = y_B + m\Delta y$, for $m = 0, \ldots, M_y - 1$. The same can be done for the time interval $[0, T]$, in this case if Δt is the time step, t_k will be the k-th temporal step, i.e. $t_k = k\Delta t$, for $k = 0, \ldots, N_{\Delta t}$.

The simulations were performed over a number of frames equal to 8640, corresponding to a final time $T = 86{,}400$s (24 h) of observations.

Besides, since experimentally it was observed that ICs leave the domain Ω, in order to manage the entrance and exit of cells and avoid numerical instabilities, we added a ghost grid to Ω_Δ, in the following defined as Ω_{Δ^*}, where the cells lie after having left the main domain. In order to extend the grid, we discretize the domain $[x_L - L_{x^*}, x_R + L_{x^*}] \times [y_B - L_{y^*}, y_T + L_{y^*}]$ with the same discretization step values Δx, Δy already introduced for Ω_Δ. In our tests we chose $L_{x^*} = L_{y^*} = 100\,\mu$m. The equations were solved on Ω_{Δ^*}.

3.1 *Discretization of the PDE (Eq. (2))*

The parabolic Eq. (2) is composed of the diffusion term, the source term, and the stiff degradation term $-\eta\varphi$.

In order to eliminate this last quantity we perform the classical exponential transformation: $\varphi(\mathbf{x}, t) = e^{-\eta t} u(\mathbf{x}, t)$, which leads to the diffusion equation with source for $u(\mathbf{x}, t)$:

$$\partial_t u = D\Delta u + e^{\eta t} \xi \sum_{j=1}^{N_T} \chi_{\mathbf{B}(\mathbf{Y}_j, R_T)}. \tag{15}$$

For this equation we apply a central difference scheme in space, i.e. the 5-point stencil for the Laplacian, and the parabolic Crank–Nicolson scheme in time.

Denoting with $u^k_{n,m}$ the approximation of u at the grid point (x_n, y_m), for any time instant t_k the numerical scheme can be written as:

$$\frac{u^{k+1}_{n,m} - u^k_{n,m}}{\Delta t} = \frac{D}{2}\left(\partial^2_x u^{k+1} + \partial^2_y u^{k+1}\right) + \frac{D}{2}\left(\partial^2_x u^k + \partial^2_y u^k\right)$$
$$+ \frac{1}{2} e^{\eta(k+1)\Delta t} \xi \sum_{j=1}^{N_T} \chi_{\mathbf{B}(\mathbf{Y}^{k+1}_j, R_T)} + \frac{1}{2} e^{\eta k \Delta t} \xi \sum_{j=1}^{N_T} \chi_{\mathbf{B}(\mathbf{Y}^k_j, R_T)},$$

where $\partial_x^2 u^k$ (and analogously $\partial_y^2 u^k$) is defined as the central difference:

$$\partial_x^2 u^k = \frac{u^k_{n-1,m} - 2u^k_{n,m} + u^k_{n+1,m}}{\Delta x^2}.$$

3.2 Boundary Conditions

Using the exponential transformation above, the associated boundary conditions rewrite as $D\frac{\partial u}{\partial \mathbf{n}} + au = e^{\eta t} b$, that we rewrite as $\frac{\partial u}{\partial \mathbf{n}} + pu = q(t)$, with $p = \frac{a}{D}$ and $q(t) = e^{\eta t}\frac{b}{D}$ eventually different on each side of Ω. For the discretization of the boundary conditions we use a central finite difference scheme. Following [31], on $y = y_B$ and $y = y_T$, we have:

$$\partial_y u^k + p_l u^k - q^k{}_l = \frac{u^k_{n,m+1} - u^k_{n,m-1}}{2\Delta y} + r_l u^k{}_{n,m} - h_l, \tag{16}$$

with $l = S, N$, and we proceed analogously for the vertical conditions, with the signs of r_l and h_l depending on the incoming/outgoing flow.

3.3 Discretization of the ODE

As in [4], the equation of motion (3) is reduced to a first order system with

$$\begin{aligned}
\dot{\mathbf{V}}_i = & \frac{\gamma}{\mathcal{W}} \int_{\mathbf{B}(\mathbf{X}_i,\mathbf{R}_0)} \chi(\varphi(\mathbf{x},t)) \nabla\varphi(\mathbf{x},t) w_i(\mathbf{x}) d\mathbf{x} + \sum_{j:\mathbf{Y}_j \in \mathbf{B}(\mathbf{X}_i,\mathbf{R}_1)\setminus\{\mathbf{X}_i\}} K(\mathbf{Y}_j - \mathbf{X}_i) \\
& + \sum_{j:\mathbf{X}_j \in \mathbf{B}(\mathbf{X}_i,\mathbf{R}_4)\setminus\{\mathbf{X}_i\}} K(\mathbf{X}_j - \mathbf{X}_i) \\
& + \frac{\beta}{N_i} \sum_{j:\mathbf{X}_j \in B(\mathbf{X}_i,R_3)\setminus\{\mathbf{X}_i\}} \frac{(\mathbf{V}_j - \mathbf{V}_i)}{\left(1 + \frac{\|\mathbf{X}_j - \mathbf{X}_i\|^2}{R_3^2}\right)^{\sigma}} - \mu \mathbf{V}_i,
\end{aligned} \tag{17}$$

$$\dot{\mathbf{X}}_i = \mathbf{V}_i, \tag{18}$$

for $i = 1, \ldots, N_I$. Equation (17) is discretized with a one step IMEX method, putting in implicit the term containing $\mathbf{V}_i$ and in explicit the other addends, and Eq. (18) is solved with forward Euler method. The two-dimensional integral in (17) can be computed by a 2D quadrature formula, which due to the truncated Gaussian weight function $w_i(\mathbf{x})$ given in (7), is reduced to a sum of the discretized integrand functions on the grid points belonging to the ball $\mathbf{B}(\mathbf{X}_i, \mathbf{R}_0)$. The two-dimensional integral in $\mathcal{W}$ defined in (6) is approximated by $\widetilde{\mathcal{W}} := \sum_{n,m \text{s.t.} (x_n, x_m) \in \mathbf{B}(\mathbf{X}_i^k, \mathbf{R}_0)} (w_i)_{n,m}^{(k)}$. The gradients of Eq. (17) are approximated with first order differences $\nabla_{n,m}\varphi^k \approx \left(\frac{\varphi_{n+1,m}^k - \varphi_{n,m}^k}{\Delta x}, \frac{\varphi_{n,m+1}^k - \varphi_{n,m}^k}{\Delta y} \right)$, then Eqs. (17)–(18) are discretized as follows:

$$
\begin{aligned}
\frac{\mathbf{V}_i^{k+1} - \mathbf{V}_i^k}{\Delta t} &= \frac{\gamma}{\widetilde{\mathcal{W}}} \sum_{n,m \text{s.t.} (x_n, x_m) \in \mathbf{B}(\mathbf{X}_i^k, \mathbf{R}_0)} \chi(\varphi^k)(\nabla_{n,m}\varphi^k)(w_i)_{n,m}^{(k)} \\
&+ \sum_{j: \mathbf{X}_j^k \in \mathbf{B}(\mathbf{X}_i^k, \mathbf{R}_1) \setminus \{\mathbf{X}_i^k\}} K(\mathbf{X}_j^k - \mathbf{X}_i^k) + \sum_{j: \mathbf{X}_j^k \in \mathbf{B}(\mathbf{Y}_i^k, \mathbf{R}_4) \setminus \{\mathbf{Y}_i^k\}} K(\mathbf{Y}_j^k - \mathbf{X}_i^k) \\
&+ \frac{\beta}{N_i} \sum_{j: \mathbf{X}_j^k \in B(\mathbf{X}_i^k, R_3) \setminus \{\mathbf{X}_i^k\}} \frac{(\mathbf{V}_j^k - \mathbf{V}_i^{k+1})}{\left(1 + \frac{\|\mathbf{X}_j^k - \mathbf{X}_i^k\|^2}{R_3^2}\right)^{\sigma}} - \mu \mathbf{V}_i^{k+1}, \\
\frac{\mathbf{X}_i^{k+1} - \mathbf{X}_i^k}{\Delta t} &= \mathbf{V}_i^{k+1}.
\end{aligned}
$$

Finally, using the differential form of Brownian motion $dW(t) = \psi(t)dt$, Eq. (1) can be written as $d\mathbf{Y}_j(t) = \sigma_W dW(t)$, where $\mathbf{Y}_j(t)$ is a one dimensional Wiener process with diffusion σ_W. This equation is discretized with the Euler-Maruyama scheme [17], the stochastic version of the deterministic Euler scheme. The increments of the Wiener process are defined as $\Delta W = W^{k+1} - W^k$, with $0 \le k \le N_{\Delta t} - 1$ and ΔW is a random variable with zero mean and variance equal to Δt, i.e. $\Delta W \sim \mathcal{N}(0, \Delta t)$. Then we have

$$
\mathbf{Y}_j^{k+1} = \mathbf{Y}_j^k + \sigma_W \sqrt{\Delta t} Z^k, \tag{19}
$$

with Z^k being standard normal variables with mean 0 and variance 1 for all k.

4 Tests

In the following, we will present a scenario-based analysis showing the effects of some model parameters, whose values cannot be obtained by laboratory experiments. First, we present the numerical results obtained with default values of model parameters used as a prototype simulation. Then, in order to study the effect on the simulated dynamics of cell alignment, growth rate and inflow of chemoattractant, we perform numerical tests with increased/decreased values of β, ξ and b, respectively. All simulations start with the same initial configuration, shown in Fig. 3: $N_T = 30$

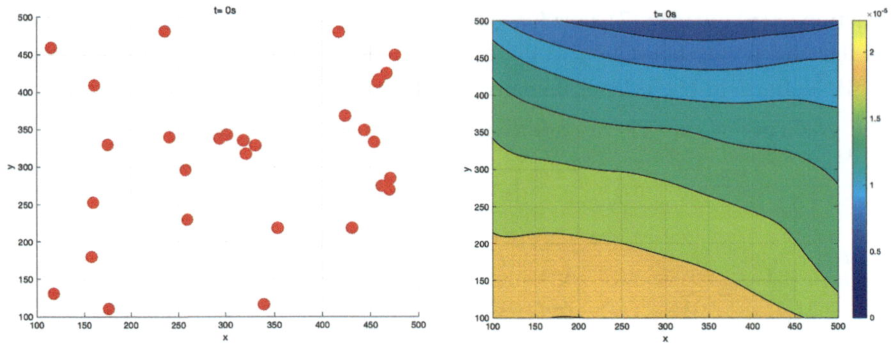

Fig. 3 Initial condition. Left: $N_T = 30$ TCs are randomly located in the square $[100, 500] \times [100, 500]$. Right: chemoattractant concentration

TCs are randomly located in the square Ω, and chemoattractant initial datum is given by the chemical concentration produced by TCs at time $t = 0$. We run the numerical code on a laptop equipped with an Intel Core i7-1060NG7 processor and 16 GB RAM and each simulation takes around 6000 s.

4.1 Test 1: Default Parameters

Figure 4 shows six snapshots of a simulation obtained with default parameters, see Table 1. The considered values are obtained from experimental settings, as well as biological assumptions and previous research works. At this stage, the majority of parameters have been chosen from a qualitative perspective, in order to reproduce the dynamics emerging in the biological experiment.

At the initial time, only TCs are present in the investigated domain. ICs randomly enter the domain from the top and right corner as observed from the video footage of laboratory experiments. After 3 h, a relevant number of ICs reaches the monitored area, moving towards the tumor. The total number of ICs travelling in the domain across time is $N_I = 100$.

Some of them are captured by groups of TCs on the right side and in the middle of the domain Ω, where higher concentrations of chemoattractant are released, as shown in the bottom panels of Fig. 4. However, the majority move downwards as in the real experiment. This is due to the sensing of the strong chemical signal produced by large amount of TCs located in wells connected to chip compartments.

It is worth noting that, in order to reproduce the experimentally observed diagonal-downward movement of ICs, the choice of parameters in Robin boundary conditions (5) is crucial to enable the model to reproduce it. In this respect, the regulation of inflow of chemoattractant from the outside towards the monitored region represents a quite delicate aspect, and it can be done by balancing the effects

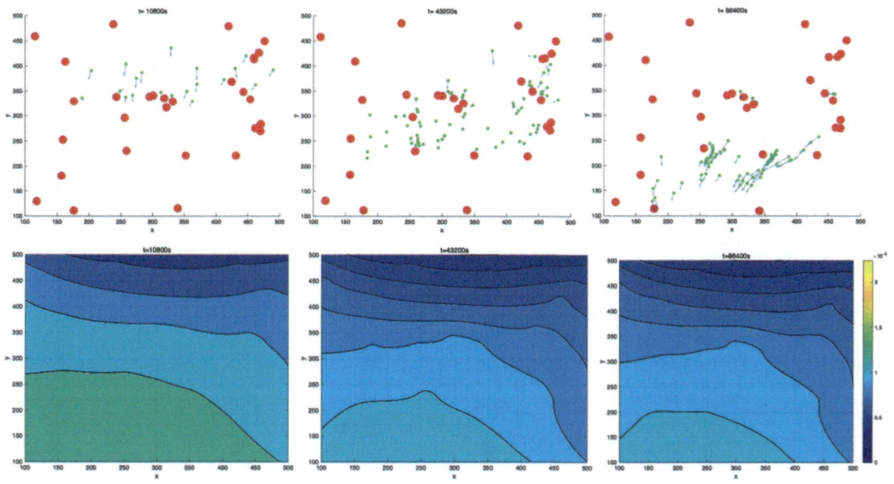

Fig. 4 Test 1: Numerical simulation of model (14) for the reference case with parameters reported in Table 1. Top Panels: dynamics of ICs plotted at T=3 h (Left), T= 12 h (Middle) and 24 h (Right). Bottom Panels: Concentration of chemoattractant plotted in 2D with contour lines at T=3 h (Left), T= 12 h (Middle) and 24 h (Right)

of parameters a and b in (5) as debated in greater details in the following, see Remark 1.

4.2 Test 2: The Effect of Alignment Parameter β

One of the main novelties introduced in the presented model relies on alignment effects among ICs. In order to investigate its role, we run numerical simulations for different values of the alignment parameter $\beta \geq 0$.

With respect to radial effect due to attraction-repulsion, the additional alignment term reflects in a tangential effect among cells. Several numerical tests (data not shown) have been performed in order to establish a threshold value $\bar{\beta}$, used in the prototype simulation: for $\beta < \bar{\beta}$, the effect of alignment is almost negligible.

Figure 5 (first column) shows different snapshots of the reference simulation performed neglecting alignment, e.g. setting $\beta = 0$. In absence of alignment, repulsion prevents overlapping phenomena, and adhesion leads to the formation of highly congested groups that are not exhibited by the experiments.

On the contrary, for higher values of β, as soon as ICs start interacting they align with the nearest ones. Indeed, as shown in Fig. 5 (second column), for $\beta = 10 \cdot \bar{\beta}$, many ICs align with the nearest in the surroundings of TCs. For greater values, e.g. $\beta = 100 \cdot \bar{\beta}$ (see Fig. 5 (third column)), cells move downwards in small aligned groups, which again is not a feature observed experimentally.

Table 1 Default values of parameters used in Test 1

Parameter	Description	Units	Value	Reference
N_I	Number of ICs in the domain	–	100	Exp. setting
N_T	Number of TCs in the domain	–	30	Exp. setting
L_x	Horizontal size of the domain	μm	400	Exp. setting
L_y	Vertical size of the domain	μm	400	Exp. setting
R_I	IC radius	μm	4	[33]
R_T	TC radius	μm	10	[3]
R_0	Detection radius of chemicals	μm	7	Biol. assumption
R_1	Radius of action of repulsion between ICs and TCs	μm	14	Biol. assumption
R_3	Radius of action of repulsion between ICs	μm	8	[13]
R_4	Radius of action of adhesion/interaction between ICs	μm	10	Biol. assumption
D	Diffusion Coeff.	$\mu m^2 s^{-1}$	900	[26]
$\bar{\xi}$	Default growth rate of φ	μm^{-2} mol s^{-1}	$1.6 \cdot 10^{-6}$	Biol. assumption
η	Consumption rate of φ	s^{-1}	10^{-4}	[11]
ω_{adh}	Coeff. of adhesion between ICs	s^{-2}	10^{-7}	Biol. assumption
ω_{rep}	Coeff. of repulsion between ICs	$\mu m^2 s^{-2}$	$5 \cdot 10^{-4}$	Biol. assumption
ω_{repTI}	Coeff. of repulsion between ICs and TCs	$\mu m^2 s^{-2}$	$8.5 \cdot 10^{-3}$	Biol. assumption
μ	Damping coeff.	s^{-1}	$2.1 \cdot 10^{-3}$	Biol. assumption
k_1	Cellular drift velocity	mol s^{-1}	$3.9 \cdot 10^{-9}$	[26]
k_2	Receptor dissociation constant	mol μm^{-2}	$5 \cdot 10^{-14}$	[26]
γ	Coeff. of chemotactic effect	μm^{-1}	$2 \cdot 10$	Biol. assumption
β	Alignment coeff.	s^{-1}	10^{-1}	[13]
σ	Exponent of alignment term		0.5	[13]
a	Rate of exchange of chemoattractant with ext. environment	μm s^{-1}	10^5	Biol. assumption
$\bar{b}$	Default flux parameters in (5)	μm^{-2} mol s^{-1}	(2.2, 1.2, 0, 1.8)	Biol. assumption
ω_W	Diffusion coefficient of Wiener process	$\mu m^2 s^{-1}$	0.01	Biol. assumption

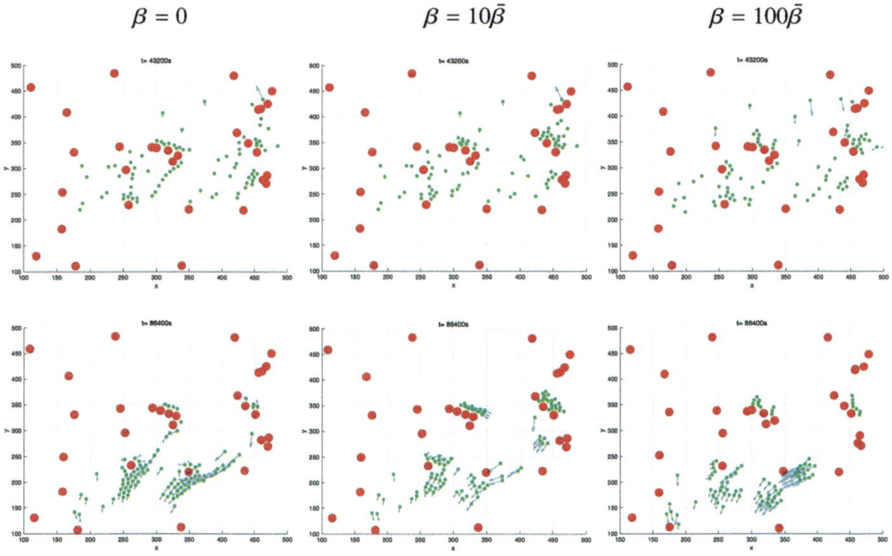

Fig. 5 Test 2: Numerical simulation of model (14) with (First column) $\beta = 0$, (Second column) $\beta = 10 \cdot \bar{\beta}$, (Third column) $\beta = 100 \cdot \bar{\beta}$

4.3 *Test 3: The Effect of Chemoattractant Growth Rate Parameter* ξ

In this test we investigate the role of parameter ξ, regulating the growth rate of chemoattractant φ, by increasing and decreasing its value of one order of magnitude with respect to the nominal one (see Table 1).

Figure 6 shows three snapshots of the reference simulation run with $\xi = 10 \cdot \bar{\xi}$. As expected, increased values of ξ result in higher chemical concentrations, and this effect is particularly evident in regions containing small groups of TCs. Consequently, the ICs dynamics becomes trivial: the effect of the chemical signal, expressed by TCs, becomes predominant with respect to the external chemical signal produced by TCs wells. As shown in Fig. 6 (first line) the downward movement of ICs is no longer simulated, and an extremely high congested scenario around TCs appears. From a mathematical point of view, high values of ξ also imply that small displacements in spatial coordinates in the domain may induce considerable variations in the values of function φ and, consequently, the gradient $\nabla\varphi$ can be very high. As a consequence of the coupled structure of the model, this leads to a considerable increase of the average speed of ICs, which rapidly reach and surround TCs.

Conversely, weakening the effect of source term in the parabolic equation, the attraction effect induced by TCs wells outside Ω plays a crucial role. Some interesting clues can be highlighted. In this case, the direction of ICs is mostly

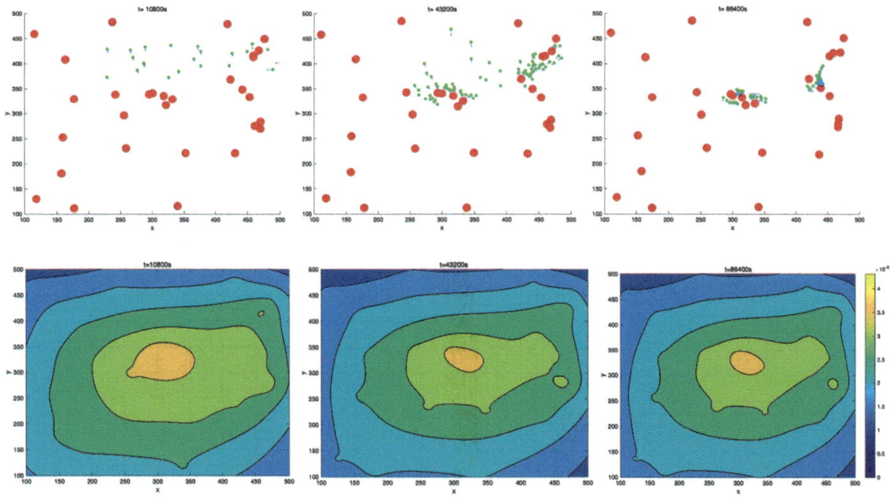

Fig. 6 Numerical simulation of model (14) with $\xi = 10 \cdot \bar{\xi}$. Top Panels: dynamics of ICs plotted at T=3 h (Left), T= 12 h (Middle) and 24 h (Right). Bottom Panels: Concentration of chemoattractant plotted in 2D with contour lines at T=3 h (Left), T= 12 h (Middle) and 24 h (Right)

influenced by the choice of flux parameter b in boundary conditions (5) rather than the chemical signal produced by TCs, and this determines the diagonally-downward directed movement observed experimentally. Moreover, the chemotactic function χ, being proportional to $1/\varphi^2$, assumes high values due to the low concentration of the chemical signal in the domain. As a consequence, ICs speed increases and they rapidly move downwards, eventually remaining in the surroundings of the TCs, as observable in Fig. 7 showing three snapshots of the simulation performed assuming $\xi = 10^{-1} \cdot \bar{\xi}$. In fact TCs are no longer strongly attractive, as exhibited by the last frame, where almost no cells are present in the domain Ω.

4.4 Test 4: The Effect of Chemoattractant Inflow Parameters $b = (b_S, b_E, b_N, b_W)$

Here we investigate numerically the effect of variations on the value of parameter b, regulating the inflow/outflow of signal through the border of Ω. More precisely, we consider the variation with respect to default values of its components in the four cardinal directions, see Fig. 2 for a schematization of the computational domain reporting the boundary parameters $b = (b_S, b_E, b_N, b_W)$.

Figure 8 shows three time instants of a numerical simulation performed with $b = 10 \cdot \bar{b}$. At a first glance, the effect seems qualitatively similar to the results of Test 3 with low ξ: the chemotactic signal produced by TCs is overwhelmed by external signals, and ICs tend to move downwards. However, two important features

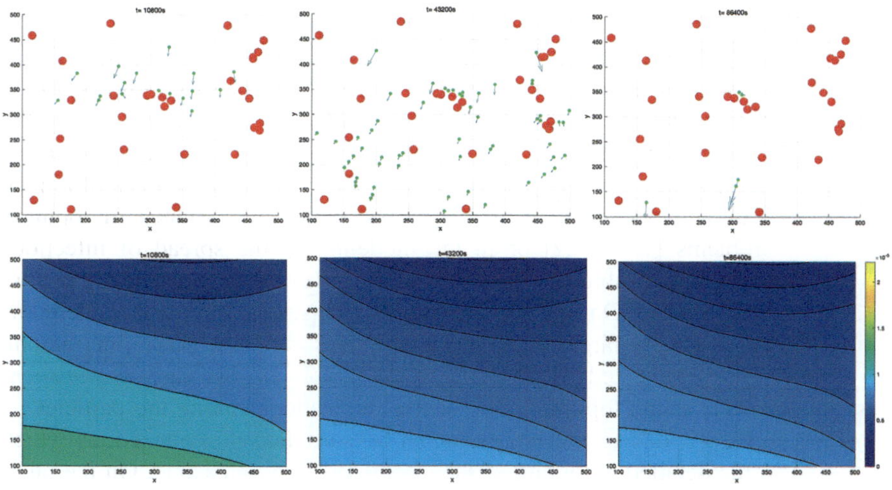

Fig. 7 Numerical simulation of model (14) with $\xi = 10^{-1} \cdot \bar{\xi}$. Top Panels: dynamics of ICs plotted at T=3 h (Left), T= 12 h (Middle) and 24 h (Right). Bottom Panels: Concentration of chemoattractant plotted in 2D with contour lines at T=3 h (Left), T= 12 h (Middle) and 24 h (Right)

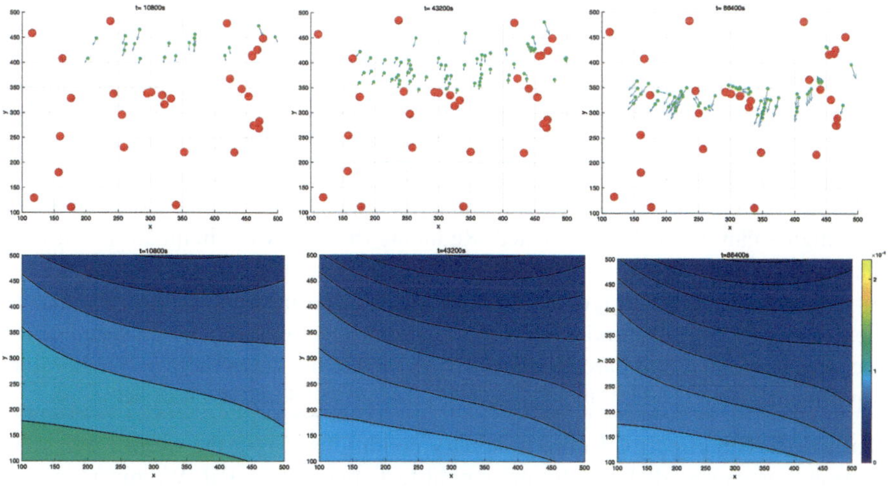

Fig. 8 Numerical simulation of model (14) with $b = 10 \cdot \bar{b}$. Top Panels: dynamics of ICs plotted at T=3 h (Left), T= 12 h (Middle) and 24 h (Right). Bottom Panels: Concentration of chemoattractant plotted in 2D with contour lines at T=3 h (Left), T= 12 h (Middle) and 24 h (Right)

can be discussed. Concerning the evolution of chemical signal in the domain, we observe that high values of b imply higher values of φ in the domain. With respect to Test 1, the values of φ are two orders of magnitude higher (from 10^{-6} to 10^{-4}). Furthermore, it is evident that ICs are moving diagonally-downwards, but their

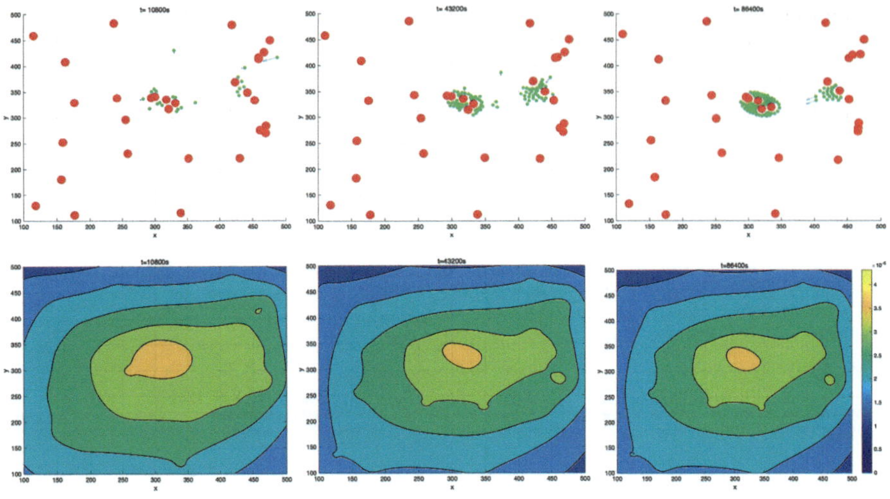

Fig. 9 Numerical simulation of model (14) with $b = 10^{-1} \cdot \bar{b}$. Top panels: dynamics of ICs plotted at T=3 h (Left), T= 12 h (Middle) and 24 h (Right). Bottom panels: Concentration of chemoattractant plotted in 2D with contour lines at T=3 h (Left), T= 12 h (Middle) and 24 h (Right)

speeds are not so high as in Fig. 7. This is in agreement with the structure of function χ, which is proportional to $1/\varphi^2$.

In order to complete our analysis, we evaluate the effect of decreased value of b on the simulated dynamics. Figure 9 shows three different snapshots of the simulation performed with $b = 10^{-1} \cdot \bar{b}$ and all the other parameters fixed as in Table 1. As in the previous test, a qualitative comparison with the results of Test 3 with high ξ can be easily established. Reducing the inflow of chemoattractant from the border, the chemical signal produced by TCs located in Ω biases ICs dynamics. As a consequence, the diagonally-downward directed movement observed experimentally is no longer reproduced by the model. Moreover, the average speed of ICs becomes too high if compared to the one observed in the video footage of laboratory experiment.

Remark 1 The effect of Robin boundary conditions (5) on the overall dynamics is strong, as can be deduced by numerical tests here presented. We must say that current boundary conditions do not properly correspond to reality. However, since from laboratory experiment we do not have any data about the space distribution and the time evolution of the chemoattractant in the chip environment, we assigned the values of parameter s a, b of Robin conditions in such a way to qualitatively reproduce the ICs dynamics observed in experimental movies. In particular, in (5) we assumed a fixed for all tests and let b vary, but, in principle, the inflow of chemoattractant can be changed letting also parameter a vary as well, taking care of the following aspects. If a increases by three orders of magnitude (i.e. $a \sim 10^8$) or more with b fixed, the effect consists in a strong decrease of inflow from outer boundaries. Then, in order to balance it and avoid having all ICs aggregated around TCs, it is necessary to let the ratio $b/a \sim 10^{-5}$. For a smaller of three order s of magnitude, i.e. $a \sim 10^2$, we started observing an uncontrolled growth of the gradient $\nabla\varphi$ and this should be contrasted by modifying parameters regulating chemotactic terms. For the mentioned reasons, for the sake of simplicity, we only acted on parameter b and this allowed us to observe the interesting features discussed above.

5 Conclusions and Perspectives

The work presented here represents a scenario-based numerical study on the effect of model parameters on the overall dynamics of COC experiments. The mathematical model is essentially the one proposed in [4], with some new elements, such as the Brownian motion of TCs and the addition of an alignment term to ICs dynamics. Starting from a reference test, we investigate from a numerical point of view, the effect of variations of some key parameters of the models.

In the future we will conduct a rigorous sensitivity analysis to assess the effects of variations in model parameters on the overall dynamics. Moreover, a quantitative calibration for fine tuning model parameters against available experimental data will be carried out. We also aim at further enriching the proposed model with new features, such as death of TCs during the time interval of interest, and this will be a relevant aspect influencing ICs travelling time. In addition, having experimental data about the space-time distribution of chemoattractant field will allow to consider conditions at the boundaries linked more closely to experiment (i.e. Dirichlet conditions) for the inflow of chemical substances at the boundaries.

Acknowledgments The authors thank Maria Patrizia Stoppelli (IGB-CNR) for the interesting discussions about biological quantities. G. Bretti, E. Campanile and M. Menci are members of GNCS-INDAM research group, R. Natalini is member of GNAMPA-INDAM research group.

References

1. Altrock, P.M., Liu, L.L., Michor, F.: The mathematics of cancer: integrating quantitative models. Nat. Rev. Cancer **15**(12), 730–745 (2015)
2. An, G., Mi, Q., Dutta-Moscato, J., Vodovotz, Y.: Agent-based models in translational systems biology. Wiley Interdiscip. Rev. Syst. Biol. Med. **1**, 159–71 (2009)
3. Boulter, E., Grall, D., Cagnol, S., Van Obberghen-Schilling, E.: Regulation of cell-matrix adhesion dynamics and Rac-1 by integrin linked kinase. FASEB J. **20**(9), 1489–1491 (2006). https://doi.org/10.1096/fj.05-4579fje
4. Bretti, G., De Ninno, A., Natalini, R., Peri, D., Roselli, N.: Estimation algorithm for a Hybrid PDE–ODE model inspired by immunocompetent cancer-on-chip experiment. Axioms **10**(4), 243 (2021)
5. Braun, E.C., Bretti, G., Natalini, R.: Mass-preserving approximation of a chemotaxis multi-domain transmission model for microfluidic chips. Mathematics **9**, 688 (2021). https://doi.org/10.3390/math9060688

6. Braun, E.C., Bretti, G., Natalini, R.: Parameter estimation techniques for a chemotaxis model inspired by Cancer-on-Chip (COC) experiments. Int. J. -Non-Linear Mech. **140**, 103895 (2022). https://doi.org/10.1016/j.ijnonlinmec.2021.103895
7. Bretti, G., De Gaetano, A.: An agent-based interpretation of leukocyte chemotaxis in cancer-on-chip experiments. Mathematics **10**, 1338 (2022). https://doi.org/10.3390/math10081338
8. Bretti, G., Natalini, R.: Numerical approximation of nonhomogeneous boundary conditions on networks for a hyperbolic system of chemotaxis modeling the physarum dynamics. J. Comput. Methods Sci. Eng. **18**, 85–115 (2018)
9. Businaro, L., De Ninno, A., Schiavoni, G., Lucarini, V., Ciasca, G., Gerardino, A., Belardelli, F., Gabriele, L., Mattei, F., Cross, F.: Talk between cancer and immune cells: exploring complex dynamics in a microfluidic environment. Lab Chip **13**, 229–239 (2013)
10. Cucker, F., Smale, S.: Emergent behavior in flocks. IEEE Trans. Autom. Control **52**, 852–862 (2007)
11. Curk, T., Marenduzzo, D., Dobnikar, J.: Chemotactic sensing towards ambient and secreted attractant drives collective behaviour of E. coli. PLoS ONE **8**, e74878 (2013). https://doi.org/10.1371/journal.pone.0074878
12. De Ninno, A., Bertani, F.R., Gerardino, A., Schiavoni, G., Musella, M., Galassi, C., Mattei, F., Sistigu, A., Businaro, L.: Microfluidic co-culture models for dissecting the immune response in in vitro tumor microenvironments. J. Vis. Exp. **170**, e61895 (2021). https://doi.org/10.1161/10.3791/61895
13. Di Costanzo, E., Natalini, R., Preziosi, L.: A hybrid mathematical model for self-organizing cell migration in the zebrafish lateral line. J. Math. Biol. **71**(1), 171–214 (2015)
14. Di Costanzo, E., Menci, M., Messina, E., Natalini, R., Vecchio, A.: A hybrid model of collective motion of discrete particles under alignment and continuum chemotaxis. Discrete Contin. Dyn. Syst., Ser. B **25**(1), 443–472 (2020)
15. Gori, M., Simonelli, M.C., Giannitelli, S.M., Businaro, L., Trombetta, M., Rainer, A.: Investigating nonalcoholic fatty liver disease in a liver-on-a-chip microfluidic device. PLoS ONE **11**(7), e0159729 (2016)
16. Greenberg, J.M., Alt, W.: Stability results for a diffusion equation with functional drift approximating a chemotaxis model. Trans. Am. Math. Soc. **300**, 235–258 (1987)
17. Higham, D.J.: An algorithmic introduction to numerical simulation of stochastic differential equations. SIAM Rev. **43**(3), 525–546 (2001). https://doi.org/10.1137/S0036144500378302
18. Hillen, T., Painter, K.J.: A user's guide to PDE models for chemotaxis. J. Math. Biol. **58**, 183–217 (2009). https://doi.org/10.1007/s00285-008-0201-3
19. Keller, E.F., Segel, L.A.: Initiation of slime mold aggregation viewed as an instability. J. Theor. Biol. **26**, 399–415 (1970)
20. Kim, H.J., Huh, D., Hamilton, G., Ingber, D.E.: Human gut-on-a-chip inhabited by microbial flora that experiences intestinal peristalsis-like motions and flow. Lab Chip **12**(12), 2165–2174 (2012)
21. Kirschner, D., Panetta, J.C.: Modeling immunotherapy of the tumor-immune interaction. J. Math. Biol. **37**(3), 235–252 (1998). https://doi.org/10.1007/s002850050127. PMID: 9785481
22. Lapidus, I.R., Schiller, R.: Model for the chemotactic response of a bacterial population. Biophys. J. **16**(7), 779–789 (1976). https://doi.org/10.1016/S0006-3495(76)85728-1
23. Lee, S.W.L., Seager, R.J., Litvak, F., Spill, F., Sieow, J.L., Leong, P.H., Kumar, D., Tan, A.S.M., Wong, S.C., Adriani,G., Zaman, M.H., Kamm, A.R.D.: Integrated in silico and 3D in vitro model of macrophagemigration in response to physical and chemical factors in the tumor microenvironment. IntegrBiol (Camb) **12**(4), 90–108 (2020). https://doi.org/10.1093/intbio/zyaa007. PMID: 32248236; PMCID:PMC7167463
24. Low, L.A., Mummery, C., Berridge, B.R., Austin, C.P., Tagle, D.A.: Organs-on-chips: Into the next decade. Nat. Rev. Drug Discov. 1–17 (2020)
25. Menci, M., Papi, M.: Existence of solutions for hybrid systems of differential equations under exogenous information with discontinuous source term. Nonlinear Anal. **221** (2022)
26. Murray, J.D.: Mathematical Biology II Spatial Models and Biomedical Applications, vol. 3. Springer, New York (2001)

27. Natalini, R., Paul, T.: The mean-field limit for hybrid models of collective motions with chemotaxis. Preprint (2021). arXiv:2107.14645
28. Othmer, H.G.: Cell-based, continuum and hybrid models of tissue dynamics. Mathematical Models and Methods for Living Systems. Lecture Notes in Mathematics, vol. 2167, pp. 1–72. Springer, Cham (2016)
29. Perfahl, H., Hughes, B.D., Alarcón, T., Maini, P.K., Lloyd, M.C., Reuss, M., Byrne, H.M.: 3D hybrid modelling of vascular network formation. J. Theor. Biol. **414**, 254–268 (2017)
30. Perthame, B.: Transport Equations in Biology. Frontiers in Mathematics. Birkhäuser, Basel (2007). https://doi.org/10.1007/978-3-7643-7842-4
31. Strikwerda, J.C.: Finite difference schemes and partial differential equations. Society for Industrial and Applied Mathematics (2004). https://doi.org/10.1137/1.9780898717938
32. Vacchelli, E. et al.: Chemotherapy-induced antitumor immunity requires formyl peptide receptor 1. Science **350**, 972–978 (2015). https://doi.org/10.1126/science.aad0779
33. Wiśniewski, J.R., Hein, M.Y., Cox, J., Mann, M.: A "proteomic ruler" for protein copy number and concentration estimation without spike-in standards. Mol. Cell. Proteomics. **13**(12), 3497–3506 (2014). https://doi.org/10.1074/mcp.M113.037309
34. Yang, T.D., Park, J.-S., Choi, Y., Choi, W., Ko, T.-W., Lee, K.J.: Zigzag turning preference of freely crawling cells. PLoS ONE **6**, e20255 (2011). https://doi.org/10.1371/journal.pone.0020255
35. Yesil-Celiktas, O., Hassan, S., Miri, A.K., Maharjan, S., Al-kharboosh, R., Quinones-Hinojosa, A., Shrike Zhang, Y.: Pathology-on-a-chip: mimicking human pathophysiology in organ-on-chip devices. Adv. Biosyst. 1800109 (2018)

The Effect of Substratum Stiffness and Stochasticity on Cell Reorientation over a Stretched Substratum

Annachiara Colombi, Nadia Loy, and Luigi Preziosi

Abstract It is observed that when cells are seeded on a substrate subject to a cyclic deformation, they tend to reorient their major axis either perpendicularly or forming an oblique angle with the main stretching direction. However, this does not occur any longer if the substrate is too soft, or if the frequency and amplitude of imposed deformation are too small. In order to explain these transitions, in this chapter we respectively schematize the cell as a set of discrete elements and deal with the random behaviour of cell response deducing a Fokker-Planck equation, that can be also related to an optimal control strategy the cell may wish to pursue.

1 Introduction

It is nowadays clear that cells react to mechanical stresses exerted on them by activating mechanotransduction pathways that link mechanical cues and chemical processes inside the cell. In particular, in this chapter we are interested in describing what happens to cells that are seeded on a substrate that is stretched periodically. In fact, it was observed that, as in many other situations, cells actively respond to the exerted stretch by an internal adaptation. Specifically, in this experimental situation they develop families of stress fibres (SFs) distributed according to a peaked distribution function around a certain direction that evolves in time until it reaches a stationary distribution. The main function of SFs is to confer the cell structural stability, as well as to develop active contractile forces. Such fibres are long bundles of actin filaments, that are cross-linked by several types of molecules, such as α-actinin, fimbrin, and fascin (see Fig. 1). In addition, other molecules link fibres, holding them at specific rest angles. For instance, filamin tends to link actin fibres perpendicularly. New (daughter) actin fibres can also originate from a mother filament with a well-defined branching angle of about 70°, due to the presence at

A. Colombi · N. Loy · L. Preziosi (✉)
Department of Mathematical Sciences "G.L. Lagrange", Politecnico di Torino, Turin, Italy
e-mail: annachiara.colombi@polito.it; nadia.loy@polito.it; luigi.preziosi@polito.it

A. d'Onofrio et al. (eds.), *Problems in Mathematical Biophysics*, SEMA SIMAI Springer Series 38, https://doi.org/10.1007/978-3-031-60773-8_4

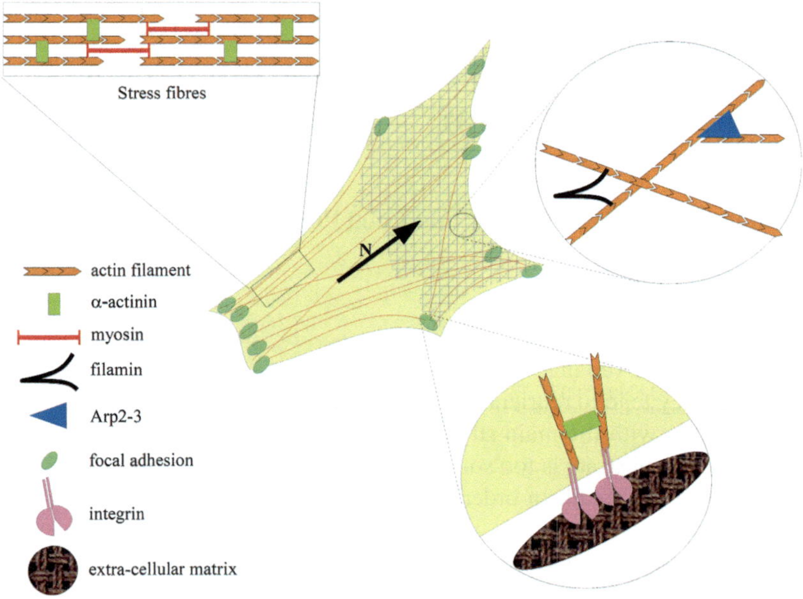

Fig. 1 Sketch of the inner structure of a typical cell and of its adhesion to the substrate

the diramation point of a specific molecular complex, called ARP2/3, which has a rest conformation leading to the formation of such an angle. Looking at these cytoskeletal structures it is then clear that, even considering the filaments as rigid and not capable of remodelling, a stretching of the network would cause a change in the rest angles. Therefore, the cross-linking molecules would like to return to their stress-free configuration.

The connection between the internal cytoskeleton and the external environment is provided by focal adhesions complexes. These are clusters of transmembrane proteins that are attached on one side on cytoskeletal elements and on the other side on the substrate. So, when stretched, deformations are transferred inside the cell deforming the cytoskeleton and triggering a cellular response. In particular, it seems that, when submitted to an external stretch, the cell reorganizes its internal structure, in order to relieve the state of stress, with a central role in such a response played by the cytoskeleton.

As discussed in detail in the review [5], it seems that this reorientation behaviour is cell-type independent. To our knowledge, only Matsumoto et al. [16] report that rat macrophages were indifferent to the mechanical stimulus. On the other hand, an alignment slightly parallel to the direction of stretch is reported for human monocytes [16]. This is in contrast with the common behaviour for all other cell types, which exhibited an avoidance of cyclic strain and reoriented obliquely or perpendicularly to the stretch. This difference is probably due to the fact that immune cells do not possess a strong cytoskeleton and do not adhere so strongly to

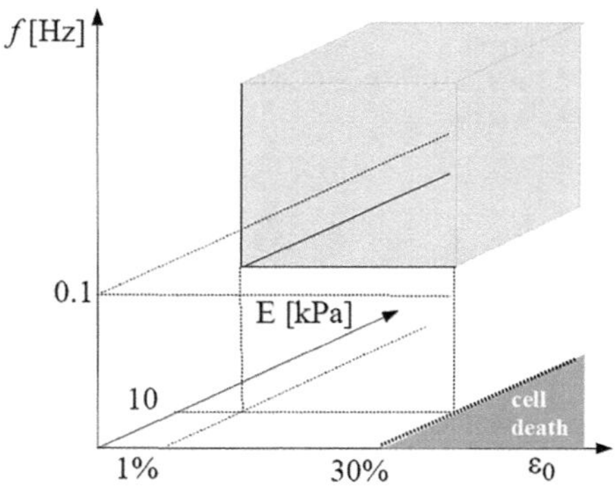

Fig. 2 Range of main parameters leading to cell re-orientation. ϵ_0 is the maximum strain, f is the frequency, and E is the substratum stiffness

the substrate. Their type of motion is more of amoeboid-type with little involvement of adhesion molecules. In fact, their role is to move fast to search for microbes, viruses, cellular debris, foreign substances, and in order to do that they need to squeeze through the extracellular network, basal membranes, cellular layers, and capillary walls.

Conversely, fibroblasts crawl in the environment by a mesenchymal motion that involves the formation of firm adhesion bonds and a strong involvement of the cytoskeleton. They pull the extracellular matrix to move or to repair wounded regions. As trivially known, also the role of muscle cells is to activate their acto-myosin machinery to exert pulling forces or to resist to stresses. Similarly, epithelial and endothelial cells, form cellular sheets and are generally attached to each other, for instance, by cadherin bonds adhering to basal membranes, for instance, via the expression of integrins.

Amazingly, it seems that for all types of cells the phenomenon is also nearly independent from the applied frequency and amplitude of the deformation, as well as from the stiffness of the substrate, as long as they are large enough. In fact, transitions are observed when each of the corresponding parameters above gets larger that some thresholds, as shown in Fig. 2. In addition, the characteristic response time and the variance of the final distribution of orientations both decrease when these parameters individually increase, as shown in Fig. 3.

The aim of this Chapter is to present some discrete and kinetic models aimed at giving a possible explanation to these transitions, namely,

- why too soft substrates inhibit cell alignment;
- why, sometimes, over soft substrates cells align parallel to the main stretching direction, rather than perpendicularly or obliquely;
- why cell alignment is not observed for too small amplitudes and frequencies;
- why an increase in the stretch amplitude and frequency leads to a peaking up of the distribution function of the orientation angles, as shown in Fig.3, but the most probable orientation remains unchanged.

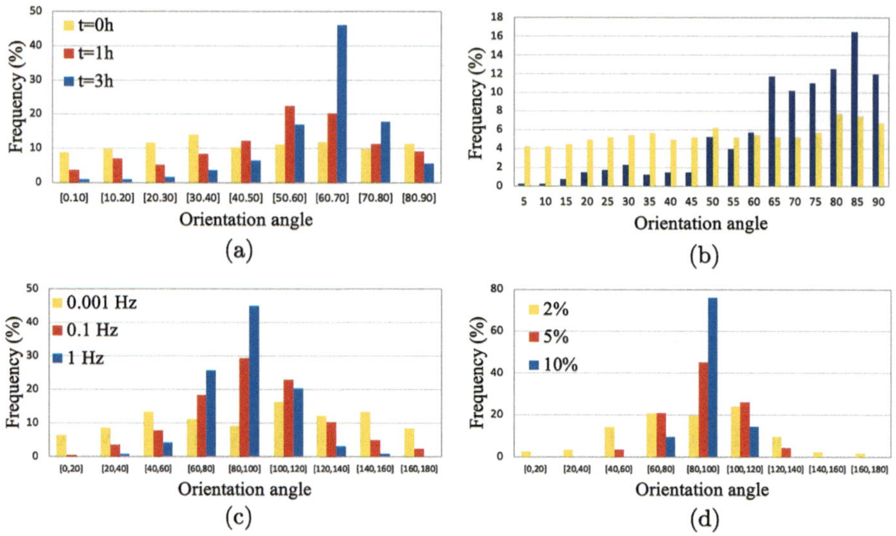

Fig. 3 (**a**) Temporal evolution of orientation angles (data from Hayakawa et al. [6]). (**b**) Distribution of final orientation for a maximum strain of 11.8% compared with the control case in yellow (data from Faust et al. [4]). (**c**) Distribution of final orientation for a maximum strain of 5% at different frequencies (data from Mao et al. [15]). (**d**) Distribution of final orientation for a frequency of 1 Hz and varying maximum strain (data from Mao et al. [15])

In more detail, Sect. 2 provides an overview on some models that schematize the cytoskeleton as a simple mechanical structure, which also allows to take into account the mechanical characteristics of the substratum and to show that, when the latter is too soft, a stability switch occurs from supercritical to subcritical. This might lead to a change of the stable orientation from oblique to parallel or perpendicular. Finally, Sects. 3 and 4 present models that look at the cell orientation behaviour from a kinetic and an optimal control points of view, respectively, focusing on the effect of randomness. In particular, it is shown that in low frequency regimes and for small amplitudes, stochastic effects get so strong that a clear orientation is no longer evident. This approach also explains the peaking up of the distributions in Fig. 3 both when the frequency of the imposed oscillation and its amplitude increases.

2 Tensegrity-Like Models

Starting from the seminal work by Ingber [8], a simplified way to describe the cytoskeletal structure of a cell is to use tensegrity-like models. These consist in simplifying the complex structure of the cytoskeleton in an ensemble of few connected elastic and possibly viscous elements. For instance, Xu et al. [19] used

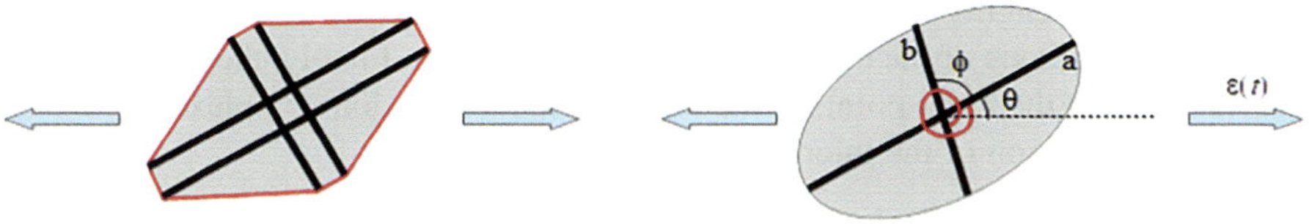

Fig. 4 Tensegrity models with two pairs of perpendicular struts (thick lines) used in [15, 19, 20] (on the left) and with two struts connected by a torsional spring between the two struts used in [3] (on the right). Arrows indicate the main stretching directions

this approach modeling the cell as formed by two parallel elastic struts along the orientation axis and two other struts perpendicular to them, as shown in the left panel of Fig. 4. A string connecting the ends of the struts is also considered, but then neglected in the computations. The mechanical energy of the system of elements is then evaluated, but unfortunately it presents some flaws when the work done by the applied force is computed. In addition, there is some confusion with some signs (e.g., Eqs. (13) and (18) in [19]). Luckily, most of the signs are a posteriori adjusted, so that the energy eventually discussed has the same structure of the one that will be discussed below (with suitable parameter changes). The same schematization was then used in [1, 15, 19, 20]. Here, we summarize the approach used in Colombi et al. [3], who had the aim of studying the effect of the interactions with the substratum and the influence of its stiffness on the final orientation of the cell, i.e., the first two questions at the end of the introduction.

The deduction of their model is based on the observation that the overall phenomenon can be split in two phases characterized by different time scales. The first one occurs at the time scale of the imposed periodic oscillations, i.e., few seconds or less. It consists in a quick stretching of the cell that immediately adapts to the deformation of the substrate resisting to it, due to its mechanical characteristics. The second one take several minutes and involves a slow internal restructuring of the actin cytoskeleton in order to minimize the stored internal energy, denoted below as U_{in}. As a result of this remodelling process, the cell main orientation evolves in time according to

$$\tau_\theta \frac{d\theta}{dt} = -\frac{\partial U_{\text{in}}}{\partial \theta}, \tag{1}$$

where τ_θ is related to the characteristic time for stress fibres and focal adhesion dynamics.

Then, as shown in the right panel of Fig. 4, the cell cytoskeleton is schematized by two elastic elements, say a and b, with elastic moduli k_{a} and k_{b}. They are in charge of describing the response of the cell to the external mechanical input: the element a along the main orientation of SFs (denoted by θ) and the element b along a transversal direction (i.e., rotated by an angle ϕ). These two elements are hinged at their centers and linked by a torsional spring, with modulus equal to k_ϕ, that is at rest when the two elements are orthogonal. The two cytoskeletal elements a and

b have relaxed lengths equal to $2L_\mathrm{a}$ and $2L_\mathrm{b}$, respectively, which can be interpreted as mean dimensions of the cell in the absence of stresses. Finally, they are linked to the substrate at their end points, which models focal adhesion complexes.

The computation of the internal energy then gives

$$U_\mathrm{in} = \frac{1}{2}\varepsilon^2 k_\mathrm{s}^2 \left\{ \frac{k_\mathrm{a} L_\mathrm{a}^2}{(k_\mathrm{a}+k_\mathrm{s})^2}(\cos^2\theta - r\sin^2\theta)^2 + \frac{k_\mathrm{b} L_\mathrm{b}^2}{(k_\mathrm{b}+k_\mathrm{s})^2}(r\cos^2\theta - \sin^2\theta)^2 + \right.$$
$$\left. + \frac{4k_\phi}{\left[\left(\frac{1}{L_\mathrm{a}^2}+\frac{1}{L_\mathrm{b}^2}\right)k_\phi + k_\mathrm{s}\right]^2}(1+r)^2\sin^2\theta\cos^2\theta \right\}, \tag{2}$$

where $\varepsilon(t)$ is the imposed strain along the x direction and $r = \max_t \varepsilon_{yy}(t)/\max_t \varepsilon(t)$ is the biaxiality ratio that in experiments is found to be constant in time (ε_{yy} is the strain along the direction perpendicular to the main stretching direction). Equation (1) therefore rewrites as

$$\frac{\tau_\theta}{k_a L_a^2}\frac{d\theta}{dt}(t) = \frac{2\hat{k}_s^2}{(1+\hat{k}_s)^2}\,\varepsilon^2(t)\,(1+r)\Big\{[(1+r)\cos^2\theta(t) - r]$$
$$+ K_\perp^s[(1+r)\cos^2\theta(t) - 1] - K_\phi^s(1+r)[2\cos^2\theta(t) - 1]\Big\}\sin\theta(t)\cos\theta(t)\,. \tag{3}$$

It is noteworthy to observe that the structure of the equation is the same as the one in [2, 11, 13, 14] (also reviewed in another Chapter of this volume) where the effect of substratum compliance is neglected. The effect of the substratum is included in the parameters

$$K_\perp^s = \lambda\hat{k}_b\left(\frac{1+\hat{k}_s}{\hat{k}_b+\hat{k}_s}\right)^2, \qquad K_\phi^s = \hat{k}_\phi\left(\frac{1+\hat{k}_s}{\frac{\lambda+1}{2\lambda}\hat{k}_\phi+\hat{k}_s}\right)^2, \tag{4}$$

where

$$\hat{k}_b = \frac{k_b}{k_a}, \qquad \hat{k}_s = \frac{k_s}{k_a}, \qquad \text{and} \qquad \hat{k}_\phi = \frac{2k_\phi}{k_a L_a^2}, \tag{5}$$

compare the elastic constant k_a of the element along the main cell axis, with the perpendicular one k_b, with the torsional spring k_ϕ (i.e., the response to shear), and with the substratum elasticity k_s, respectively, while $\lambda = \frac{L_b^2}{L_a^2}$ is a measure of the aspect ratio of the cell, usually smaller than 1. These parameters turn out to be crucial in the dynamics of the system and in determining the stability of the equilibrium orientations.

If we formally consider a time independent strain, a stability analysis similar to that performed in [13, 14] can be repeated. In particular, the r.h.s. of Eq. (3) vanishes when the representative cell orients along the principal strain directions $\theta_{\parallel}^{\text{eq}} = 0$, or $\theta_{\perp}^{\text{eq}} = \frac{\pi}{2}$, or along an oblique direction $\theta_{\text{obl}}^{\text{eq}}$ given by

$$\cos^2 \theta_{\text{obl}}^{\text{eq}} = \frac{1}{2} + \frac{1}{\mathcal{K}} \left(\frac{1}{2} - \frac{1}{r+1} \right) , \tag{6}$$

where

$$\mathcal{K} = \frac{1 + K_{\perp}^{s} - 2K_{\phi}^{s}}{1 - K_{\perp}^{s}} . \tag{7}$$

Before proceeding with the stability analysis, we observe that, taking separately the three terms on the r.h.s. of Eq. (3), for a time independent stretch, the first two tend to destabilize the equilibrium orientations $\theta_{\parallel}^{\text{eq}}$ and $\theta_{\perp}^{\text{eq}}$, while the last tends to stabilize them. The opposite holds true for $\theta_{\text{obl}}^{\text{eq}}$. As we will see in Sects. 2.1 and 2.2, this will lead to important stability switches also seen in some experiments (see, for instance, [17, 18]).

We also recall that the range of r usually tested in experiments is $r \in [0, 1]$, where $r = 0$ corresponds to a substrate clamped along the sides parallel to the main stretching direction, while $r = 1$ implements an isochoric planar deformation (i.e., $\varepsilon_{yy} = -\varepsilon_{xx}$). Very few results are reported for negative values of r, which corresponds to stretching in both directions (see, for instance, [10]). However, the fact that the x-axis is assumed parallel to the maximum stretching direction implies that $r \geq -1$. The other extremum $r > 1$ would correspond to an untested strong compression of the substrate along the y-axis. For this reason, we hereafter focus on the range $r \in [-1, 1]$, but, of course, the stability analysis also holds outside this interval.

Regarding the other parameters, from the mechanical point of view, it is also reported that the response of cells to deformations along the main orientation is stronger than along the perpendicular one (even much stronger., We can therefore focus on the range $\hat{k}_{\text{b}} \in (0, 1)$ (i.e., $0 < k_{\text{b}} < k_{\text{a}}$). Similarly, we expect $\hat{k}_{\phi} \in \left(0, \frac{1}{2}\right)$. This can be realized comparing, for the same deformation involving the opening of the angle in O, the energetic terms related to the torsional spring (proportional to k_{ϕ}) with the energy that would be stored by a spring with rigidity k_{a} connecting the extrema of the elements a and b. So, we will focus the following discussion on the subspace of parameters defined by

$$r \in [-1, 1], \quad \lambda \in (0, 1], \quad \hat{k}_{\text{b}} \in (0, 1), \quad \text{and} \quad \hat{k}_{\phi} \in \left(0, \frac{1}{2}\right) , \tag{8}$$

having in mind the biologically sound constraints above.

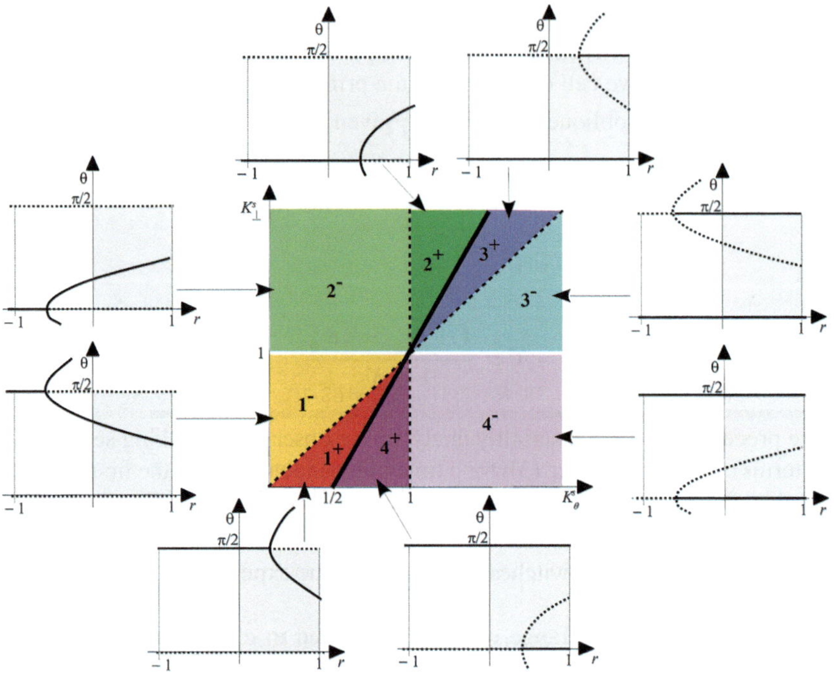

Fig. 5 Bifurcation scenarios in the $(K^s_\perp, K^s_\phi)$ plane. The black line corresponds to a vanishing numerator $\mathcal{N}$ of $\mathcal{K}$ in Eq. (7) (i.e., $K^s_\perp = 2K^s_\phi - 1$) with positive values above it. It corresponds to the switch between supercritical bifurcations with stable oblique configuration to subcritical bifurcations. The white line corresponds to a vanishing denominator of $\mathcal{K}$ (i.e., $K^s_\perp = 1$). $\mathcal{K}$ is thus positive in regions 1 and 3. The dashed black lines $K^s_\perp = K^s_\phi$ and $K^s_\phi = 1$ identify the regions where the bifurcation point is negative or positive. The shaded regions in the (r, θ) planes is the usually tested one

2.1 Discussion in Terms of $K^s_\perp$ and K^s_ϕ

The stability of the equilibria mentioned above depends on the sign of both the numerator $\mathcal{N}$ and the denominator $\mathcal{D}$ of $\mathcal{K}$ in Eq. (7). In particular, a positive numerator (see the regions above the black line in Fig. 5) leads to supercritical bifurcations, while a negative numerator to subcritical ones. On the other hand, Eq. (7) implies that if $\mathcal{K} > 0$ then $\theta^{\text{eq}}_{\text{obl}} \in \left(\frac{\pi}{4}, \frac{\pi}{2}\right)$ (see regions 1 and 3 in Fig. 5), otherwise $\theta^{\text{eq}}_{\text{obl}} \in \left(0, \frac{\pi}{4}\right]$.

More precisely, proceeding clockwise in Fig. 5, we have the following four scenarios (see also Table 1):

1. In Region 1, i.e., if $K^s_\perp < 1$ and $K^s_\phi < \frac{1}{2}(K^s_\perp + 1)$, both $\mathcal{N}$ and $\mathcal{K}$ are positive. Then the stable equilibria are the oblique orientation $\theta^{\text{eq}}_{\text{obl}} \in (\frac{\pi}{4}, \frac{\pi}{2})$

Table 1 Bifurcation scenarios

	$\theta_{\parallel}^{eq}$	$\theta_{\perp}^{eq}$	θ_{obl}^{eq}	Bifurcation
Region 1	Unstable	Stable	Stable	
$\mathcal{N} > 0, \mathcal{D} > 0$	$\forall r \in [-1, 1]$	if $r < \frac{K_\phi^s - K_\perp^s}{1 - K_\phi^s}$	if $r > \frac{K_\phi^s - K_\perp^s}{1 - K_\phi^s}$ $\theta_{obl}^{eq} \in (\frac{\pi}{4}, \frac{\pi}{2})$	Supercritical
Region 2	Stable	Unstable	Stable	
$\mathcal{N} > 0, \mathcal{D} < 0$	if $r < \frac{1 - K_\phi^s}{K_\phi^s - K_\perp^s}$	$\forall r \in [-1, 1]$	if $r > \frac{1 - K_\phi^s}{K_\phi^s - K_\perp^s}$ $\theta_{obl}^{eq} \in (0, \frac{\pi}{4})$	Supercritical
Region 3	Stable	Stable	Unstable	
$\mathcal{N} < 0, \mathcal{D} < 0$	$\forall r \in [-1, 1]$	if $r > \frac{K_\phi^s - K_\perp^s}{1 - K_\phi^s}$	$\forall r \in [-1, 1]$ $\theta_{obl}^{eq} \in (\frac{\pi}{4}, \frac{\pi}{2})$	Subcritical
Region 4	Stable	Stable	Unstable	
$\mathcal{N} < 0, \mathcal{D} > 0$	if $r > \frac{1 - K_\phi^s}{K_\phi^s - K_\perp^s}$	$\forall r \in [-1, 1]$	$\forall r \in [-1, 1]$ $\theta_{obl}^{eq} \in (0, \frac{\pi}{4})$	Subcritical

for $r > \frac{K_\phi^s - K_\perp^s}{1 - K_\phi^s}$ and the perpendicular one for lower values of r. Bifurcation is supercritical.

2. In Region 2, i.e., if $K_\perp^s > 1$ and $K_\phi^s < \frac{1}{2}(K_\perp^s + 1)$, $\mathcal{N}$ is positive, while $\mathcal{K}$ is negative. Then the stable equilibria are the oblique orientation $\theta_{obl}^{eq} \in (0, \frac{\pi}{4})$ for $r > \frac{1 - K_\phi^s}{K_\phi^s - K_\perp^s}$ and the parallel one for lower values of r. Bifurcation is supercritical.
3. In Region 3, i.e., if $K_\perp^s > 1$ and $K_\phi^s > \frac{1}{2}(K_\perp^s + 1)$, $\mathcal{N}$ is negative, while $\mathcal{K}$ is positive. Then the stable equilibria are the parallel one (always) and the perpendicular one for $r > \frac{K_\phi^s - K_\perp^s}{1 - K_\phi^s}$; while the oblique equilibrium $\theta_{obl}^{eq} \in (\frac{\pi}{4}, \frac{\pi}{2})$ is always unstable. Bifurcation is subcritical.
4. In Region 4, i.e., if $K_\perp^s < 1$ and $K_\phi^s > \frac{1}{2}(K_\perp^s + 1)$, both $\mathcal{N}$ and $\mathcal{K}$ are negative. Then the stable equilibria are the perpendicular one (always) and the parallel one for $r > \frac{1 - K_\phi^s}{K_\phi^s - K_\perp^s}$; while the oblique equilibrium $\theta_{obl}^{eq} \in (0, \frac{\pi}{4})$ is always unstable. Bifurcation is subcritical.

In Fig. 5, each region is further subdivided in two subregions denoted either by the apex "+" or "−", according to whether the bifurcation value is positive or negative. It follows that, if the range of values of r is further restricted to the usually tested values [0, 1], in Region 1^- only the oblique equilibrium with orientation larger than $\frac{\pi}{4}$ is stable; in Region 2^- only the oblique equilibrium with orientation smaller than $\frac{\pi}{4}$ is stable; in Region 3^- and Region 4^-, both the parallel and the perpendicular equilibria are always stable.

Summarizing, when the black thick straight line in Fig. 5 is overcome from left to right, there is a stability switch from a configuration allowing for a stable oblique equilibrium $\theta_{\text{obl}}^{\text{eq}}$ to a configuration having only stable equilibria along the stretching direction $\theta_{\parallel}^{\text{eq}}$ and/or the perpendicular one $\theta_{\perp}^{\text{eq}}$. In addition, when the horizontal white line or the black thick straight line are overcome, there is a switch from an oblique orientation $\theta_{\text{obl}}^{\text{eq}}$ above $\frac{\pi}{4}$ (either stable or unstable) to one below it, or vice versa. For instance, focusing on a value of r close to 1 (so that r is higher than both $\frac{K_\phi^s - K_\perp^s}{1-K_\phi^s}$ and $\frac{1-K_\phi^s}{K_\phi^s - K_\perp^s}$), i.e., a setting that corresponds to a nearly isochoric stretch in the plane, then the equilibrium orientation close to $\frac{\pi}{4}$ is stable for all positive values of the numerator of $\mathcal{K}$ and becomes suddenly unstable for negative values.

2.2 Discussion in Terms of Substratum Stiffness $\hat{k}_s$

As $K_\perp^s$ and K_ϕ^s depend on $\hat{k}_s$, we are here interested in translating the discussion above in terms of substrate stiffness to prove that it can determine a stability switch. In particular, we want to show that decreasing the substrate stiffness, the bifurcation landscape can change from a supercritical one as in [13, 14] to a subcritical one with an unstable oblique orientation and stable parallel and/or perpendicular orientations, as observed in [17, 18]. Actually, the inclusion of substratum compliance has also the effect of slowing down the reorientation process, as evidenced by the presence of the $\hat{k}_s^2$ factor multiplying the r.h.s. of Eq. (3). This might explain why for very soft substrates no re-orientation occurs during the usual duration of the experiments (typically, nearly a day). Actually, in this case, very little deformation is transferred to the cell.

Referring to Fig. 6, the most important thing to notice, however, is that, if the substratum is soft enough, equilibrium orientations with the cell axis parallel to the main stretching direction might become stable. In fact, for a fixed biaxiality ratio r, stable oblique orientations are favoured by a sufficiently large substratum stiffness, corresponding to the region above the upper orange surface in Fig. 6a. Instead, the subcritical condition, characterized by the stability of parallel and perpendicular orientations, is preferential for softer substrates, corresponding to the region between the orange surfaces in Fig. 6a.

As an example, referring to Fig. 6b obtained fixing $\hat{k}_\theta = 0.4$ in addition to the biaxiality ratio $r = 0.5$, focusing to the region $\hat{k}_b > \hat{k}_\theta = 0.4$, the stable equilibrium angle first decreases from a value slightly above $\frac{\pi}{4}$ to a value below $\frac{\pi}{4}$. Decreasing further $\hat{k}_s$, the stable equilibrium angle is the one along the principal strain direction, with the oblique equilibrium that becomes unstable. A further softening of the substrate then leads to the stability of both the parallel and perpendicular configurations, still with an unstable oblique equilibrium.

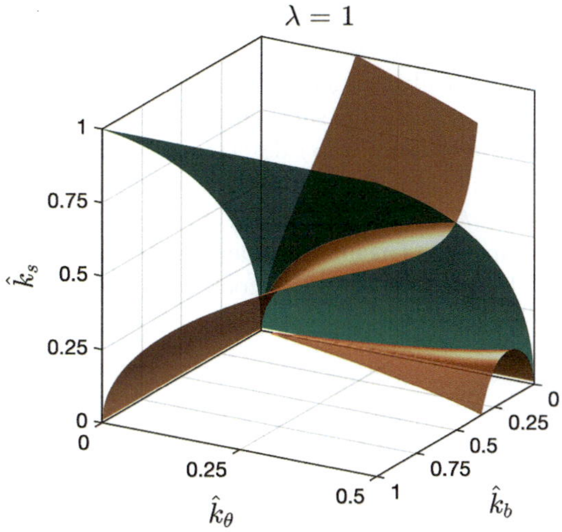

(a)

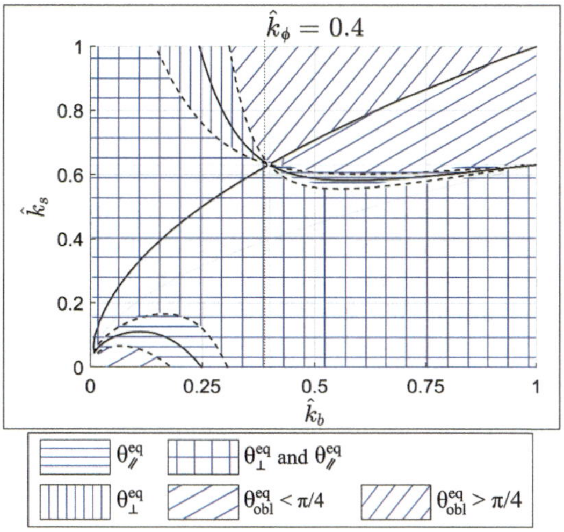

(b)

Fig. 6 (**a**) Surfaces delimiting the different bifurcation scenarios. In particular, the oblique equilibrium orientation is stable above the upper orange surface and below the lower orange surface and unstable in between them, corresponding to a supercritical bifurcation in the former region and subcritical in the latter. (**b**) Cross section of the space of parameters for $\hat{k}_\theta = 0.4$. The stable configurations for $r = 0.5$ are sketched

3 Kinetic Modelling and Fokker-Planck Equation

In this Section, we briefly mention some works based on classical tools of statistical mechanics that allow to describe the statistical distribution of the orientations of the cells. Indeed, as shown in Fig. 3, the biological phenomenon under study is intrinsically affected by a certain degree of randomness and, because of the already mentioned possible symmetries of the system, measurements of the frequencies of the orientations are reported over $\left[0, \frac{\pi}{2}\right]$ or $[0, \pi]$. Therefore, an interesting issue is also the determination of the probability density function $f = f(\theta)$

(over $\left[0, \frac{\pi}{2}\right]$ or $[0, \pi]$, respectively) of the orientations of the cells, that allows for a complete statistical description. This is typically carried out, in the field of statistical mechanics, by defining a microscopic dynamics for the orientation θ and by deriving subsequently an evolution equation for a probability density function $f = f(t, \theta)$ describing the statistical distribution of the orientations of cells following the aforementioned microscopic process. We also highlight the fact that, if f is considered on the symmetry interval according to the given experimental set-up, then f is unimodal and the linear and circular average coincide.

The starting point is a stochastic differential equation obtained by adding a noise to Eq. (1), i.e.

$$d\theta = -\frac{\varepsilon^2}{\tau_\theta}\frac{\partial U}{\partial \theta}dt + \sqrt{\frac{\sigma^2}{\tau_\theta}}dW_t \tag{9}$$

where $dW_t = \sqrt{t}\xi$, being ξ a Gaussian noise with zero mean and unitary variance. The forward equation of $f = f(t, \theta)$ is

$$\frac{\partial}{\partial t}f(t,\theta) = \frac{\varepsilon^2}{\tau_\theta}\frac{\partial}{\partial \theta}\left(\frac{\partial \bar{U}}{\partial \theta}(\theta)f(t,\theta)\right) + \frac{1}{2\tau_\theta}\frac{\partial^2}{\partial \theta^2}\left(\sigma^2 f(\theta,t)\right), \tag{10}$$

where s takes into account the stochastic fluctuations related to the randomness in cell motion, τ_θ is the characteristic time, and $\bar{U} = \frac{U}{K_{\parallel}^s \varepsilon^2}$ is the strain independent dimensionless energy, where U is for instance the energy defined in Eq. (2). From Eq. (10) it can be readily observed that an increase in the applied strain ε would increase the importance of the drift term with respect to diffusion. Vice versa, decreasing excessively ε would lead to a diffusion-dominated problem, in which the random noise covers the re-orientation tendency expressed by the drift term. This is also evident by looking at the stationary state of Eq. (10) that writes

$$f^{\infty}(\theta) = C\exp\left[-\frac{2\varepsilon^2}{\sigma^2}\bar{U}(\theta)\right], \tag{11}$$

where C is a normalization constant. In fact, it presents maxima (resp. minima) corresponding to minima (resp. maxima) of $\bar{U}$ and peaks up when ε increases.

The kinetic theory approach is then developed by considering a Boltzmann-like kinetic equation for the probability density function in which the microscopic dynamics for the evolution of the orientation angle θ is described by specifying the new angle θ' after transition as follows:

$$\theta' = \theta - \gamma\varepsilon^2\frac{\partial \bar{U}}{\partial \theta} + \sqrt{\gamma\sigma^2}\xi \quad \mathrm{mod}(\pi), \tag{12}$$

where ξ is a Gaussian random variable with zero mean and unitary variance, and $\gamma \ll 1$ is a small parameter describing how much cells reorient at each transition. The application of the quasi-invariant direction limit (that is a classical

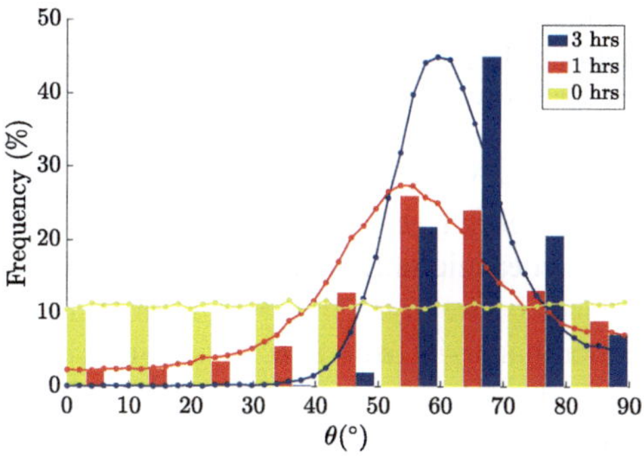

Fig. 7 Comparison of the evolution of the probability density function obtained by performing a Montecarlo simulation of Eq. (10) with the experimental data reported in [7]. In particular, $\varepsilon = 20\%$ and $r = 0.4$. Bars refer to experimental data at $t = 0, 1, 3$ h, respectively in yellow, red and blue. Curves are the recovered probability density functions at $t = 0, 1, 3$ h. Solution for $\theta_{eq} \approx 61°$, $s \approx 0.04$, and $\tau_\theta \approx 0.18$ s

tool of kinetic theory) to the Boltzmann-like integro-differential equation joined with Eq. (12) allows to derive Eq. (10).

Coming to a comparison of the model (10) with the experimental results, Fig. 7a reports the result of a Montecarlo simulation of the microscopic process (12) in the quasi-invariant regime allowing to approximate (10). In Fig. 7b the results are also compared with the experimental data reported in [7] for $\varepsilon = 20\%$, $r = 0.4$ and $\omega = 1$ Hz. In these experiments it was found that at $t = 1$ h the average orientation is 52.8°, while at $t = 3$ h, when more than the 80% of the cells are oriented at angles of 50°-80°, the average orientation is 62.02°. By using Eq. (6) and the value of $\mathcal{K}$ suggested in [11], the minimum of the elastic energy is achieved at $\theta_{eq} \approx 61°$. The value of σ is set to $\sigma \approx 0.04$ so that the mean of the distribution function over $\left[0, \frac{\pi}{2}\right]$, i.e. $\bar{\theta}_\ell = 62.2°$. After 1 h the average orientation is 54.6° and after 3 h it is 62.04° and 85% of the cells are oriented at angles of 50°–80°.

4 Reorientation as an Optimal Control Problem

In [12] it is assumed that in the reorientation process, the cell rotates in an optimal way, i.e. in order to minimize a certain energy functional. This is described as a control problem, assuming that the dynamics of the cell is governed by the energy functional that tends to be minimized and the cell changes its reorientation by an angle ψ_{opt} that minimizes a certain cost functional $\mathcal{J}$. In particular, this is is implemented in the Boltzmann kinetic framework introduced in Sect. 3, describing the evolution of f for orientation angles by performing the following dynamics

$$\theta' = \theta + \nu\psi_{opt}, \quad \psi_{opt} = \text{argmin}_\psi \mathcal{J}(\psi), \tag{13}$$

where $\mathcal{J}$ is an energy functional given by a kinetic contribution related to the control and by an energy function related to the reorientation process

$$\mathcal{J}(\psi) = \nu \frac{\psi^2}{2} + \langle h(\theta') \rangle ,$$

where ν is a penalization coefficient. It is then proved that Eq. (10) is recovered if

$$h = \varepsilon^2 \bar{U}. \tag{14}$$

Instead, if, for instance, for $\theta_{eq} \in \left(0, \frac{\pi}{2}\right)$ we

1. define

$$\hat{\theta}(\theta) = \theta_{eq} p(\theta) + (1 - p(\theta))(\pi - \theta_{eq}) \tag{15}$$

   with $p(\theta)$ a non negative and continuous function defined on $[0, \pi)$ that satisfies

$$p(\theta^1_{eq}) = 1 \quad p(\pi - \theta^1_{eq}) = 0, \quad p(0) = p(\pi/2) = p(\pi) = 1/2,$$
$$p'(\theta^1_{eq}) = p'(\pi - \theta^1_{eq}) = 0,$$

   modelling, essentially, that if θ is already close to an equilibrium orientation, then it is more likely to stay there;
2. use the more classical quadratic rule

$$h(\theta') = \frac{\varepsilon^2}{2} [\theta' - \hat{\theta}(\theta)]^2 , \tag{16}$$

   with $\hat{\theta}(\theta)$ defined in (15);

then it is possible to explicitly determine the control as

$$\psi_{opt} = - \frac{\varepsilon^2}{1 + \nu\varepsilon^2} (\theta - \hat{\theta}) .$$

The corresponding reorientation rule (with the addition of a stochastic fluctuation weighted by σ_c) becomes

$$\theta' = \theta + \gamma\varepsilon^2 (\hat{\theta}(\theta) - \theta) + \sqrt{\gamma\sigma_c^2}\xi \quad \mathrm{mod}(\pi) . \tag{17}$$

This rule implies the fact that at each reorientation the cell will activate a control to reach a better orientation that is given by a rotation of $\gamma\varepsilon^2(\hat{\theta}(\theta) - \theta)$ (plus a white noise). This process will stop when the cell has oriented along the stable equilibria. In the symmetry points $\theta = 0, \pi/2, \pi$ the cell has the same probability ($= 1/2$) of re-orienting towards θ_{eq} or $\pi - \theta^1_{eq}$.

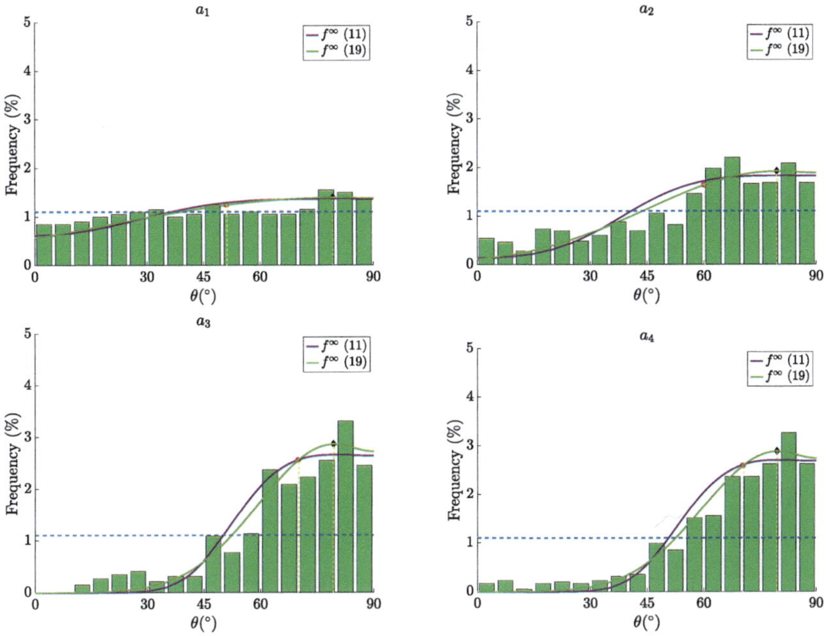

Fig. 8 Equilibrium distributions (11) and (19) for $\varepsilon = 4.9, 8.4, 11.8, 14\%$ and data from [4]. The red circles represent the average circular orientation and the black diamonds represent θ_{eq}

In this case the quasi-invariant direction limit procedure leads to the following Fokker-Planck equation

$$\frac{\partial}{\partial \tau} f(\tau, \theta) = -\frac{\varepsilon^2}{\tau_\theta} \frac{\partial}{\partial \theta} \left[(\hat{\theta}(\theta) - \theta) f(\tau, \theta) \right] + \frac{1}{2\tau_\theta} \frac{\partial^2}{\partial \theta^2} \left[\sigma_c^2 f(\tau, \theta) \right]. \tag{18}$$

whose equilibrium distribution is given by

$$f^\infty(\theta) = C \exp\left(\int_0^\theta \frac{2\varepsilon^2}{\sigma_c^2} [\hat{\theta}(\theta) - \theta] d\theta \right), \tag{19}$$

where C is the normalization constant.

Figure 8 gives a comparison of the stationary distributions (11) and (19) with the experimental data by Faust et al. [4]. They perform the experiment applying different stretching amplitudes, namely 4.9% (denoted as Case a_1), 8.4% (Case a_2), 11.8% (Case a_3), and 14% (Case a_4) and measuring a biaxiality ratio $r = 0.15$. Taking again $\mathcal{K} = 1.26$, as suggested in [11], the minimum elastic energy and therefore the mode is obtained at $\theta_{eq} \approx 79°$. The value of σ_c is set in such a way that the mean orientation evaluated in $\left[0, \frac{\pi}{2}\right]$ is the same as in [4]. We also remark that (19) gives a better approximation than (11).

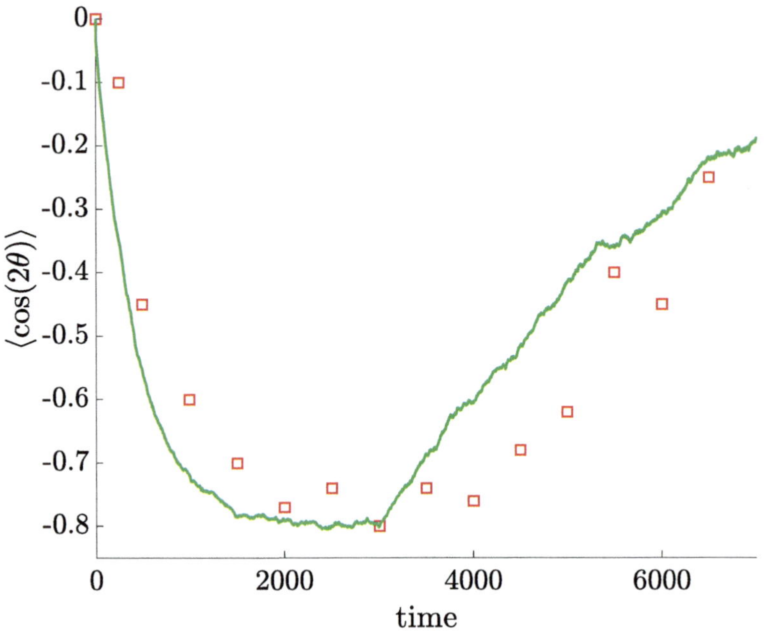

Fig. 9 Temporal evolution of the mean of the orientation distribution (reported in terms of $\cos 2\theta$) compared with the one in [9]. After 3000 s stretching stops and cells tend to re-orient uniformly over the angle

Figure 9 compares the results obtained by Jungbauer [9] with the mean of the distribution function obtained integrating (18). In the experiment stretching with $\omega = 2\,\text{Hz}$, $r = 0.194$ and $\varepsilon = 8\%$ is stopped after 3000 s. Hence, the contribution of the drift term vanishes and only the random term is active, leading to a recovery phase toward a uniform distribution. In Fig. 9, $\mathcal{K} = 1.26$ and the re-orientation time $\tau_\theta = 6.6\,\text{s}$, as suggested in [11].

In conclusion, the above models show that that taking randomness into account explains why increasing either frequency or amplitude cell reorientation becomes more evident, while the drift term, related to an energy minimization trend, governs the mode of the distribution function. On the opposite a decrease in the imposed frequency or amplitude leads to the fact that the random behaviour of the cell hides the reorientation trend.

5 Concluding Remarks

In this chapter we have shown how a discrete and a kinetic model were able to describe different features of the cell reorientation dynamics when they are seeded on a periodically stretched substratum. They both are able to catch the evolution toward an asymptotic orientation that can be identified minimizing a very general elastic energy. In addition, the former was particularly fit to include the effect of substrate stiffness, the latter, starting from the definition of suitable microscopic rules, the effect of stochasticity.

Regarding the effect of substrate stiffness, coherently with the experiments, it was found that the supercritical scenario characterized by the presence of a stable oblique configuration is favoured by sufficiently large substrate stiffness, while the subcritical scenario characterized by the stability of parallel and perpendicular orientations is favoured when the for softer substrates. In addition, the region characterized by subcriticality (i) increases as $\hat{k}_b$ (measuring the ratio of the cell response along the direction perpendicular to the main cell orientation axis versus the latter) decreases and (ii) increases as $\hat{k}_\theta$ (measuring the ratio of the cell response to shear versus the one along the main cell orientation axis) increases.

We also showed that the r.h.s. of the equation governing cell reorientation (3) presents a factor $\frac{\hat{k}_s^2}{(1+\hat{k}_s)^2}$ that leads to a considerable increase of the reorganization time as the substrate softens. Specifically, if the stiffness decreases by one order of magnitude (e.g., from about 1 MPa for silicon elastomers and polydimetylsiloxane, to about 100 KPa of collagen gels, or even less), we can expect the reorientation time to increase by two orders of magnitude (i.e., from a couple of hours to more than a week). This practically implies that over very soft substrates reorientation is not observed in the time of the experiment and might be easily hidden by random behaviours, similar to the dynamics shown by the stochastic models presented in Sects. 3 and 4. In fact, in these two sections it was shown that when the stretch amplitude or the stretching frequency were too small, randomness takes over and reorientation is hardly visible from a macroscopic point of view.

Acknowledgments This work was partially supported by the Italian National Group of Mathematical Physics (GNFM–INdAM), the Italian Ministry of Education, Universities and Research through the MIUR grant Dipartimenti di Eccellenza 2018–2022, Project No. E11G18000350001, and the Scientific Research Programmes of Relevant National Interest Project No. 2017KL4EF3.

Competing Interests The authors have no conflicts of interest to declare that are relevant to the content of this chapter.

References

1. Benvenuti, E., Reho, G.A., Palumbo, S., Fraldi, M.: Pre-strains and buckling in mechanosensitivity of contractile cells and focal adhesions: a tensegrity model. J. Mech. Behav. Biomed. Mater. **135**, 105413 (2022)
2. Ciambella, J., Lucci, G., Nardinocchi, P., Preziosi, L.: Passive and active fiber reorientation in anisotropic materials. Int. J. Eng. Sci. **176**, 103688 (2022)
3. Colombi, A., Preziosi, L., Scianna, M.: Modelling cell orientation under stretch: The effect of substrate elasticity. Bull. Math. Biol. **85**, 79 (2023)
4. Faust, U., Hampe, N., Rubner, W., Kirchgessner, N., Safran, S., Hoffmann, B., Merkel, R.: Cyclic stress at mHz frequencies aligns fibroblasts in direction of zero strain. PLoS ONE **6**, e28963 (2011)

5. Giverso, C., Loy, N., Lucci, G., Preziosi, L.: Cell orientation under stretch: A review of experimental findings and mathematical modelling. J. Theor. Biol. **572**, 111564 (2023)
6. Hayakawa, K., Hosokawa, A., Yabusaki, K., Obinata, T.: Orientation of smooth muscle-derived A10 cells in culture by cyclic stretching: relationship between stress fiber rearrangement and cell reorientation. Zool. Sci. **17**, 617–624 (2000)
7. Hayakawa, K., Sato, N., Obinata, T.: Dynamic reorientation of cultured cells and stress fibers under mechanical stress from periodic stretching. Exp. Cell Res. **268**, 104–114 (2001)
8. Ingber, D.E.: Cellular tensegrity: defining new rules of biological design that govern the cytoskeleton. J. Cell Sci. **104**(3), 613–627 (1993)
9. Jungbauer, S., Gao, H., Spatz, J.P., Kemkemer, R.: Two characteristic regimes in frequency-dependent dynamic reorientation of fibroblasts on cyclically stretched substrates. Biophys. J. **95**, 3470–3478 (2008)
10. Li, Z., Lee, H., Zhu, C.: Molecular mechanisms of mechanotransduction in integrin-mediated cell-matrix adhesion. Exp. Cell Res. **349**, 85–94 (2016)
11. Livne, A., Bouchbinder, E., Geiger, B.: Cell reorientation under cyclic stretching. Nat. Commun. **5**, 3938 (2014)
12. Loy, N., Preziosi, L.: A statistical mechanics approach to describe cell reorientation under stretch. Bull. Math. Biol. **85**, 60 (2023)
13. Lucci, G., Preziosi, L.: A nonlinear elastic description of cell preferential orientations over a stretched substrate. Biomech. Model. Mechanobiol. **20**, 631–649 (2021)
14. Lucci, G., Giverso, C., Preziosi, L.: Cell orientation under stretch: Stability of a linear viscoelastic model. Math. Biosci. **337**, 108630 (2021)
15. Mao, T., He, Y., Gu, Y., Yang, Y., Yu, Y., Wang, X., Ding, J.: Critical frequency and critical stretching rate for reorientation of cells on a cyclically stretched polymer in a microfluidic chip. AMS Appl. Mater. Interfaces **13**, 13934–13948 (2021)
16. Matsumoto, T., Delafontaine, P., Schnetzer, K.J., Tong, B.C., Nerem, R.M.: Effect of uniaxial cyclic stretch on the morphology of monocytes/macrophages in culture. J. Biomech. Eng. **118**, 420–422 (1996)
17. Terracio, L., Miller, B., Borg, T.K.: Effects of cyclic mechanical stimulation of the cellular components of the heart: in vitro. In Vitro Cell. Dev. Biol. **24**(1), 53–58 (1988)
18. Thodeti, C.K., Matthews, B., Ravi, A., Mammoto, A., Ghosh, K., Bracha, A.L., Ingber, D.E.: TRPV4 channels mediate cyclic strain-induced endothelial cell reorientation through integrin-to-integrin signaling. Circ. Res. **104**, 1123–1130 (2009)
19. Xu, G., Li, B., Feng, X., Gao, H.: A tensegrity model of cell reorientation on cyclically stretched substrates. Biophys. J. **111**, 1478–1486 (2016)
20. Xu, G., Feng, X., Gao, H.: Orientations of cells on compliant substrates under biaxial stretches: a theoretical study. Biophys. J. **114**, 701–710 (2018)

Kinetic and Macroscopic Equations for Action Potential in Neural Networks

Martina Conte, Maria Groppi, and Andrea Tosin

Abstract Starting from the concept of binary interactions between pairs of particles, a kinetic framework for the description of the action potential dynamics on a neural network is proposed. It consists of two coupled levels: the description of a single brain region dynamics and the interactions among different regions. On one side, the pairwise interaction between neurons exchanging membrane potential is statistically described to account for the unmanageable number of neuron synapses within a single brain region. On the other, the network connections accounting for the brain region topology are represented and studied using concepts of the graph theory. Equilibria and stability of the obtained macroscopic systems are analyzed as well as numerical simulations of the system dynamics are performed in different scenarios. In particular, the latter allows us to observe the influence of the discrete network topology on the membrane potential propagation and synchronization through the different regions, in terms of its spiking characteristics.

1 Introduction

Collective behaviors and network dynamics have proven to play a key role in the modeling of complex systems in various fields. In particular, their importance is recognized in the mathematical description of information processing in the brain, both at macroscopic and cellular levels, and in related studies of synchronization phenomena. It is conjectured that synchronous brain activity is the most likely mechanism for many cognitive functions, such as attention and feature binding, as well as learning and memory formation. However, synchronization is not always useful, because brain disorders, such as schizophrenia, epilepsy, Alzheimer's and

M. Conte · A. Tosin (✉)
Department of Mathematical Sciences "G. L. Lagrange", Politecnico di Torino, Torino, Italy
e-mail: martina.conte@polito.it; andrea.tosin@polito.it

M. Groppi
Department of Mathematical, Physical, and Computer Sciences, University of Parma, Parma, Italy
e-mail: maria.groppi@unipr.it

A. d'Onofrio et al. (eds.), *Problems in Mathematical Biophysics*, SEMA SIMAI Springer Series 38, https://doi.org/10.1007/978-3-031-60773-8_5

Parkinson's diseases, have been linked to high levels of synchronization in the neuronal activities [5, 14].

Emerging collective behaviors like synchronization are strongly influenced by the microscopic interactions of the individual agents involved in the process. For this reason, in recent years, methodological approaches rooted in the statistical mechanics have gained a lot of momentum in the modeling and analysis of biophysical phenomena. Among them, we refer in particular to those which revisit methods of the classical kinetic theory to tackle systems that, at the microscopic scale, may be described as large ensembles of interacting particles. It is worth pointing out that, while in classical kinetic theory the word "particle" refers essentially to a gas molecule, in the modern kinetic theory for multi-agent systems it may indicate any kind of fundamental microscopic entity, whose interactions shape the dynamics of the entire system at hand. For instance, in this paper, a particle will be a brain cell, i.e., a neuron. The reason why the kinetic theory may be profitably exploited to investigate mathematically biophysical phenomena is that it provides a sound and flexible framework to bridge the often stochastic particle level, where elementary cell dynamics take place, and the aggregate level of tissue or organ dynamics. Without pretending to be exhaustive, we mention that, in the recent literature, examples of the use of kinetic theory methods for biophysical applications are found e.g. in the modeling of the onset and propagation of Alzheimer's disease [2, 3] as well as of tumor growth and therapeutic control strategies [8, 9, 19].

In addition to the revisitation of kinetic theory methods, the contemporary literature about interacting multi-agent systems is increasingly focussed on *networked* interactions. This means that, unlike gas molecules, the interactions among the considered particles are mediated by a background *graph structure* discriminating which interactions may actually take place and at which rate. Examples of kinetic descriptions of networked interactions can be found, for instance, in opinion formation problems [7, 16, 22] or in the modeling of the spread of infectious diseases [15]. In these problems, the background network may enter the big picture in essentially two ways. On one hand, particles themselves may be considered as the vertices of a usually large graph and interactions may take place only among particles directly connected by an edge of the graph. In this case, a *statistical* description of the graph topology, viz. of the connections among the particles, is typically necessary, due to the large number of particles and connections which would make the detailed modeling of the graph structure unfeasible. On the other hand, particles may populate the vertices of a relatively small graph, which describes the structure of the connections among different particle communities. Within each node particles typically experience "all-to-all" interactions, namely interactions not mediated by a network structure, whereas interactions among particles located in different nodes are possible only if an edge exists between the corresponding nodes. Alternatively, interactions may be limited to particles belonging to the same node but particles may migrate across the nodes following the edges of the graph. In this case, a statistical description of the graph is not necessary due to the small number of nodes and vertices, which are then modeled one by one in detail.

In this paper, we adopt the methodological ideas just outlined to address action potential dynamics in neural networks. In particular, we aim to take into account both levels of the network description discussed above to build a model which, starting from simple neuron-to-neuron exchanges of electrical stimuli, is able to represent the activation/deactivation and synchronization of entire brain regions. For this, we combine a statistical description of the (large) network of neuron synapses within each brain region with a detailed description of the (small) network connecting the brain regions themselves. This allows us to recover, as a particular case, classical macroscopic neural network models featuring a coupling among brain regions described by a discrete Laplacian. In standard approaches to neural networks, these models are usually postulated directly at the macroscopic scale of the brain regions [5, 14]. Instead, with our approach, they are obtained consistently with elementary interactions among single neurons and can be possibly enriched with aggregate features resulting from the statistical description of the in-region network of neuron synapses. At a macroscopic level, we get at the end a complex network in which the nodes interact pairwise through a set of links, that encode the network topology. The evolution of the whole network is described by a system of coupled differential equations, and we investigate in this paper the complete synchronization of the resulting neural networks of dynamical systems. This is the simplest form of synchronization and consists in a perfect convergence of the trajectories of the single node systems in the course of time [4], when some kind of coupling between them is introduced. The dynamical network is said to achieve globally (locally) asymptotic synchronization if the synchronous state, namely the solution in which the corresponding state variables are equal in each node, is globally (locally) asymptotically stable. The stability will be investigated by borrowing ideas from the Master Stability Function (MSF) approach [18], taking advantage of the simple linear structure of the obtained network model. The MSF extends Lyapunov stability methods to complex networks of identical (or almost identical [21]) oscillators and provides conditions for the local stability of the synchronous state, depending on the graph structure.

The paper is organized as follows. In Sect. 2, we describe and derive the spatially homogenous model for the action potential dynamics in a single node and its generalization to the spatially inhomogeneous case, accounting for the partition of the brain into different macro-areas. Then, in Sect. 3 we specify the microscopic rule of neuron-to-neuron interaction in the previously derived settings to describe the propagation of the action potential among neurons and we study the equilibria and their stability for both the spatially homogenous and inhomogeneous case. Section 4 collects four sets of numerical tests aimed at showing the influence of the brain region topology on the propagation and synchronization of the action potential, its spiking characteristics, and its timing. Finally, in Sect. 5 we draw some conclusions and illustrate the relevance and the possible outcomes of this work.

2 Model Set Up

We propose a kinetic-based model for the description of the spread of an "information" on a neural network. We use the word "information" to refer to any kind of mechanism whose dynamics are characterized by a neuron-to-neuron transmission. In this specific, we focus here on the propagation of the action potential among neurons. Considering the unmanageable number of nodes and links constituting the neuronal network that describes the connections among neurons, we propose a statistical description of the neuron connections. Precisely, to model the degree of connections of a single neuron we introduce the parameter c, which ranges between 0 and $+\infty$. Although this parameter c should be described by a discrete variable, at this stage, we approximate it as a continuous one. This allows us to define the function $g(c)$, representing the distribution of the neuron connections, such that the quantity

$$\mathbb{P}(c^*) := \int_0^{c^*} g(c)dc$$

describes the probability that a neuron has a number of connections between 0 and c^*.

In the following, we first assume that there is no spatial dependence on the function $g(c)$. This leads to a spatially homogenous model where each neuron has the same distribution of connections, independently of its spatial position. This choice is motivated by the fact the human brain can be divided into different macro-regions [20], each of them controlling a particular neural activity and whose neurons have similar characteristics. Therefore, the corresponding spatially homogenous model would describe the neuron dynamics inside each macro-area. Since the connections among macro-areas have been well described [6, 20], we then generalize the derived setting to the spatially inhomogeneous case, where $g = g(c, x)$. This second description allows us to take into account the variability of the distribution of neuron connections in the different brain regions. In this case, the system is properly described by a classical discrete network, where each node is a brain macro-area and each edge represents the connection between two macro-areas. Moreover, inside each node, the description of the connections would be given in statistical terms, following the spatially homogenous model, while the coupling terms between different macro-areas are deduced from the microscopic level. Indeed, we derive the macroscopic network model from a detailed description of the microscopic dynamics. This determines a novel justification and an original integration between a microscopic continuous and the macroscopic network descriptions.

2.1 Spatially Homogenous Case

Let $\mathcal{V}(t)$ be the dimensionless *membrane potential* of a single neuron [13], which we assume to evolve according to a neuron-to-neuron transmission mechanism, and $\mathcal{W}(t)$ the *recovery variable* associate to it. Precisely, the membrane potential is defined as the difference in electric potential between the interior and the exterior of a biological cell, depending on the concentration of different types of ions, among which sodium (Na^+) and potassium (K^+) play the major role. Its dynamics is driven by the opening of the Na^+ channels and the consequent emergence of a Na^+ electrochemical gradient. A typical dynamics of the membrane potential is the so called *action potential*, which is defined as a sudden, fast, transitory, and propagating change of the resting membrane potential. Only neurons and muscle cells are capable of generating an action potential and this property is called excitability. The recovery variable, instead, is introduced to acts as a negative feedback on the membrane potential, accounting for the activation of the K^+ ionic current and the inactivation of the Na^+ ionic current. Let $t \geq 0$ be the time variable and $C(t)$ the *neuron degree of connections*, the microscopic state of a single neuron is characterized by the triplet $(\mathcal{V}, \mathcal{W}, C)$ in the state space $[0, 1] \times [0, 1] \times [0, +\infty)$. As such, the microscopic variables $\mathcal{V}$ and $\mathcal{W}$ have to be regarded as independent of C, in the sense that their evolution is not a function of C and vice versa, and all variables are functions of time.

Assumption 1 *As it is reasonable to assume that no relevant changes of the cerebral tissue happen in the temporal window in which the action potential evolves, we assume that the number of connections C of a single neuron does not evolve in time.*

In order to lighten the notation, we set $\mathcal{V}_t := \mathcal{V}(t)$ and $\mathcal{W}_t := \mathcal{W}(t)$. We rely on the concept of binary interactions between pairs of neurons to derive a Boltzmann-type mesoscopic model to describe the evolution of the microscopic state $(\mathcal{V}_t, \mathcal{W}_t, C)$ depending on repeated individual interactions with other neurons. Considering a generic pair of neurons with microscopic states $(\mathcal{V}_t, \mathcal{W}_t, C)$, $(\mathcal{V}_t^*, \mathcal{W}_t^*, C^*) \in [0, 1] \times [0, 1] \times [0, +\infty)$, the effect of a binary interaction is modeled by

$$L(\mathcal{V}_t, \mathcal{W}_t, \mathcal{V}_t^*, \mathcal{W}_t^*) : [0, 1]^4 \to [0, 1]^2 ,$$

a prescribed function describing the interaction rule between the two pairs of neurons and accounting for the dynamics of the membrane potential and the recovery variable. To introduce a variation rule for $(\mathcal{V}_t, \mathcal{W}_t)$, we assume that in a short time interval $\Delta t > 0$ there is a probability proportional to Δt that a neuron undergoes the transmission mechanism. Precisely, we introduce the Bernoulli random variable $T_p \in \{0, 1\}$ of parameter p, such that

$$\mathbb{P}(\mathbb{T}_p = 1) = p, \quad \mathbb{P}(\mathbb{T}_p = 0) = 1 - p .$$

This random variable describes whether an interaction leading to a change of microscopic state takes place ($\mathbb{T}_p = 1$) or not ($\mathbb{T}_p = 0$). To include the statistical description of the neuron connections, we assume that $p = \Delta t\, G(C, C^*)$, where the function $G(C, C^*)$ takes into account the probability that a neuron with microscopic states $(\mathcal{V}, \mathcal{W}, C)$ is connected by a synapsis with a neuron with microscopic state $(\mathcal{V}^*, \mathcal{W}^*, C^*)$. The function $G(C, C^*)$ influences the frequency of interactions in the collision operator. Several different expressions for the function G can be proposed, each of them taking into account a different description of the probability of two neurons being connected. We derive the model in a general case without specifying the expression of G. Then, in the following sections, we would assume that $G(C, C^*) = 1$, indicating that all the neurons within a single brain region are connected. However, we remark that the general derivation here proposed allows for several different choices of G.

Considering $(\mathcal{V}, \mathcal{W})$ the membrane potential related variables, the microscopic law for the pair $(\mathcal{V}_t, \mathcal{W}_t, C)$, $(\mathcal{V}_t^*, \mathcal{W}_t^*, C^*)$ can be written as

$$\begin{cases} (\mathcal{V}_{t+\Delta t}, \mathcal{W}_{t+\Delta t}) = (1 - \mathbb{T}_p)(\mathcal{V}_t, \mathcal{W}_t) + \mathbb{T}_p(\mathcal{V}_t', \mathcal{W}_t') \\ (\mathcal{V}_{t+\Delta t}^*, \mathcal{W}_{t+\Delta t}^*) = (1 - \mathbb{T}_p)(\mathcal{V}_t^*, \mathcal{W}_t^*) + \mathbb{T}_p(\mathcal{V}_t^{*'}, \mathcal{W}_t^{*'}) . \end{cases} \tag{1}$$

We indicate with $(\mathcal{V}_t', \mathcal{W}_t') := (\mathcal{V}_t, \mathcal{W}_t) + L(\mathcal{V}_t, \mathcal{W}_t, \mathcal{V}_t^*, \mathcal{W}_t^*)$. To derive the Boltzmann-type equations for the evolution of $\mathcal{V}_t$ and $\mathcal{W}_t$, let us consider

$$\phi : [0, 1] \times [0, 1] \times [0, +\infty) \to \mathbb{R}$$

a test function representing any observable quantity that can be computed out of the microscopic state $(\mathcal{V}_t, \mathcal{W}_t, C)$ of a neuron. It holds

$$\phi(\mathcal{V}_{t+\Delta t}, \mathcal{W}_{t+\Delta t}, C) = \phi((\mathcal{V}_t, \mathcal{W}_t) + \mathbb{T}_p L(\mathcal{V}_t, \mathcal{W}_t, \mathcal{V}_t^*, \mathcal{W}_t^*), C) .$$

For simplicity of notation in the following calculations, we set $\mathcal{A} := (\mathcal{V}, \mathcal{W})$. Considering the average of both sides and computing the mean with respect to $\mathbb{T}_p$ we get

$$\begin{aligned} \langle \phi(\mathcal{A}_{t+\Delta t}, C) \rangle &= \langle (1 - p)\phi(\mathcal{A}_t, C) \rangle + \langle p\phi(\mathcal{A}_t', C) \rangle \\ &= \langle \phi(\mathcal{A}_t, C) \rangle + \langle p\phi(L(\mathcal{A}_t, \mathcal{A}_t^*) + \mathcal{A}_t, C) \rangle - \langle p\phi(\mathcal{A}_t, C) \rangle . \end{aligned} \tag{2}$$

Recalling that $p = \Delta t\, G(C, C^*)$, we can write

$$\begin{aligned} \langle \phi(\mathcal{A}_{t+\Delta t}, C) \rangle =& \langle \phi(\mathcal{A}_t, C) \rangle + \Delta t \, \langle G(C, C^*)\phi(\mathcal{A}_t + L(\mathcal{A}_t, \mathcal{A}_t^*), C) \rangle \\ &- \Delta t \, \langle G(C, C^*)\phi(\mathcal{A}_t, C) \rangle \end{aligned} \tag{3}$$

i.e.,

$$\frac{\langle\phi(\mathcal{A}_{t+\Delta t},C)\rangle-\langle\phi(\mathcal{A}_t,C)\rangle}{\Delta t}=\langle G(C,C^*)\left[\phi(\mathcal{A}_t+L(\mathcal{A}_t,\mathcal{A}_t^*),C)-\phi(\mathcal{A}_t,C)\right]\rangle . \tag{4}$$

In the limit $\Delta t\to 0^+$, we get

$$\frac{d}{dt}\langle\phi(\mathcal{A}_t,C)\rangle=\langle G(C,C^*)\left[\phi(\mathcal{A}'_t,C)-\phi(\mathcal{A}_t,C)\right]\rangle . \tag{5}$$

Going back to the previous notation, we have

$$\frac{d}{dt}\langle\phi(\mathcal{V}_t,\mathcal{W}_t,C)\rangle=\langle G(C,C^*)\left[\phi(\mathcal{V}'_t,\mathcal{W}'_t,C)-\phi(\mathcal{V}_t,\mathcal{W}_t,C)\right]\rangle . \tag{6}$$

Let us now define $f(v,w,c,t):[0,1]\times[0,1]\times[0,+\infty)\times\mathbb{R}_+\to\mathbb{R}$ the kinetic distribution function such that $f(v,w,c,t)=f(\mathcal{V}_t=v,\mathcal{W}_t=w,C_t=c,t)$. Precisely, $f(v,w,c,t)dvdwdc$ represents the fraction of neurons that at time t have membrane potential in $[v,v+dv]$, recovery variable in $[w,w+dw]$, and degree of connections in $[c,c+dc]$. We assume that the pair $(\mathcal{V}_t,\mathcal{W}_t,C)$, $(\mathcal{V}_t^*,\mathcal{W}_t^*,C^*)$ consists in independent processes, such that their joint law is given by $f(v,w,c,t)f(v_*,w_*,c_*,t)$. We can, thus, rewrite (6) as

$$\begin{aligned}\frac{d}{dt}&\int_0^1\int_0^1\int_0^{+\infty}\phi(v,w,c)f(v,w,c,t)dcdvdw\\ &=\iint_0^1\iint_0^1\iint_0^{+\infty}G(c,c_*)[\phi(v',w',c)-\phi(v,w,c)]\times\\ &\qquad f(v,w,c,t)f(v_*,w_*,c_*,t)dcdc_*dvdv_*dwdw_*\end{aligned} \tag{7}$$

where $(v',w'):=(v,w)+L(v,w,v_*,w_*)$ denotes the new state that a neuron with previous state (v,w) acquires after an interaction with another neuron with state (v_*,w_*). We assume that the number of connections is not affected by the interactions. Moreover, it has to hold

$$\int_0^1\int_0^1 f(v,w,c,t)dvdw=g(c)\quad\forall t>0 .$$

Considering Eq. (7), we can extract some information about the average macroscopic trends of our system. In particular, we can analyze the evolution of the average membrane potential $V(t)$ and the average recovery variable $W(t)$, defined as

$$V(t) := \int_0^{+\infty}\int_0^1\int_0^1 v f(v, w, c, t) dv dw dc\,,$$

$$W(t) := \int_0^{+\infty}\int_0^1\int_0^1 w f(v, w, c, t) dv dw dc\,, \tag{8}$$

respectively. Considering Eq. (7) and choosing the test function as $\phi(v, w, c) := v$ $\forall c \in [0, +\infty),\ \forall w \in [0, 1]$, we get

$$\frac{dV(t)}{dt} = \iint_0^{+\infty}\iint_0^1\iint_0^1 G(c, c_*)(v' - v) f(v, w, c, t) \times f(v_*, w_*, c_*, t) dv dv_* dw dw_* dc dc_*\,, \tag{9}$$

while, choosing $\phi(v, w, c) := w$ $\forall c \in [0, +\infty),\ \forall v \in [0, 1]$, we obtain

$$\frac{dW(t)}{dt} = \iint_0^{+\infty}\iint_0^1\iint_0^1 G(c, c_*)(w' - w) f(v, w, c, t) \times f(v_*, w_*, c_*, t) dv dv_* dw dw_* dc dc_*. \tag{10}$$

2.2 *Spatially Inhomogeneous Case*

The above introduced spatially homogeneous model can be generalized to the spatially inhomogeneous case to account for the partition of the brain into different macro-areas, each of them with a (possibly) different distribution of the neuronal connections [6]. Precisely, in this case, each neuron is identified by (x, v, w, c), where v indicates the membrane potential, w the recovery variable, c its degree of connections, and x the macro-areas to which it belongs. In particular, x is a discrete variable such that $x \in \{1, \ldots, N\} = \mathbb{I}$, with N the number of macro-areas. In this context, we define the distribution function $f : \mathbb{I}\times[0, 1]\times[0, 1]\times[0, +\infty)\times\mathbb{R}_+\to\mathbb{R}$ as a combination of the different distribution functions f_i defined in each of the macro-areas, i.e.,

$$f = f(x, v, w, c, t) := \sum_{i=1}^{N} f_i(v, w, c, t)\delta(x - i)\,. \tag{11}$$

In this case, it holds

$$\int_0^1 \int_0^1 f_i(v, w, c, t)dvdw = g_i(c) \quad \forall t > 0 \text{ and } \forall i \in \mathbb{I}\,,$$

where $g_i(c)$ is the probability distribution of the neuronal connections in the i-th macro-area. The kinetic equation reads

$$\begin{aligned}
&\frac{d}{dt}\int_{\mathbb{I}}\int_0^1\int_0^1\int_0^{+\infty} \phi(x, v, w, c) f(x, v, w, c, t)dcdwdvdx \\
&= \iint_{\mathbb{I}} \iint_0^1 \iint_0^1 \iint_0^{+\infty} B(x, x_*)G(c, c_*)\left[\phi(x, v', w', c) - \phi(x, v, w, c)\right] \times \\
&\qquad f(x, v, w, c, t) f(x_*, v_*, w_*, c_*, t)dcdc_*dwdw_*dvdv_*dxdx_*\,.
\end{aligned} \tag{12}$$

Here, $B(x, x_*)G(c, c_*)$ represents the collision kernel. Even though, in general, the collision kernel can be given as a unique function of (x, c, x_*, c_*), here we assume that it can be factorized in such a way that $B(x, x_*)$ describes the influence of the neuron location on the interactions, while $G(c, c_*)$ refers to their degree of connections. Substituting (11) into (12) we get

$$\begin{aligned}
&\frac{d}{dt}\sum_{i=1}^{N}\int_0^1\int_0^1\int_0^{+\infty} \phi_i(v, w, c) f_i(v, w, c, t)dcdwdv \\
&= \sum_{i,j=1}^{N} \iint_0^1 \iint_0^1 \iint_0^{+\infty} B_{ij}G(c, c_*)[\phi_i(v', w', c) - \phi_i(v, w, c)] \times \\
&\qquad f_i(v, w, c, t) f_j(v_*, w_*, c_*, t)dcdc_*dwdw_*dvdv_*
\end{aligned} \tag{13}$$

where $\phi_i(v, w, c) := \phi(x = i, v, w, c)$ and $B_{ij} := B(x = i, x_* = j)$. Fixing a certain index i and assuming that $\phi(x, v, w, c) = \tilde{\phi}(x)\Phi(v, w, c)$, with $\tilde{\phi}(x) = 1$

for $x = i$, while $\tilde{\phi}(x) = 0$ for $x \neq i$, we get

$$\frac{d}{dt}\int_0^1\int_0^1\int_0^{+\infty}\Phi(v,w,c)f_i(v,w,c,t)dcdwdv$$

$$=\sum_{j=1}^{N}\iint_0^1\iint_0^1\iint_0^{+\infty}B_{ij}G(c,c_*)[\Phi(v',w',c)-\Phi(v,w,c)]\times \tag{14}$$

$$f_i(v,w,c,t)f_j(v_*,w_*,c_*,t)dcdc_*dwdw_*dvdv_*.$$

Defining the average membrane potential and recovery variables for the i-th macro-areas as

$$V_i(t) := \int_0^{+\infty}\int_0^1\int_0^1 vf_i(v,w,c,t)dvdwdc\,,$$

$$W_i(t) := \int_0^{+\infty}\int_0^1\int_0^1 wf_i(v,w,c,t)dvdwdc\,, \tag{15}$$

respectively, we can derive the equations for the evolution of $V_i(t)$ and $W_i(t)$ in the spatially inhomogeneous case. Precisely, choosing $\Phi(v,w,c) := v\ \forall c \in [0,+\infty)$, $\forall w \in [0,1]$, we obtain

$$\frac{dV_i(t)}{dt}=\sum_{j=1}^{N}\iint_0^{+\infty}\iint_0^1\iint_0^1 B_{ij}G(c,c_*)(v'-v)f_i(v,w,c,t)\times \tag{16}$$

$$f_j(v_*,w_*,c_*,t)dvdv_*dwdw_*dcdc_*\,,$$

while, choosing $\Phi(v,w,c) := w\ \forall c \in [0,+\infty)$, $\forall v \in [0,1]$, we obtain

$$\frac{dW_i(t)}{dt}=\sum_{j=1}^{N}\iint_0^{+\infty}\iint_0^1\iint_0^1 B_{ij}G(c,c_*)(w'-w)f_i(v,w,c,t)\times \tag{17}$$

$$f_j(v_*,w_*,c_*,t)dvdv_*dwdw_*dcdc_*.$$

In the following, we refer to this spatially inhomogeneous case as network system.

3 Microscopic Rule for Action Potential Propagation

To derive the explicit equations for the average quantities $V(t)$ and $W(t)$, we need to specify the microscopic rule which would describe the propagation of the action potential between two neurons. Inspired by the nondimensional Morris–Lecar neuron model [17] and the FitzHugh–Nagumo model [10], we define

$$L(v, w, v_*, w_*) := (i_{ext} + \gamma(\bar{v} - v) + (v_* - v) - w, v - aw)$$

with i_{ext} representing all the external current sources stimulating the neuron, $\bar{v}$ the relaxation value for the membrane potential, while $\gamma > 0$ and $a > 0$ are dimensionless parameters. Here, the term $(v_* - v)$ is used to take into account at the microscopic level the coupling between two different neurons. Thus, the microscopic rule reads

$$\begin{cases} v' = v + i_{ext} + \gamma(\bar{v} - v) + (v_* - v) - w \\ w' = w + v - aw. \end{cases} \tag{18}$$

We recall that, depending on the choice of the function $G(c, c_*)$, it is possible to obtain different macroscopic spatially homogenous and inhomogeneous systems. As above stated, in these notes we choose $G(c, c_*) = 1$. Thus, from Eqs. (9) and (10), for the spatially homogenous case we obtain the following system

$$\begin{cases} \dfrac{dV(t)}{dt} = i_{ext} + \gamma(\bar{v} - V(t)) - W(t) \\ \dfrac{dW(t)}{dt} = V(t) - aW(t). \end{cases} \tag{19}$$

Instead, from Eqs. (16) and (17), for the network system we obtain

$$\begin{cases} \dfrac{dV_i(t)}{dt} = (i_{ext} + \gamma(\bar{v} - V_i(t)) - W_i(t)) \sum_{j=1}^{N} B_{ij} + \sum_{j=1}^{N} B_{ij}(V_j(t) - V_i(t)) \\ \dfrac{dW_i(t)}{dt} = (V_i(t) - aW_i(t)) \sum_{j=1}^{N} B_{ij} \qquad \forall i \in \mathbb{I}. \end{cases} \tag{20}$$

We notice that system (20) represents a simpler form of the Morris–Lecar model defined on a graph and the derived term $\sum_{j=1}^{N} B_{ij}(V_j(t) - V_i(t))$ relates directly to the definition of the Laplacian on a graph. This term, typical of electrical synapses modeling [5] and known as electric coupling, has not been added directly to the single-node system at the macroscopic level, but it has been derived from a more detailed definition of the microscopic dynamics. In particular, given the graph $\mathcal{G} = (\mathbb{I}, \mathbb{E})$ containing a set $\mathbb{I} = \{1, 2, \ldots, N\}$ of nodes (brain macro-area) and a set $\mathbb{E}$ of arcs (i, j) leading from initial node i to terminal node j, we indicate with B its adjacent matrix, taking into account the network topology into the system. For $i = j$, we set $B_{ij} = 1$, while for $i \neq j$ we set

$$B_{ij} = \begin{cases} b_{ij} > 0 & \text{if } (i, j) \in \mathbb{E} \\ 0 & \text{otherwise} . \end{cases} \tag{21}$$

We recall that an undirected graph is a graph where all the edges are bidirectional (namely, B is symmetric), while a directed graph refers to a graph in which the edges have a direction. For non-weighted undirected and directed graph, when $(i, j) \in \mathbb{E}$, we set $b_{ij} = 1$. Instead, the values b_{ij} could be lower or greater than 1 if the graph is weighted, i.e., when each edge has an associated numerical value.

3.1 Equilibria and Linear Stability of the Macroscopic Systems

Here, we study equilibria and stability of the spatially homogenous (19) and the network (20) systems. Starting from (19), we observe that the configuration

$$\begin{cases} V^* = \dfrac{i_{ext} + \gamma \bar{v}}{\gamma + \dfrac{1}{a}} \\ W^* = \dfrac{1}{a} V^* \end{cases} \tag{22}$$

is the unique equilibrium of the single node system. We can prove that it is asymptotically stable by studying the eigenvalues of matrix A associated with the single node dynamics, which is given by

$$A = \begin{pmatrix} -\gamma & -1 \\ 1 & -a \end{pmatrix} .$$

We notice that $det(A) = \gamma a + 1 > 0$, while $Tr(A) = -(\gamma + a) < 0$, thus the real part of the corresponding eigenvalues is negative. This means that the equilibrium is asymptotically stable. Moreover, we observe that

$$\begin{aligned} p(A-\lambda I)=0 &\Longleftrightarrow (-\gamma-\lambda)(-a-\lambda)+1=0 \\ &\Longleftrightarrow \lambda^2+\lambda(\gamma+a)+1+a\gamma=0 \\ &\Longleftrightarrow \lambda_{1,2}=\frac{-(a+\gamma)\pm\sqrt{(a-\gamma)^2-4}}{2}. \end{aligned} \tag{23}$$

Thus,

$$\lambda_{1,2}\in\mathbb{R} \quad \Longleftrightarrow \quad |a-\gamma|\geq 2\,.$$

If this condition is satisfied, the equilibrium is an asymptotically stable node, otherwise $\lambda_{1,2}\in\mathbb{C}$ and the equilibrium would be an asymptotically stable focus.

Referring to the network system (20), we can study its equilibria and stability using the framework of the Master Stability Function (MSF) approach, proposed in [18, 21]. Precisely, starting from (20) we observe that

$$\frac{dW_i(t)}{dt}=0 \Longleftrightarrow (V_i(t)-aW_i(t))=0 \Longleftrightarrow W_i(t)=\frac{1}{a}V_i(t) \quad \forall i=1\dots N$$

and

$$\frac{dV_i(t)}{dt}=0$$

$$\Longleftrightarrow (i_{ext}+\gamma(\bar{v}-V_i(t))-W_i(t))\sum_{j=1}^{N}B_{ij}+\sum_{j=1}^{N}B_{ij}(V_j(t)-V_i(t))=0$$

$$\Longleftrightarrow (i_{ext}+\gamma\bar{v})\sum_{j=1}^{N}B_{ij}-\left(\gamma+\frac{1}{a}\right)V_i(t)\sum_{j=1}^{N}B_{ij}+\sum_{j=1}^{N}B_{ij}(V_j(t)-V_i(t))=0$$

$$\Longleftrightarrow V_i(t)=\frac{i_{ext}+\gamma\bar{v}}{\gamma+\frac{1}{a}}+\frac{\sum_{j=1}^{N}B_{ij}(V_j(t)-V_i(t))}{\left(\gamma+\frac{1}{a}\right)\sum_{j=1}^{N}B_{ij}}.$$

Thus, the configuration

$$
\begin{cases}
V_i^* = \dfrac{i_{ext} + \gamma \bar{v}}{\gamma + \dfrac{1}{a}} & \forall i = 1 \dots N \\
\\
W_i^* = \dfrac{1}{a} V_i^* & \forall i = 1 \dots N \, .
\end{cases}
\tag{24}
$$

is the synchronization manifold for our system. In particular we denote by $\mathbf{x}^*$ this synchronization manifold, whose i-th component, indicated by $\mathbf{x}_i^*$, is the vector of the equilibrium state variables of the i-th region, i.e., $\mathbf{x}_i^* = (V_i^*, W_i^*)$, and it is given by

$$
\mathbf{x}_i^* = \begin{pmatrix} \dfrac{i_{ext} + \gamma \bar{v}}{\gamma + \dfrac{1}{a}} \\ \\ \dfrac{i_{ext} + \gamma \bar{v}}{a\gamma + 1} \end{pmatrix} \qquad \forall i = 1 \dots N.
$$

In order to use the MSF method for the stability analysis, we first rewrite system (20) using the classic notation of the graph theory as

$$
\dot{\mathbf{x}}_i(t) = F_i(\mathbf{x}_i) + M_i(\mathbf{x}_1, \dots, \mathbf{x}_N) \qquad \forall i = 1 \dots N. \tag{25}
$$

For each i-th node, $F_i(\mathbf{x}_i) := A_i \mathbf{x}_i + b_i$ is the vector field of single node dynamics, with

$$
A_i = B_i \begin{pmatrix} -\gamma & -1 \\ 1 & -a \end{pmatrix}, \qquad b_i = B_i \begin{pmatrix} i_{ext} + \gamma \bar{v} \\ 0 \end{pmatrix},
$$

and $B_i = \sum_{j=1}^N B_{ij}$ the degree of the i-th node. One difference with respect to the classical notation used in graph theory is that the vector field $F_i(\mathbf{x}_i)$ in this case depends on the i-th node. However, among the nodes this vector differs only for the multiplicative constant B_i, which does not influence the synchronization manifold or its stability. We are, thus, in the frame of a network of nearly identical dynamical systems [21]. Instead, $M_i(\mathbf{x}_1, \dots, \mathbf{x}_N)$ is the vector field taking into account the coupling of i-th node with all the other nodes to which it is connected, i.e.,

$$
M_i(\mathbf{x}_1, \dots, \mathbf{x}_N) = \begin{pmatrix} \sum_{j=1}^N B_{ij}(V_j - V_i) \\ 0 \end{pmatrix}.
$$

We recall the definition of the Laplacian of a graph, indicated by $L = (L_{ij})$, and the matrix for the diffusion profile, indicated by H. Precisely, the Laplacian on a graph is defined as

$$L_{ij} = \begin{cases} -B_{ij} & \text{for } i \neq j \\ B_i & \text{for } i = j \end{cases} \tag{26}$$

while, for the microscopic rule introduced in (18), the matrix describing the diffusion profile reads

$$H = \begin{pmatrix} 1 & 0 \\ 0 & 0 \end{pmatrix}.$$

Thus, system (25) can be rewritten as

$$\dot{\mathbf{x}}_i(t) = F_i(\mathbf{x}_i) - \sum_{j=1}^{N} L_{ij} H \mathbf{x}_j \qquad \forall i = 1 \ldots N. \tag{27}$$

To study the stability of the synchronization manifold $\mathbf{x}^*$, for each node we consider the perturbation $\boldsymbol{\delta}_i := \mathbf{x}_i - \mathbf{x}_i^*$ which satisfies the equation

$$\dot{\boldsymbol{\delta}}_i(t) = A_i \boldsymbol{\delta}_i - \sum_{j=1}^{N} L_{ij} H \boldsymbol{\delta}_j \qquad \forall i = 1 \ldots N. \tag{28}$$

Considering the Laplacian matrix L, if it is diagonalizable in $\mathbb{R}$, then exists a matrix T such that $TL = DT$, with D diagonal matrix such that $D_{ii} = \epsilon_i$, $\forall i = 1 \ldots N$, and $\epsilon_i \in \mathbb{R}$ eigenvalues of L. Thus, multiplying (28) by T and defining $\boldsymbol{\xi}_i = \sum_{k=1}^{N} T_{ik} \boldsymbol{\delta}_i^k$, we obtain the expression of system (28) in the canonical basis of the Laplacian eigenvalues

$$\dot{\boldsymbol{\xi}}_i(t) = (A_i - \epsilon_i H)\boldsymbol{\xi}_i \qquad \forall i = 1 \ldots N, \tag{29}$$

which represents the system of the normal modes. Here, the equilibrium $\boldsymbol{\xi}_i^* = \mathbf{0}$, $\forall i = 1 \cdots N$, is asymptotically stable if and only if the eigenvalues of the associated matrix

$$A_i - \epsilon_i H = B_i \begin{pmatrix} -\gamma - \dfrac{\epsilon_i}{B_i} & -1 \\ 1 & -a \end{pmatrix}$$

have a negative real part. We observe that

$$Tr(A_i - \epsilon_i H) = B_i(-a - \gamma - \frac{\epsilon_i}{B_i}) < 0 \Longleftrightarrow \epsilon_i > -B_i(a + \gamma)$$

$$det(A_i - \epsilon_i H) = B_i^2 \left(1 + a\left(\gamma + \frac{\epsilon_i}{B_i}\right)\right) > 0 \Longleftrightarrow \epsilon_i > -B_i\left(\gamma + \frac{1}{a}\right). \tag{30}$$

These two conditions can be written as

$$\epsilon_i > -B_i \min\left\{(a + \gamma), \left(\frac{1}{a} + \gamma\right)\right\}. \tag{31}$$

Recalling that ϵ_i are the eigenvalues of the Laplacian matrix L, we can observe that this condition is always satisfied in two particular cases.

- If we consider an undirected graph (weighted or not), then both B and L are symmetric. Therefore L is diagonalizable with real and positive eigenvalues, i.e., $\epsilon_i \in \mathbb{R}$ and $\epsilon_i > 0\ \forall i = 1 \cdots N$. Thus, condition (31) is always satisfied and $\mathbf{x}^*$ is an asymptotically stable manifold for system (20). It is worth noticing that the electrical coupling is bidirectional by construction, and this leads to biophysically meaningful undirected graphs for the network topology [5].
- If B is a triangular matrix, then also L would have the same structure. Thus, it is diagonalizable with real and positive eigenvalues that allow to always satisfy condition (31). This would be, for instance, the case of any directed graph for which $B_{ij} = 0$ for any $i > j$.

For a generic directed graph, it is not possible to obtain the same conclusion and the stability of the synchronization manifold would depend on the topology of the network described by B. As a final observation, we notice that, in the case of system (20), the condition that distinguishes the case of $\mathbf{x}^*$ to be a node or a focus, meaning the eigenvalues of $(A_i - \epsilon_i H)$ being real, reads

$$\left|a - \gamma - \frac{\epsilon_i}{B_i}\right| > 2.$$

4 Numerical Simulations

In this section, we present some numerical tests obtained by numerical integration of both the spatially homogenous model (9)–(10) with the microscopic rule (18), summarized in (19), and the corresponding network system (16)–(17) with the microscopic rule (18), summarized in (20). For the numerical simulations, we use a fourth order Runge–Kutta method. The simulation of the spatially homogenous

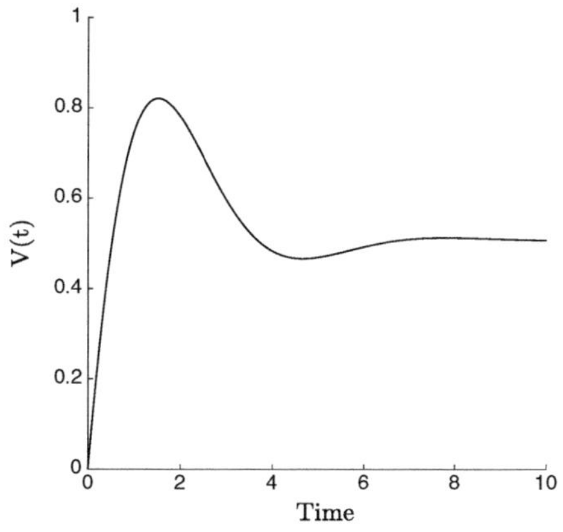

Fig. 1 Dynamics of the membrane potential variable $V(t)$ emerging from the system (19)

model is used for a matter of comparison to analyze how the underlying network introduced in the spatially inhomogeneous model affects the single node dynamics, in terms of timing of the spike, amplitude, and width. Setting $\bar{v} = 1$, $\gamma = 0.7$, $a = 0.6$, and $i_{ext} = 0.5$, in Fig. 1 we show the dynamics of the action potential in a single macro-area for the initial condition $V(0) = 0$ and $W(0) = 0$.
This result shows how the developed model is able to replicate the emergence of a spike in the evolution of the action potential, representing the depolarizing and repolarizing phases of the considered area, as well as the undershoot phenomenon and the convergence towards the resting state. From the observation done in Sect. 3.1 and considering the value of the parameters, we know that the equilibrium is an asymptotically stable focus.

To show the capability of our approach to capture a large variety of scenarios, we shall present four main numerical tests performed with the network system (20).

Test 1: in Sect. 4.1, we consider undirected and non-weighted graphs. We first analyze the case of a complete undirected graph, in order to compare the effects of this topology with respect to the single node dynamics. Then, we analyze a configuration built on five nodes differently connected, studying the effect of the underlying network on the spiking characteristics of each region with respect to the others.

Test 2: in Sect. 4.2, we specify the previously discussed example for a directed and non-weighted graph, looking at the differences between the undirected case and the role of the network in this directed example. In particular, this test is motivated by the fact that there can be two type of communication between neurons, meaning electrical connections, which is bidirectional, and excitatory/inhibitory connections, which is unidirectional, and a directed network allows us to model this second scenario.

Test 3: in Sect. 4.3, we specify the previous example to the case of an undirected and weighted graph. In particular, this test is motivated by the fact the physical distance between brain areas has an impact on the propagation of signals, such as the action potential, and a weighted network could be used to take this aspect into account.

Test 4: in Sect. 4.4 we consider the special case of a ring network in both the undirected and non-weighted case, directed and non-weighted case, and undirected and weighted case.

4.1 Test 1: Undirected and Non-Weighted Graph

We analyze here two examples of an undirected and non-weighted graph consisting of $N = 5$ macro-areas differently connected. The different configurations of nodes and edges are described through the adjacency matrix B_{ij}, for $i, j = 1 \dots N$ and, for all tests, we set the initial conditions $V_i(t) = W_i(t) = 0\ \forall i = 1 \dots N$. Moreover, we set $i_{ext} = 0.5$, $\bar{v} = 1$, $\gamma = 0.7$, and $a = 0.6$, as done for the simulation of Fig. 1. From the results of Sect. 3.1, we already know that the equilibrium configuration is asymptotically stable, independently of the specific graph topology.

Firstly, as **Example 1**, we consider a complete graph, meaning that we set $B_{ij} = 1$ for each $i, j = 1 \dots N$. In this case, the evolution of the potentials $V_i(t)$ is described by the left plot of Fig. 2, while on the right plot we compare the dynamics of a single node and of the 5 nodes network.

From the left plot of Fig. 2, we observe that considering a complete graph allows us to recover a perfect synchronization of the action potentials in all the nodes and for all times. The curves related to each of them are, in fact, perfectly overlapping and no differences can be grasped in either the spike time, width, or amplitude. Observing the comparison between the single node dynamics and the complete

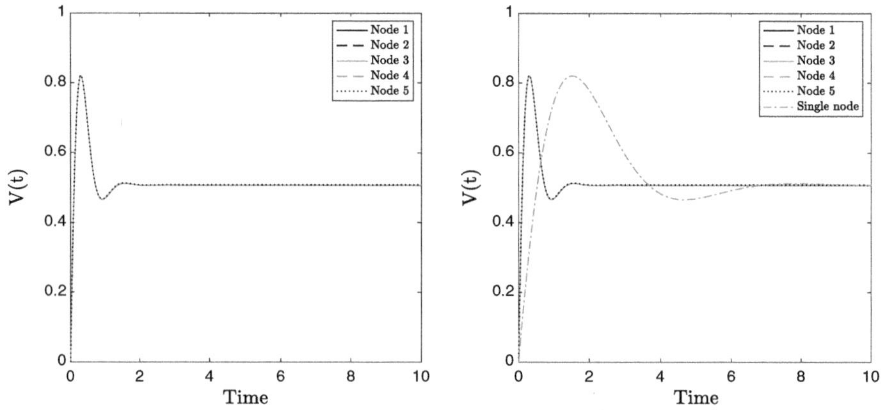

Fig. 2 Example 1. Dynamics of the potentials $V_i(t)$, for $i = 1 \cdots N$, emerging from system (20) when a complete graph is considered (left plot). Comparison between the single node and the complete graph dynamics (right plot)

graph dynamics in the right plot, we notice that the connections between the different macro-areas determine a faster activation of the membrane potential and a shorter duration of the spike with respect to the case of an isolated node, i.e., a node not connected with any other. Therefore, this allows us to hypothesize that the undirected connections among nodes support action potential activation and its faster propagation in the network.

As **Example 2**, we consider the graph described by the following symmetric adjacency matrix

$$B = \begin{pmatrix} 1 & 1 & 0 & 0 & 0 \\ 1 & 1 & 0 & 1 & 1 \\ 0 & 0 & 1 & 1 & 1 \\ 0 & 1 & 1 & 1 & 0 \\ 0 & 1 & 1 & 0 & 1 \end{pmatrix}. \tag{32}$$

The corresponding graph is schematized in the left plot of Fig. 3, while the dynamics of the potentials $V_i(t)$ are shown in the right plot of Fig. 3.

With respect to the dynamics emerging in Fig. 2, where a complete graph is considered, in Fig. 3 we observe that reducing the number of edges determines different delays in the action potential spike of each macro-area. Precisely, we can observe that the greater the number of connections of a node (meaning its degree), the earlier the spike occurs. In fact, node 2, which is the only one with three edges, shows the earliest spike; nodes 3, 4, and 5, which have the same degree, i.e., two edges, have almost identical dynamics with an intermediate spike emergence; node 1, instead, which has only one connection, is the last one to spike. Moreover, we can observe that, for the nodes with the same degree, the small differences in their action potential dynamics are determined by the characteristics of the node to which

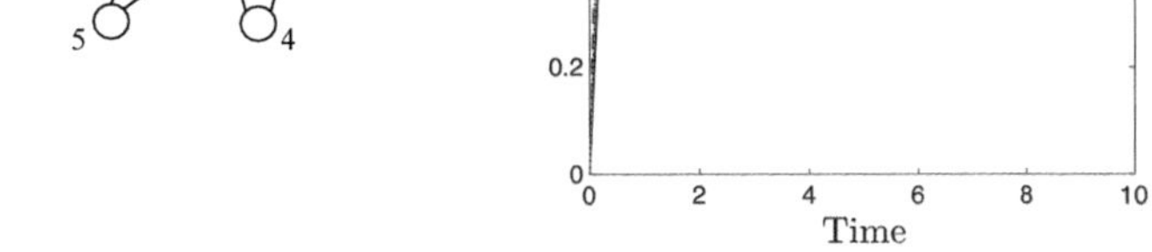

Fig. 3 **Example 2.** Schematic representation of the graph described by the adjacent matrix (32) (left plot) and the dynamics of the potentials $V_i(t)$, for $i = 1 \cdots N$, emerging from system (20) on this schematized network (right plot)

they are connected. In fact, nodes 4 and 5, which are connected with node 2 (that has the highest degree), have a slightly faster spike with respect to node 3.

4.2 Test 2: Directed and Non-Weighted Graph

Here, we analyze **Example 3**, which describes a directed and non-weighted graph consisting of $N = 5$ macro-areas differently connected. Precisely, we consider the same configuration described in **Example 2**, but choosing a specific orientation for each edge. The configuration of nodes and edges is described through the following (non symmetric) adjacency matrix

$$B = \begin{pmatrix} 1 & 1 & 0 & 0 & 0 \\ 0 & 1 & 0 & 0 & 1 \\ 0 & 0 & 1 & 0 & 0 \\ 0 & 1 & 1 & 1 & 0 \\ 0 & 0 & 1 & 0 & 1 \end{pmatrix}. \tag{33}$$

The initial conditions are set to $V_i(t) = W_i(t) = 0 \; \forall i = 1 \ldots N$, and the constant parameters are the same of **Test 1**. The graph is schematized in the left plot of Fig. 4, while the dynamics of the potentials $V_i(t)$ are shown in the right plot of the same Figure.

Comparing the right plots of Figs. 3 and 4, we can immediately grasp the differences between the emerging dynamics. When a directed graph is considered, in fact, the timing at which each node spikes is related to the net outflow of a node, meaning to the differences between the number of outgoing edges and the number of incoming edges, rather than to its degree (as in the undirected case). If we indicate

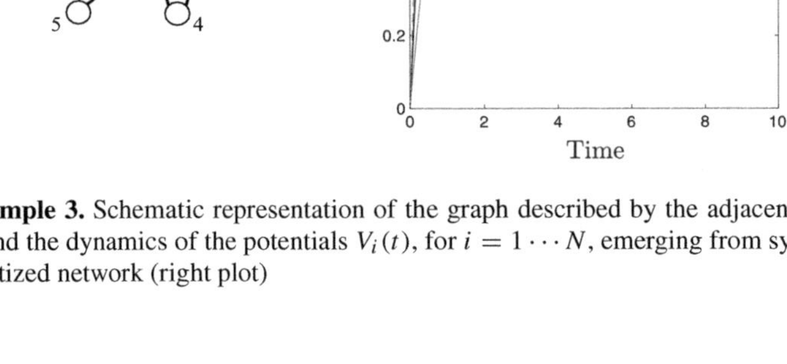

Fig. 4 **Example 3.** Schematic representation of the graph described by the adjacent matrix (33) (left plot) and the dynamics of the potentials $V_i(t)$, for $i = 1 \cdots N$, emerging from system (20) on this schematized network (right plot)

with O_i the net outflow of the i-th node, it holds

$$O_i = \sum_{j=1}^{N} (B_{ij} - B_{ji})\,.$$

Precisely, the higher the value of this net outflow, the earlier the spike. For instance, we observe the dynamics of node 4, for which $O_4 = 2$, having two outgoing edges and no incoming edges, and of node 3, which has $O_3 = -2$. Looking at the curves of the corresponding action potentials, node 4 shows the earliest spike, while node 3 is the latest. This example is clearly showing how much the network topology affects the system dynamics. In this case, we cannot use the results of Sect. 3.1 to infer about the equilibrium stability. However, we can numerically calculate the eigenvalues of the associated Laplacian matrix, which turns out to be $\epsilon_1 = \epsilon_2 = 1$, $\epsilon_3 = 3$, $\epsilon_4 = 2$, and $\epsilon_5 = 3$. Thus, condition (31) is satisfied and the equilibrium manifold is asymptotically stable.

4.3 Test 3: Undirected and Weighted Graph

For this test, we analyze the case of an undirected, but weighted graph consisting of $N = 5$ macro-areas. Precisely, we consider the same configuration described in **Example 2**, but choosing a specific weight for each edge. As for **Test 1**, the analysis in Sect. 3.1 allows us to directly conclude about the asymptotic stability of the equilibrium configuration, independently of the graph topology. The configuration of nodes and edges is described through the following symmetric adjacency matrix

$$B = \begin{pmatrix} 1 & 0.25 & 0 & 0 & 0 \\ 0.25 & 1 & 0 & 0.5 & 1 \\ 0 & 0 & 1 & 0.25 & 1 \\ 0 & 0.5 & 0.25 & 1 & 0 \\ 0 & 1 & 1 & 0 & 1 \end{pmatrix}. \tag{34}$$

We refer to this as **Example 4**, we set the initial conditions to $V_i(t) = W_i(t) = 0$ $\forall i = 1 \dots N$, and we use the same constant parameters of **Test 1** and **2**. In this case, the closer two macro-areas i and j are, the higher the value of B_{ij} is. The corresponding graph is schematized in the left plot of Fig. 5, while the dynamics of the potentials $V_i(t)$ are given in the right plot.

If we assign to each node a value corresponding to the sum of the weights on its edges, i.e, for the i-th node we define the quantity

$$T_i = \sum_{j=1}^{N} B_{ij}$$

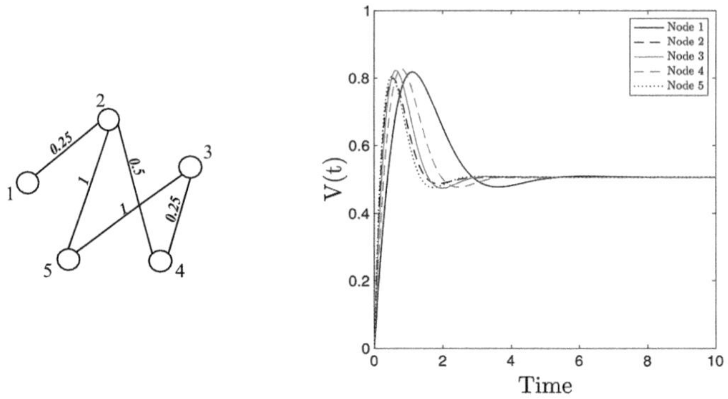

Fig. 5 **Example 4.** Schematic representation of the graph described by the adjacent matrix (34) (left plot) and the dynamics of the potentials $V_i(t)$, for $i = 1 \cdots N$ emerging from system (20) on this schematized network (right plot)

we can notice that the order of emergence of the spikes in Fig. 5 is directly related to T_i. Precisely, the greater T_i, the faster the i-th action potential spike. For instance, if we look at node 1, the related value $T_1 = 0.25$ is the smallest with respect to the other nodes and, in fact, it is the last one to spike. Instead, node 5, which has $T_5 = 2$, shows the fastest dynamics and, in fact, it is the node with the biggest value of T_i. This phenomenon means that faster dynamics are promoted in those nodes that are more connected to the others and also closer to them, with respect to less connected and/or farther nodes.

4.4 Test 4: Ring Graph

The last set of tests refers to the special case of a ring graph. We perform three sets of simulations, assuming the graph to be undirected (**Example 5**), directed (**Example 6**), or weighted (**Example 7**), respectively. Starting from the undirected case, we consider the following symmetric adjacency matrix

$$B = \begin{pmatrix} 1 & 1 & 0 & 0 & 1 \\ 1 & 1 & 1 & 0 & 0 \\ 0 & 1 & 1 & 1 & 0 \\ 0 & 0 & 1 & 1 & 1 \\ 1 & 0 & 0 & 1 & 1 \end{pmatrix}. \tag{35}$$

A scheme of this graph is given in Fig. 6 (left plot), while the right plot shows the action potential dynamics.

The particular topology of the ring net replicates the same behavior we observed in the case of a complete graph, i.e., full synchronization of the action potentials. In

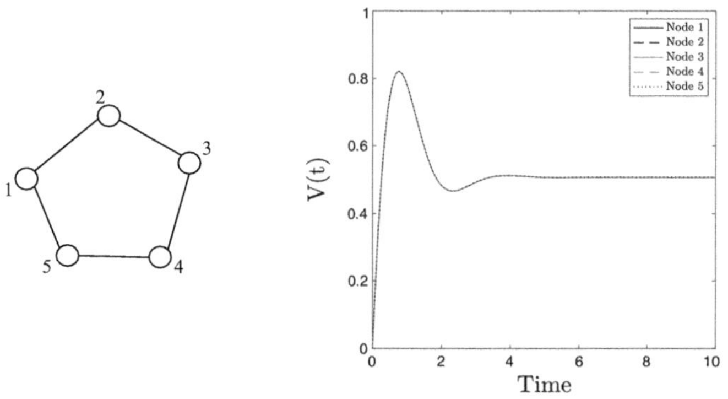

Fig. 6 Example 5. Schematic representation of the graph described by the adjacent matrix (35) (left plot). Dynamics of the potentials $V_i(t)$, for $i = 1 \cdots N$ emerging from system (20) (right plot)

fact, in this case, all the nodes are identical and it is not possible to distinguish their dynamics. Moreover, because the graph is undirected, the asymptotic stability of its equilibrium configuration is guaranteed.

Considering, instead, the case of a directed and non-weighted ring, we define two possible non-symmetric adjacency matrices, i.e.,

$$B = \begin{pmatrix} 1 & 1 & 0 & 0 & 0 \\ 0 & 1 & 1 & 0 & 0 \\ 0 & 0 & 1 & 1 & 0 \\ 0 & 0 & 0 & 1 & 1 \\ 1 & 0 & 0 & 0 & 1 \end{pmatrix} \tag{36}$$

and the analogous one in which we only change the direction of one edge, meaning $B_{15} = 1$ and $B_{51} = 0$. The dynamics of the potentials $V_i(t)$ in these two cases are shown in Fig. 7. Precisely, on the left plot we have the ring described by $B_{15} = 0$ and $B_{51} = 1$, while on the right plot $B_{15} = 1$ and $B_{51} = 0$.

The left plot of Fig. 7 highlights the fact that, when each node has exactly one incoming and one outgoing edge, the resulting action potential dynamics does not show any difference with respect to the corresponding undirected graph. In fact, the nodes still remain indistinguishable. Instead, changing the direction of at least one edge (as in the right plot of Fig. 7), drastically modifies the evolution. Node 1, which in this case has $O_1 = 2$, has the fastest dynamics. Instead, node 5, which has $O_5 = -2$, shows the slowest dynamics. The remaining nodes, which have $O_i = 0$ for $i = 2, 3, 4$, show the same timing for the spike, but different amplitudes. This effect could depend on the characteristics of the nodes to which they are connected: the more similar the connections of the adjacent nodes are, the closer to their neighbors the node dynamics are. Concerning the stability of the equilibrium, when the adjacent matrix is described by the modified version (36)

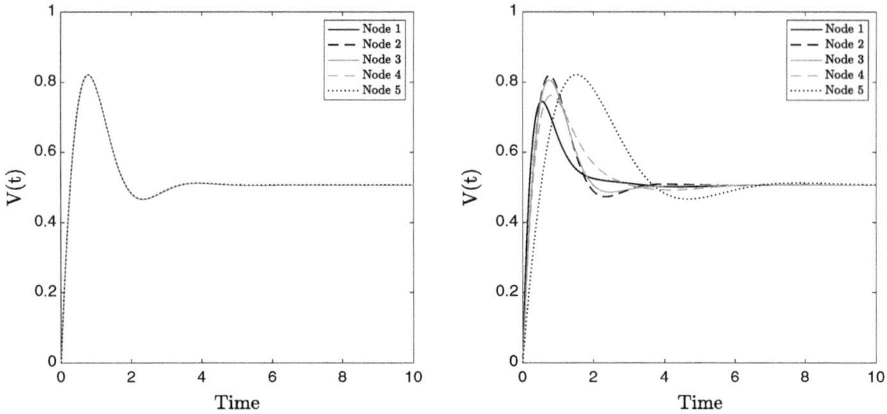

Fig. 7 **Example 6.** Dynamics of the potentials $V_i(t)$, for $i = 1 \cdots N$, emerging from system (20) when the connections are described by the adjacent matrix (36) with $B_{15} = 0$ and $B_{51} = 1$ (left plot) or with $B_{15} = 1$ and $B_{51} = 0$ (right plot)

with $B_{15} = 1$ and $B_{51} = 0$, the Laplacian matrix is triangular and, thus, from the results in Sect. 3.1, the equilibrium manifold is asymptotically stable. Instead, when $B_{15} = 0$ and $B_{51} = 1$, we can numerically calculate the eigenvalues of the Laplacian matrix that result to be complex and such that the Laplacian matrix is diagonalizable in $\mathbb{C}$. From the numerical study of the matrix associated with the system of the normal mode, it is possible to check the asymptotic stability of the equilibrium manifold.

Finally, we consider the weighted and undirected ring described by the following symmetric adjacency matrix

$$B = \begin{pmatrix} 1 & 0.25 & 0 & 0 & 0.25 \\ 0.25 & 1 & 1 & 0 & 0 \\ 0 & 1 & 1 & 1 & 0 \\ 0 & 0 & 1 & 1 & 0.5 \\ 0.25 & 0 & 0 & 0.5 & 1 \end{pmatrix}. \tag{37}$$

The dynamics of the graph and a schematic representation of it are given in Fig. 8. As for the case of the directed graph in Fig. 7 (right plot), assigning weights to the edges perturbs the synchronized dynamics shown in Fig. 6 (right plot). Thus, even if the graph is undirected, instead of having a fully synchronized network, in the case of a weighted graph we observe delays among the spikes, which are related to the value T_i assigned to each node. As for **Example 4**, the stability of the equilibrium is guaranteed by the results of Sect. 3.1.

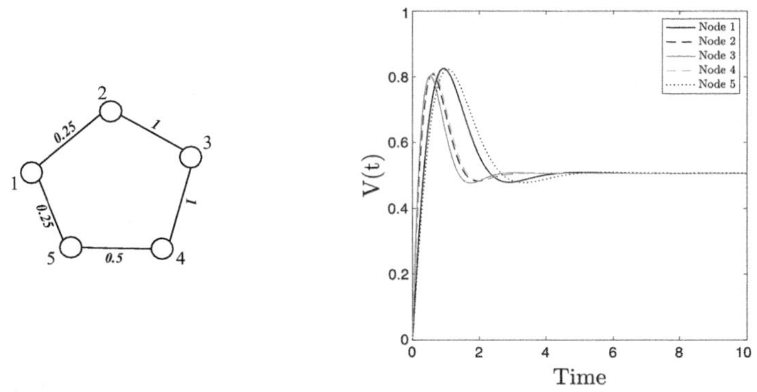

Fig. 8 **Example 7.** Schematic representation of the graph described by the adjacent matrix (37) (left plot) and the dynamics of the potentials $V_i(t)$, for $i = 1 \cdots N$, emerging from system (20) on this schematized network (right plot)

5 Discussion

In this note, we have studied the dynamics related to the spread of an information on a neural network, specializing the description to the case of the action potential propagation among neurons. On the basis of well-established studies on brain connectome [11, 20], we describe the human brain as a graph whose nodes correspond to the different brain regions and the edges to the connections among them, which can be obtained from tractography of diffusion tensor images [20]. The main focus of the proposed approach is the coupling between a local and statistical description of the interactions between neurons in a single brain region with a non-local and discrete description of the interactions between connected brain regions.

We analyze how the coupling affects the dynamics of the action potential at the macroscopic scale. Precisely, relying on the concept of binary interactions between pairs of neurons and starting from a microscopic description of them, we derive a Boltzmann-type mesoscopic equation in two specific scenarios. Firstly, considering a single brain region, we characterize each neuron by its number of connections and its membrane potential variables and we formally derived a spatially homogenous model from the microscopic description of neuron interactions. These are given in statistical terms using the function $g(c)$, which describes neuron connections. Then, we generalized the approach to the spatially inhomogeneous case, characterizing each neuron also by its discrete position in the brain connectome. This allows us to derive a macroscopic system for the average quantities describing the action potential evolution on the brain network. In particular, using this formalism we are able to obtain, in the equations at the macroscopic level, coupling terms of electric type among the different brain regions that correspond to the classical discrete Laplacian on a graph. However, here we do not add the coupling directly in the network system at the macroscopic level, but we derive it from a more detailed description of the microscopic dynamics, thus providing a microscopic justification for its macroscopic expression.

The specification of the microscopic interaction rule in the simplest and linear case allows us to explicitly study equilibria and stability of the spatially homogenous and inhomogeneous models, which result to be linear, using techniques from the Master Stability Function approach. In particular, we obtain a general result about the asymptotic stability of the synchronization manifold of the network macroscopic system for a general undirected graph, independently of the specific network topology. Instead, we specify the condition describing the dependence of this stability on the network characteristics in the case of a directed graph. The numerical simulations proposed in the last sections reinforce our observations about the influence of the specific brain region connections on the overall dynamics, showing how they affect the propagation and synchronization of the action potential, its spiking characteristics, and its timing.

As a first attempt, here we have chosen the simplest expression for the function $G(c, c_*)$ describing the probability of two neurons being connected. However, several possible choices for this can be considered. Each of them would give rise to a different macroscopic model for the average membrane potential and recovery variable in a single brain region. Future plans will involve the use of different expressions for this function and the analysis of the corresponding system, which could result to be not closed and for which only estimations of the macroscopic behavior of $V(t)$ and $W(t)$ could be studied. At the same time, similar challenges in the rigorous derivation of the macroscopic equations could emerge when other choices for the microscopic rule are considered, some of which might involve a nonlinear coupling between the microscopic variables [12]. Concerning the numerical results, we have shown here the most illustrative examples aimed at highlighting the main features of the influence of the network on action potential propagation. The choice of a simple network itself (five nodes differently connected) has been done with illustrative purposes and a more realistic description of the neural network is planned to be included in future works, as well as the application of the proposed approach to the description of specific disease progression [1, 2]. Finally, we have mainly focused here on the synchronization equilibrium of the system and on the network conditions that can alter it. However, also the asynchronous state might be of great interest. We have already obtained some preliminary results when an alternation of the input current on the different nodes is considered, and future plans include also a further analysis of asynchronization.

Acknowledgments The authors are members of INdAM-GNFM. The work has been performed in the frame of the Italian National Research Project "Integrated Mathematical Approaches to Socio–Epidemiological Dynamics" (Prin 2020JLWP23, CUP: E15F21005420006) and it was partially supported by the Ministry of University and Research through the MUR grant Dipartimento di Eccellenza 2018-2022, Project no. E11G18000350001. MC thanks the support of INdAM–GNFM Project "From kinetic to macroscopic models for tumor-immune system competition" (CUP_E53C22001930001). The research of MC has partially been performed during her research stay at the Department of Computational and Quantitative Medicine of the Beckman Research Institute of City of Hope National Medical Center (Duarte, CA, USA), which she thanks for the support and hospitality. The research of MG was granted by University of Parma through the action Bando di Ateneo 2022 per la ricerca co-funded by MUR-Italian Ministry of Universities and Research—D.M. 737/2021—PNR—PNRR—NextGenerationEU (project: "Collective and self-organised dynamics: kinetic and network approaches"). The work has been performed in the frame of the project PRIN 2022 PNRR "Mathematical Modelling for a Sustainable Circular Economy in Ecosystems" (project code P2022PSMT7, CUP D53D23018960001) funded by the European Union - NextGenerationEU and by MUR-Italian Ministry of Universities and Research.

Competing Interests The authors have no conflicts of interest to declare that are relevant to the content of this chapter.

References

1. Bertsch, M., Franchi, B., Marcello, N., Tesi, M.C., Tosin, A.: Alzheimer's disease: a mathematical model for onset and progression. Math. Med. Biol. **34**(2), 193–214 (2017)
2. Bertsch, M., Franchi, B., Tesi, M.C., Tosin, A.: Microscopic and macroscopic models for the onset and progression of Alzheimer's disease. J. Phys. A: Math. Theor. **50**(41), 1–22 (2017)
3. Bertsch, M., Franchi, B., Meschini, V., Tesi, M.C., Tosin, A.: A sensitivity analysis of a mathematical model for the synergistic interplay of amyloid beta and tau on the dynamics of Alzheimer's disease. Brain Multiphysics. **2**, 1–13 (2021)
4. Boccaletti, S., Latora, V., Moreno, Y., Chavez, M., Hwang, D.U.: Complex networks: structure and dynamics. Phys. Rep. **424**(4), 175–308 (2006)
5. Bonacini, E., Burioni, R., di Volo, M., Groppi, M., Soresina, C., Vezzani, A.: How single node dynamics enhances synchronization in neural networks with electrical coupling. Chaos Soliton Fract. **85**, 32–43 (2016)
6. Bullmore, E., Sporns, O.: Complex brain networks: graph theoretical analysis of structural and functional systems. Nat. Rev. Neurosci. **10**(3), 186–198 (2009)
7. Burger, M.: Network structured kinetic models of social interactions. Vietnam J. Math. **49**(3), 937–956 (2021)
8. Conte, M., Groppi, M., Spiga, G.: Qualitative analysis of kinetic-based models for tumor-immune system interaction. Discrete Contin. Dyn. Syst. Ser. B. **23**(9), 3663–3684 (2018)
9. Conte, M., Dzierma, Y., Knobe, S., Surulescu, C.: Mathematical modeling of glioma invasion and therapy approaches via kinetic theory of active particles. Math. Models Methods Appl. Sci. **33**(5), 1009–1051 (2023)
10. FitzHugh, R.: Impulses and physiological states in theoretical models of nerve membrane. Biophys. J. **1**(6), 445–466 (1961)
11. Fornari, S., Schäfer, A., Kuhl, E., Goriely, A.: Spatially-extended nucleation-aggregation-fragmentation models for the dynamics of prion-like neurodegenerative protein-spreading in the brain and its connectome. J. Theor. Biol. **486**, 110102 (2020)
12. Izhikevich, E.M.: Which model to use for cortical spiking neurons? IEEE Trans. Neural Netw. **15**(5), 1063–1070 (2004)
13. Izhikevich, E.M.: Dynamical Systems in Neuroscience. MIT Press, Cambridge (2007)
14. Jalili, M.: Synchronizing Hindmarsh-Rose neurons over Newman–Watts networks. Chaos. **19**(3), 033103 (2009)
15. Loy, N., Tosin, A.: A viral load-based model for epidemic spread on spatial networks. Math. Biosci. Eng. **18**(5), 5635–5663 (2021)
16. Loy, N., Raviola, M., Tosin, A.: Opinion polarization in social networks. Philos. Trans. R. Soc. A. **380**(2224), 1–15 (2022)
17. Morris, C., Lecar, H.: Voltage oscillations in the barnacle giant muscle fiber. Biophys. J. **35**(1), 193–213 (1981)

18. Pecora, L.M., Carroll, T.L.: Master stability functions for synchronized coupled systems. Phys. Rev. Lett. **80**(10), 2109 (1998)
19. Preziosi, L., Toscani, G., Zanella, M.: Control of tumor growth distributions through kinetic methods. J. Theor. Biol. **514**, 110579 (2021)
20. Sporns, O., Tononi, G., Kötter, R.: The human connectome: a structural description of the human brain. PLoS Comput. Biol. **1**(4), e42 (2005)
21. Sun, J., Bollt, E.M., Nishikawa, T.: Master stability functions for coupled nearly identical dynamical systems. Europhys. Lett. **85**(6), 60011(2009)
22. Toscani, G., Tosin, A., Zanella, M.: Opinion modeling on social media and marketing aspects. Phys. Rev. E. **98**(2), 1–15 (2018)

Mechanics-Based Models to Predict the Alignment of Cells on a Cyclically Stretched Substrate

Chiara Giverso, Giulio Lucci, and Luigi Preziosi

Abstract As pointed out by numerous experiments, the cell cytoskeleton appears to be responsive to external mechanical stimuli. In particular, experimental evidence demonstrates that a population of cells adhering to a substrate which is deformed periodically, for instance to mimic the heartbeat, responds to the strain by reorienting their stress fibres and focal adhesions. At the end of such a process, the cell achieves a specific orientation with respect to the strain direction, and such orientation depends on the characteristics of the external strain. The increasing interest in understanding mechanotransduction phenomena led to a growing attention to the cell realignment behaviour, both from the experimental and from the modelling point of view. Indeed, the contribution of mathematical models turns out to be very important to elucidate some relevant mechanisms and to suggest possible improvements in experimental assays. In this Chapter, we present an overview of mechanics-based mathematical models that have been proposed to describe cell reorientation and we highlight connections among the different research contributions within this field.

1 Phenomenological Observations

The response of cells to mechanical cues started to attract attention in the 1980s, following the study of cardiovascular pathophysiology. In particular, by observing the cellular arrangement in blood vessels, it was found that cells forming the walls of arteries were oriented along specific directions depending on their location. Endothelial cells in the innermost layer, in direct contact with blood flow, tended to be aligned in the axial direction of the vessel, while smooth muscle cells in deeper layers, e.g., in the intima and in the internal elastic lamina, exhibited an oblique or perpendicular orientation, forming helical-like structures, with cells sometimes

C. Giverso · G. Lucci · L. Preziosi (✉)
Department of Mathematical Sciences "G.L. Lagrange", Politecnico di Torino, Turin, Italy
e-mail: chiara.giverso@polito.it; giulio.lucci@polito.it; luigi.preziosi@polito.it

A. d'Onofrio et al. (eds.), *Problems in Mathematical Biophysics*, SEMA SIMAI Springer Series 38, https://doi.org/10.1007/978-3-031-60773-8_6

disposed at an angle of 20°–40° with respect to the vascular axial direction [16]. The pulsatile behaviour of the heart and the arteries, whose cells are constantly exposed to periodic deformations, stimulated the first investigations in vitro, with the aim of gaining a deeper understanding of cell mechanosensitivity and orientation in the circulatory system.

Motivated by his own studies on the orientation of cells in aortic walls, Buck was the first to examine the response of cells to mechanical cues in vitro [4], by seeding a cell population on a rubber plastic substrate which was then cyclically stretched, to mimic the periodic vessel inflation during pulsatile flow. Similarly to cells *in vivo*, he found that nearly 81% of fibroblasts tended to reorient between 45° and 90° with respect to the stretching direction. This result attracted some interest since it firstly put in evidence such an alignment behaviour of stimulated cells in vitro, paving the way for successive investigations, as reviewed in detail in [14]. Remarkably, it was observed that the reorientation behaviour was almost cell-type independent, namely, epithelial cells, endothelial cells, fibroblasts, osteoblasts, melanocytes, mesenchymal stem cells, in addition to the already cited muscle-type cells, all responded in a similar way.

Specifically, in all experiments, it was observed that cells actively responded to the exerted stretch by reorienting and generating strongly linked actin bundles that are named stress fibres (SFs) oriented along specific directions, with a consequent morphological elongation of the cell in the same direction. The SFs are then linked to the external substrate by focal adhesion complexes, so that through them the cell is able to sense the mechanical deformation, to transfer the information inside the cell and to respond accordingly. In particular, it seems that, when submitted to an external stretch, the cell reorganizes its internal structure, in order to relieve the state of stress. A confirmation of the central role played by the cytoskeleton is given by the experiments performed by Wang et al. [40] who observed that the use of an inhibitor of cellular contractility disrupted the stress fibres and blocked cell reorientation in response to cyclic stretching. For a similar reason, immune cells seem to be less responsive [26], since they do not possess a strong cytoskeleton and do not adhere firmly to the substrate. In fact, their role is to patrol the body for potential pathogens, and therefore to squeeze through endothelial linings with an amoeboid motion, with little involvement of adhesion molecules.

In addition to being a common response to several cell types, it seems that the phenomenon of cell realignment is nearly independent of the applied frequency and amplitude of the periodic deformation, as well as of the stiffness of the substrate, provided that all these factors are large enough.

Starting from these phenomenological observations, the aim of this Chapter is to summarize the mechanics-based approaches which have been adopted so far to describe the cell realignment behaviour under stretch. Indeed, in recent years, several mathematical models have been proposed to target such a problem, with the purpose of better understanding mechanotransduction phenomena, which are still not fully elucidated.

In detail, after the introduction of some basic notation in Sects. 2, 3 and 4 will examine the outcome of models based on the assumption that cells tend to minimize, respectively, either the strain or the stress acting on them. Along the same line, Sect. 5 will describe some models focusing on the minimization of a very general elastic energy, giving an explanation of why the equilibrium distribution of cells turns out to be so robust. Then, in Sect. 6 we will discuss a viscoelastic model, to give an explanation of the fact that, at low frequencies, the reorientation phenomenon is not observed.

2 Basic Notation

The two-dimensional set-ups that are typically used to test the behaviour of cells under stretch consist of a silicon or polydimetylsiloxane (PDMS) membrane, very often coated with collagen or fibronectin to favour cell attachment. Cells are then seeded on the substrate at a certain density: after that, the substrate is pulled along one or two perpendicular directions, either statically or periodically with different waveforms. In most of the experiments a cyclic stretch in only one direction is applied.

A simple pulling of two opposite sides would naturally lead to a narrowing in the central region, as shown in Fig. 1a, where the thin black lines refer to the undeformed square specimen, whereas the color map indicates the first principal strain in the deformed sample. In these cases, data are usually reported for those cells in the central region of the specimen (denoted by the dashed lines and shown magnified in the inset of Fig. 1a), where the stress and strain are almost homogeneous with principal directions along the direction of pulling and the perpendicular one.

So, assuming that the deformation applied to the substrate can be taken as purely extensional, the deformation gradient can be written as

$$\mathbb{F}(t) := \mathrm{diag}\{\lambda_x(t), \lambda_y(t), \lambda_z(t)\} = \mathrm{diag}\{1 + \delta_x(t), 1 + \delta_y(t), 1 + \delta_z(t)\},$$

where x is the direction characterized by the maximum deformation, that will also be denoted in the following as the *main stretching direction*, and z is the direction perpendicular to the substrate plane. We also introduce the quantity $R(t)$ such that $\delta_y(t) = -R(t)\delta_x(t)$, that represents, for instance in static conditions, the percentage of contraction in the y-direction with respect to the extension in the x-direction. It is observed that during cyclic deformations $R(t)$ is nearly constant, so that an important parameter is the so-called *bi-axiality ratio*

$$r := -\frac{\max\limits_t \delta_y(t)}{\max\limits_t \delta_x(t)}.$$

In order to avoid the central narrowing of the specimen, some countermeasures are sometimes adopted, like thickening the borders parallel to the main stretching

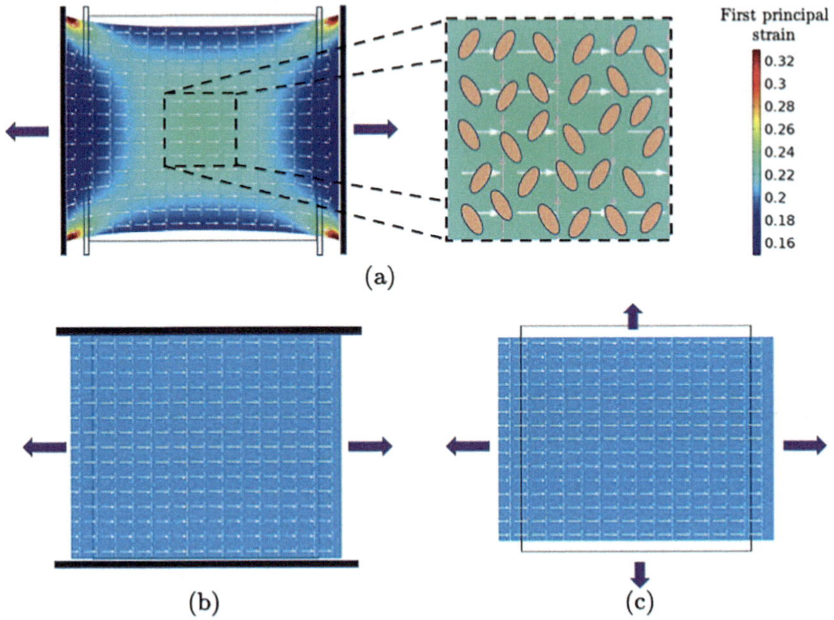

Fig. 1 Types of 2D deformations. In all panels, the thin black lines refer to the undeformed square specimen, whereas the colour plots refer to the first principal strain in the deformed sample. White and grey arrows represents the principal directions of the strain. The purple thick arrows represent the direction of the applied load/deformation. (**a**) Deformed substrate subject to a given deformation (10% strain) on two opposite sides while the others are stress-free. (**b**) Purely uniaxial stretching. (**c**) Simple elongation (when the top and bottom boundary of the specimen are stress-free) and biaxial deformation

direction (as done, for instance, in [8]) or attaching that side to a more rigid structure or substratum. This procedure allows to have a larger region with the desired stress and strain fields, if not the entire specimen. If these borders are fixed as in Fig. 1b, then $\delta_y = r = 0$ and it is said that the specimen undergoes a *pure uniaxial stretching*. If they are allowed to move inwards while staying parallel to the main stretching direction, as in Fig. 1c, then $\delta_y < 0$ and $r > 0$, and it is said that the specimen undergoes a *simple elongation*. In any case, most experiments are performed with values of $r \in [0, 1]$. A negative value of r is obtained if the substrate is pulled in both directions, as in [18, 21]. In particular, an equi-biaxial extension corresponds to $r = -1$. At the other extremum, a value larger than 1 corresponds to compression along y while pulling along x, which, to our knowledge, has not been done in experimental assays.

A cell that is oriented along a direction identified by a unit vector $\mathbf{N}$ undergoes a strain along that direction given by

$$\epsilon_N := \sqrt{(\mathbb{F}\mathbf{N}) \cdot (\mathbb{F}\mathbf{N})} - 1 = \sqrt{\mathbf{N} \cdot \mathbb{C}\mathbf{N}} - 1 := \sqrt{\mathrm{I}_4} - 1\,, \tag{1}$$

where $\mathbb{C} := \mathbb{F}^T\mathbb{F}$ is the right Cauchy-Green strain tensor, whereas $I_4 := \mathbf{N} \cdot \mathbb{C}\mathbf{N}$ is known in elasticity as the fourth invariant that quantifies the squared stretch experienced in direction $\mathbf{N}$. If we work in two dimensions, so that we can denote $\mathbf{N} = (\cos\theta, \sin\theta, 0)$, then we have

$$\begin{aligned}\epsilon_N(t) &= \sqrt{[1+\delta_y(t)]^2 + [\delta_x(t) - \delta_y(t)][2+\delta_x(t)+\delta_y(t)]\cos^2\theta} - 1 \\ &\approx \sqrt{[1 - r\delta_x(t)]^2 + [2+(1-r)\delta_x(t)](1+r)\delta_x(t)\cos^2\theta} - 1\,. \qquad (2)\end{aligned}$$

In the small deformation regime, the infinitesimal strain tensor $\mathbb{E} := \text{diag}\{\epsilon_{xx}, \epsilon_{yy}, \epsilon_{zz}\}$ is usually introduced and δ_i is replaced by ϵ_{ii}. Hence, $\mathbb{C} \approx \mathbb{I}+2\mathbb{E}$, and $r \approx -\epsilon_{yy}/\epsilon_{xx}$. In this limit

$$\epsilon_N \approx \mathbf{N}\cdot\mathbb{E}\mathbf{N} = \epsilon_{xx}\cos^2\theta + \epsilon_{yy}\sin^2\theta = [(1+r)\cos^2\theta - r]\epsilon_{xx} := \hat{I}_4\epsilon_{xx}\,, \qquad (3)$$

as can be also obtained by linearizing Eq. (2). We notice that the term $\hat{I}_4 = [(1+r)\cos^2\theta - r]$, which will appear in many models focusing on single stress fibres, is nothing else but the approximation of $I_4 - 1$ in linear elasticity.

For the sake of simplicity, in the following we will denote ϵ_{xx} simply by ϵ and the maximum strain by $\epsilon_0 = \max_t \epsilon_{xx}(t)$.

3 Strain Avoidance

Historically, the first modelling approaches to describe cell reorientation under stretch were based on a strain minimization principle, which led to naming the phenomenon as *strain avoidance*. According to this simple hypothesis, the cell preferentially reorients in the direction where the minimum strain is sensed. A direct application of this principle to Eqs. (2), or (3), readily leads to the fact that the only stable equilibrium should satisfy

$$\cos^2\theta_{eq} = \frac{r}{1+r}\,.$$

With respect to this trivial approach, Wang et al. [37] assumed that there was a maximum strain $\bar{\epsilon}_N$ that can be sustained by cells. As a consequence, the cell avoids any direction along which the perceived strain is greater than the maximum threshold. So, by considering the maximum strain δ_0 and inverting Eq. (2) they obtained the condition

$$\cos^2\theta \le \frac{2\bar{\epsilon}_N + \bar{\epsilon}_N^2 + 2r\delta_0 - r^2\delta_0^2}{[2+(1-r)\delta_0](1+r)\delta_0}\,, \qquad (4)$$

which determines a critical angle

$$\theta_{cr} = \cos^{-1}\sqrt{\frac{2\bar{\epsilon}_N + \bar{\epsilon}_N^2 + 2r\delta_0 - r^2\delta_0^2}{[2+(1-r)\delta_x](1+r)\delta_0}}.$$

Actually, in [37] the $\bar{\epsilon}_N^2$ term is not present, probably due to an approximation or to the fact that they use the Green-Lagrange tensor $\mathbb{G} = \frac{1}{2}(\mathbb{C} - \mathbb{I})$ as deformation measure. This approach was later applied in [27], where Eq. (2) is written (but with a misprinted sign and two ξ's that should be δ's).

In [38], the model by Wang was a bit more elaborated and accounted for actin filament dynamics. More in detail, he evaluated the change in the strain energy of a fibre due to the action of the axial strain ϵ_f transmitted to it. It was also assumed that only a fraction of the substrate strain was effectively transferred to the filaments, that is, $\epsilon_f = \alpha_t\epsilon_N$ in tension and $\epsilon_f = \alpha_c\epsilon_N$ in compression, with $\alpha_t < \alpha_c$ and ϵ_N as in Eq. (2). The model is then proposed under the following assumptions:

(i) in the absence of deformation, each filament of actin owns a basal strain energy given by $E_b = k\delta^2/2$, where δ is the pre-strain of the SF due to its inherent tension; then, the total energy of the fibre is $E_f = k(\delta + L\epsilon_f)^2/2$, where L is the initial length of the filament;
(ii) only the normal substrate strain ϵ_N is transmitted to individual actin filaments;
(iii) the actin filaments cannot bear compression;
(iv) the actin filaments undergo disassembly if their strain energy E_f is changed to zero or to a value greater than twice their basal strain energy E_b.

The limits in assumption (iv) are crucial because they allow to identify through Eq. (3) an interval of admissible orientations

$$\theta \in \left[\cos^{-1}\sqrt{\frac{r}{1+r} + \frac{(\sqrt{2}-1)\delta}{(1+r)\alpha_t L\epsilon}},\ \cos^{-1}\sqrt{\frac{r}{1+r} - \frac{\delta}{(1+r)\alpha_c L\epsilon}}\right]. \tag{5}$$

Then, to simplify the resulting formulas, it is conveniently assumed that the relation $\alpha_t = (\sqrt{2}-1)\alpha_c$ between coefficients holds.

In the limit $\dfrac{\delta}{\alpha_t L\epsilon}, \dfrac{\delta}{\alpha_c L\epsilon} \to 0$ the interval in Eq. (5) collapses to the point

$$\theta_{eq} = \cos^{-1}\sqrt{\frac{r}{1+r}} = \tan^{-1}\sqrt{\frac{1}{r}}, \tag{6}$$

while its width increases when the above values increase and, in particular, if ϵ decreases. Conversely, according to this model, increasing ϵ (without going into conflict with the small strain hypothesis) leads to a narrowing of the range of angles for which filament disassembly does not occur. This intriguingly well correlates with the fact that higher strain amplitudes lead to narrower ranges of measured

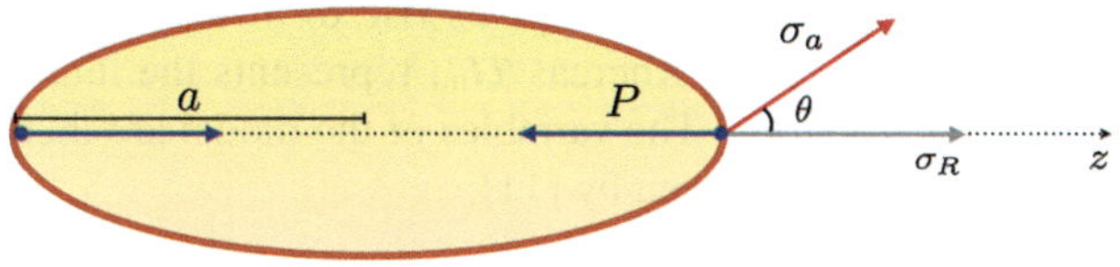

Fig. 2 Sketch of the cell dipole description used in some models of realignment. The variables of the model are the contractile dipole force $P < 0$ and the angle θ at which the external stress σ_a is applied

orientation angles, and therefore to more peaked distribution probabilities of the orientation. A similar approach was also used in [3] and [13].

A slightly different approach, though still within the strain avoidance class, was used by De and coworkers [10, 11]. In particular, using a coarse-grained model, they consider the cell as a force dipole (sketched in Fig. 2) to mimic the contraction of stress fibres along a given direction. Then, it is assumed that cell reorganisation happens with the aim of reaching an optimal level of strain ϵ_{target}. As a consequence, a reaction strain in the surrounding matrix is identified:

$$\epsilon_R = -\frac{(1+r)P}{\pi a^3 E},$$

where P is the (negative) cell contractile dipole magnitude, E is the Young modulus of the matrix and a is the cell size. Instead, the component of the applied strain along the cell axis is

$$\epsilon_a = \frac{\sigma_a}{E}\hat{\mathrm{I}}_4, \tag{7}$$

with σ_a the external uniaxial applied stress. Starting from these considerations, a free energy is built:

$$\mathcal{U}_\varepsilon = \mathcal{U}_{\text{cell},\epsilon} + \mathcal{U}_{\text{self}} + \mathcal{U}_{\text{int}},$$

where

$$\begin{aligned} \mathcal{U}_{\text{cell},\epsilon} &= \frac{1}{2}\chi\left[\frac{\sigma_a}{E}\hat{\mathrm{I}}_4 - \frac{P(1+r)}{\pi a^3 E} - \epsilon_{\text{target}}\right]^2, \\ \mathcal{U}_{\text{self}} &= (1+r)\frac{10r^2 - 14r + 9}{30(1-r)^2}\frac{P^2}{2\pi a^3 E}, \\ \mathcal{U}_{\text{int}} &= P\frac{\sigma_a}{E}\hat{\mathrm{I}}_4. \end{aligned} \tag{8}$$

In detail, the contribution $\mathcal{U}_{\text{cell},\epsilon}$ is the free energy related to the contractile activity of the cell, which is minimized when $\epsilon_a + \epsilon_R = \epsilon_{\text{target}}$, i.e. when the cell feels the optimal strain. On the other hand, $\mathcal{U}_{\text{self}}$ gives rise to a force that tends to reduce the magnitude of the dipole [10], whereas $\mathcal{U}_{\text{int}}$ represents the interaction between the dipole and the external strain. The variables of the model are therefore P and θ, with a dynamics assumed to be given by [11]:

$$\frac{dP}{dt} = -\frac{1}{\tau_P}\frac{\partial \mathcal{U}_\epsilon}{\partial P}, \qquad \frac{d\theta}{dt} = -\frac{1}{\tau_\theta}\frac{\partial \mathcal{U}_\epsilon}{\partial \theta}. \tag{9}$$

Actually, in a subsequent work [10], the equation for θ is modified by multiplying for a factor $1/P^2$, even though the results are only altered slightly. When a cyclic deformation with a high frequency is applied, in the form $\sigma_a(t) = \sigma_a(1 - \cos\omega t)$, the energy may be averaged over a period, since the cell is not able to follow the stress instantaneously. The resulting averaged energy has an additional contribution, namely,

$$\langle \mathcal{U}_\epsilon \rangle = \mathcal{U}_{\text{cell},\epsilon} + \mathcal{U}_{\text{self}} + \mathcal{U}_{\text{int}} + \mathcal{U}_{\text{av},\epsilon}\,, \tag{10}$$

in which the first three terms have the same form as Eq. (8) and

$$\mathcal{U}_{\text{av},\epsilon} = \frac{1}{4}\chi\left(\frac{\sigma_a}{E}\hat{\text{I}}_4\right)^2. \tag{11}$$

The stationary solutions can be derived by imposing that the derivatives of such an energy with respect to both P and θ are null. These steady state solutions are found to be $\theta_{eq} = 0$, $\theta_{eq} = \pi/2$ and an oblique one, which is defined in terms of its squared cosine by the implicit relation:

$$\cos^2\theta_{eq} = 1 + \frac{P_{eq}(\theta_{eq})}{\pi a^3 \sigma_a} - \left(1 + \frac{E P_{eq}(\theta_{eq})}{\chi \sigma_a} - \frac{E \epsilon_{\text{target}}}{\sigma_a}\right)\frac{1}{1+r}, \tag{12}$$

in a suitable range of parameters. In particular, if the interaction forces with the extracellular matrix are negligible, one can assume that $P \approx 0$ and, if $\epsilon_{\text{target}} = 0$ is taken, Eq. (6) is recovered from Eq. (12).

As we will discuss in Sect. 4, the same authors also explored the possibility that cell remodelling aims at reaching a target stress rather than a target strain.

As a final remark, we mention that the dipole approach has been generalized to the case of several cells in [43], while in [9], still in the same framework, more attention is paid to the dynamics of focal adhesions. A generalization of the dipole approach is also proposed in [29] to investigate the cell response to point loads.

4 Stress Avoidance

Within the same framework as the one discussed in Sect. 3, in [11, 12, 34] the alternative hypothesis of *stress avoidance* is examined. Specifically, it is assumed that the cell attempts to reach a target stress rather than a target strain by means of its contractile activity. In this case, the reaction stress in the matrix writes as

$$\sigma_R = -\frac{2-r}{1-r}\frac{P}{2\pi a^3},$$

with the same notation as above. The force component due to the applied stress on the surrounding matrix along the direction of the cell axis is given by $\sigma_a \cos^2\theta$. The free energy of the system, which includes the active contribution from the cell cytoskeleton and the forces acting on the cellular dipole due to the elasticity of the matrix, becomes

$$\mathcal{U}_\sigma = \mathcal{U}_{\text{cell},\sigma} + \mathcal{U}_{\text{self}} + \mathcal{U}_{\text{int}},$$

where

$$\mathcal{U}_{\text{cell},\sigma} = \frac{1}{2}\chi\left(\sigma_a \cos^2\theta - \frac{2-r}{1-r}\frac{P}{2\pi a^3} - \sigma_{\text{target}}\right)^2$$

is the active energy of the cell, whereas the contributions due to the interactions of the dipole with the surrounding matrix and the external stress are the same as in Eq. (8). Instead, σ_{target} identifies the target stress of the cell, so that $\mathcal{U}_{\text{cell},\sigma}$ is minimized if the sum of the reactive stress σ_R and the applied stress σ_a is equal to the target. If an averaging over a cycle is performed, the energy becomes

$$\langle\mathcal{U}_\sigma\rangle = \mathcal{U}_{\text{cell},\sigma} + \mathcal{U}_{\text{self}} + \mathcal{U}_{\text{int}} + \mathcal{U}_{\text{av},\sigma} \qquad \text{with} \qquad \mathcal{U}_{\text{av},\sigma} = \frac{1}{4}\chi\sigma_a^2\cos^4\theta\,.$$

The equilibrium orientations of the system, i.e. the solutions of Eq. (9), are then identified with $\theta_{eq} = 0$, $\theta_{eq} = \pi/2$ and an oblique orientation defined in general by the implicit relation

$$\cos^2\theta_{eq} = \frac{2}{3}\left[\frac{2-r}{1-r}\frac{P_{eq}(\theta_{eq})}{2\pi a^3\sigma_a} + \frac{\sigma_{\text{target}}}{\sigma_a} - \frac{P_{eq}(\theta_{eq})}{\chi E\sigma_a}(1+r)\right]. \tag{13}$$

It should be noticed that there are strong constraints, especially on r, for the existence of such an oblique equilibrium. For instance, it might not exist for values of r suitably close to 1, which correspond to a 2D incompressible situation. It is also worth to point out that, if solely the energy of the cell is considered, i.e. $\mathcal{U}_{\text{self}}$ and $\mathcal{U}_{\text{int}}$ are neglected, the only equilibrium orientations are $\theta_{eq} = 0$ and $\theta_{eq} = \pi/2$. The latter is the stable one and corresponds to the direction of minimal stress, if a uniaxial applied stress is considered. The authors then study the response of such a dipole model to external stresses, both in static and cyclic conditions [10, 11].

5 Elastic Energy Approach

Both the approaches based on strain avoidance and stress avoidance have been questioned by experimental data obtained by Livne et al. [22], who performed a wide series of experiments by carefully controlling the biaxiality ratio. It was shown in particular that neither the strain nor the stress avoidance-based laws for the preferential angle of the cell were able to reproduce the experimental data found in [22]. Instead, a change of perspective was proposed in their work, suggesting that the cell attempts to minimize the stored elastic energy, rather than the strain. This choice allowed to obtain a better fitting of the orientations found experimentally.

Before discussing further this approach, it is useful to observe that, since a cell does not have a real polarization given by a head and a tail during the reorientation process, configurations with cells aligned along θ and $\theta + \pi$ are geometrically indistinguishable and therefore also equivalent from the energetic point of view, i.e. $\mathcal{U}(\theta+\pi) = \mathcal{U}(\theta)$, where $\mathcal{U}$ is the elastic energy. In addition, also the orientation of the axis is equivalent and, as a consequence, $\mathcal{U}(\pi - \theta) = \mathcal{U}(\theta)$. So, in conclusion, $\mathcal{U}(\theta)$ is an even π-periodic function, so that

$$\mathcal{U}(\theta) = \mathcal{U}(2\pi - \theta) = \mathcal{U}(\pi - \theta) = \mathcal{U}(\pi + \theta), \quad \forall\theta. \tag{14}$$

Let us then recall that, generally speaking, denoting by $\mathbf{N}_\perp = (-\sin\theta, \cos\theta, 0)$ the orientation perpendicular to $\mathbf{N} = (\cos\theta, \sin\theta, 0)$ on the substratum, a general elastic energy density for an orthotropic material can depend on the invariants [28, 35]

$$\begin{aligned}
\mathrm{I}_4 &:= \mathbf{N}\cdot\mathbb{C}\mathbf{N} = (\lambda_x^2 - \lambda_y^2)\cos^2\theta + \lambda_y^2\,,\\
\mathrm{I}_5 &:= \mathbf{N}\cdot\mathbb{C}^2\mathbf{N} = (\lambda_x^4 - \lambda_y^4)\cos^2\theta + \lambda_y^4\,,\\
\mathrm{I}_6 &:= \mathbf{N}_\perp\cdot\mathbb{C}\mathbf{N}_\perp = \lambda_x^2 - (\lambda_x^2 - \lambda_y^2)\cos^2\theta\,,\\
\mathrm{I}_7 &:= \mathbf{N}_\perp\cdot\mathbb{C}^2\mathbf{N}_\perp = \lambda_x^4 - (\lambda_x^4 - \lambda_y^4)\cos^2\theta\,,\\
\mathrm{I}_8 &:= \mathbf{N}_\perp\cdot\mathbb{C}\mathbf{N} = -(\lambda_x^2 - \lambda_y^2)\sin\theta\cos\theta\,,
\end{aligned} \tag{15}$$

in addition to the usual invariants characterizing an isotropic material

$$\mathrm{I}_1 := \mathrm{tr}\,\mathbb{C}, \qquad \mathrm{I}_2 := \frac{1}{2}\left[(\mathrm{tr}\,\mathbb{C})^2 - \mathrm{tr}\,\mathbb{C}^2\right], \qquad \text{and} \qquad \mathrm{I}_3 := \det\mathbb{C}\,,$$

that do not depend on the angle θ. We notice that all the invariants in Eq. (15) except I_8 satisfy a priori the symmetry conditions in Eq. (14). Therefore, as an additional requirement on $\mathcal{U}$, we assume it to be an even function of I_8, e.g. a function of $I_8^2 = (\lambda_x^2 - \lambda_y^2)^2(1 - \cos^2\theta)\cos^2\theta$. Under this assumption, the energy becomes a function of the angle only through its squared cosine, i.e. $\mathcal{U} = U(\cos^2\theta)$, and the stationary points are trivially identified by

$$U'(\cos^2\theta)\sin\theta\cos\theta = 0\,. \tag{16}$$

Therefore, one always has the trivial equilibrium orientations $\theta_{eq} = 0$ and $\theta_{eq} = \pi/2$, whereas further equilibria might exist, depending on the specific energy considered, for the values of $\theta_{eq} = \hat{\theta}$ such that $U'\left(\cos^2\hat{\theta}\right) = 0$.

Stability of these configurations is achieved if they correspond to minima of the elastic energy, and therefore it depends on the positivity of $U''(\cos^2\theta)\sin^2\theta\cos^2\theta - U'(\cos^2\theta)(\cos^2\theta - \sin^2\theta)$ computed at equilibrium. So, one has the following general stability conditions:

$$\begin{aligned}
\theta = 0:&\ \text{stable} && \Longleftrightarrow U'(1) < 0\,,\\
\theta = \frac{\pi}{2}:&\ \text{stable} && \Longleftrightarrow U'(0) > 0\,,\\
\theta = \hat{\theta}:&\ \text{stable} && \Longleftrightarrow U''\left(\cos 2\hat{\theta}\right) > 0\,,
\end{aligned} \tag{17}$$

whenever the last position exists. In the next Subsections, we will specialize the elastic energy approach in the framework of both linear (see Sect. 5.1) and nonlinear elasticity (see Sect. 5.2). Finally, we will look at the evolution in time of the angle θ in Sect. 5.3.

5.1 Linear Elasticity

If linear elasticity is used, as done by Livne et al. [22], the most general elastic energy satisfying the symmetry properties in Eq. (14) can be written as (see [23])

$$\begin{aligned}
U(\mathbb{E}, \mathbf{N}, \mathbf{N}_\perp) = {}& 2K'_\parallel(\mathrm{tr}\,\mathbb{E})\mathbf{N}\cdot\mathbb{E}\mathbf{N} + 2K'_\perp(\mathrm{tr}\,\mathbb{E})\mathbf{N}_\perp\cdot\mathbb{E}\mathbf{N}_\perp + \frac{1}{2}K_\parallel(\mathbf{N}\cdot\mathbb{E}\mathbf{N})^2\\
&+\frac{1}{2}K_\perp(\mathbf{N}_\perp\cdot\mathbb{E}\mathbf{N}_\perp)^2 + \frac{1}{2}K_\phi(\mathbf{N}_\perp\cdot\mathbb{E}\mathbf{N})^2 + K_{\parallel\perp}(\mathbf{N}\cdot\mathbb{E}\mathbf{N})(\mathbf{N}_\perp\cdot\mathbb{E}\mathbf{N}_\perp)\,,
\end{aligned} \tag{18}$$

where $K_\parallel$ is a coefficient related to the cell mechanical response to stretching in the direction of its main stress fibre orientation, $K_\perp$ to the one in the orthogonal direction, and K_ϕ to the one related to shear. The other coefficients are due to mixing effects.

In terms of θ, Eq. (18) then writes

$$U(\epsilon,\theta) = 2\epsilon^2(1-r)\Big[(K'_{\parallel} - K'_{\perp})\xi(\theta) + (K'_{\perp} - rK'_{\parallel})\Big] + \frac{1}{2}\epsilon^2\Big\{K_{\parallel}[\xi(\theta)-r]^2$$
$$+K_{\perp}[1-\xi(\theta)]^2 + K_{\phi}\,\xi(\theta)[r+1-\xi(\theta)] + 2K_{\parallel\perp}[\xi(\theta)-r][1-\xi(\theta)]\Big\}, \tag{19}$$

where $\xi(\theta) := (r+1)\cos^2\theta$.

We remark that the use of discrete models focusing on the behaviour of stress fibres aligned along a direction θ, as done for instance in [11, 12, 34], is equivalent to considering an energy that only depends on I_4 (see Eqs. (1) and (3)).

Recalling the general observations for the stationary points of the energy given by Eq. (16), for an elastic energy in the form of Eq. (19), the equilibrium orientations are found to be $\theta_{eq} = 0, \frac{\pi}{2}$ and the oblique ones satisfying

$$\cos^2\theta_{eq} = \frac{1}{2} + \mathcal{K}\left(\frac{1}{2} - \frac{1}{1+r}\right), \tag{20}$$

where

$$\mathcal{K} := \frac{K_{\parallel} - K_{\perp} + 4(K'_{\parallel} - K'_{\perp})}{K_{\parallel} + K_{\perp} - K_{\phi} - 2K_{\parallel\perp}}. \tag{21}$$

This is a generalisation of the linear relation firstly found by Livne et al. [22] starting from a simple linear elastic energy:

$$\cos^2\theta_{eq} = b + \frac{1-2b}{1+r}, \tag{22}$$

where b is a material parameter. Therefore, we can argue that the linear relation defined in Eq. (22) holds true even if a very general linear elastic orthotropic model is considered. The only difference is that more coefficients contribute to the slope of the line defined by Eq. (20).

Moreover, the stability of the equilibrium orientations of the cell can be reported on bifurcation diagrams as the ones shown in Fig. 3. In particular, for $\mathcal{K} > 0$ we find two supercritical bifurcation points at

$$\frac{1}{1+r} = \rho_{\parallel} := \frac{1}{2}\left(1 - \frac{1}{\mathcal{K}}\right) \quad \text{and} \quad \frac{1}{1+r} = \rho_{\perp} := \frac{1}{2}\left(1 + \frac{1}{\mathcal{K}}\right).$$

In particular, as $(1+r)^{-1}$ crosses $\rho_{\parallel}$, the oblique orientation (blue line) appears and becomes the only stable one. Instead, for $(1+r)^{-1} > \rho_{\perp}$, the oblique orientation does not exist and the stable alignment is perpendicular to the main stretch direction (green line). We remark that $\rho_{\parallel}$ is negative if $0 < \mathcal{K} < 1$.

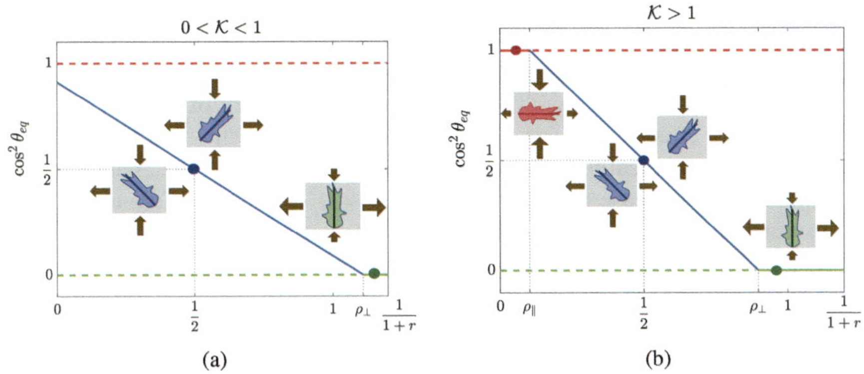

Fig. 3 Bifurcation diagrams for $\mathcal{K} > 0$, with $0 < \mathcal{K} < 1$ in (**a**) and $\mathcal{K} > 1$ in (**b**). Solid lines indicate stable orientations, whereas dashed lines identify unstable positions. The insets and dots in (**a**) and (**b**) show representative cellular orientations: perpendicular (green), oblique (blue) and parallel (red). The arrows indicate the directions of stretching or compression

Conversely, from Eq. (20), for an imposed $r \neq 1$ and for a measured $\theta_{eq} \neq 0$, $\theta_{eq} \neq \frac{\pi}{2}$, one can estimate the material parameter $\mathcal{K}$ by using

$$\mathcal{K} = -\frac{1+r}{1-r} \cos 2\theta_{eq} \,. \tag{23}$$

Data obtained using fibroblasts reported in [22] suggest that $\mathcal{K} \approx 1.26 \pm 0.08$. One could then ask whether such a value is universal, that is, independent of the cell type and the substratum of the specific experiment in sub-confluent conditions (since the confluent case might lead to differences [15]). In this respect, the bifurcation diagram in Fig. 3 puts in evidence that, if $\mathcal{K} > 1$, the only stable solution for $\frac{1}{1+r} > \rho_\perp$, and therefore $r > \frac{1}{\rho_\perp} - 1$, is $\theta_{eq} = \frac{\pi}{2}$. This includes the case in which $\epsilon_{yy} = r = 0$. Now, some pure uniaxial experiments (for which $r = 0$) seem instead to put in evidence that $\theta_{eq} \neq \frac{\pi}{2}$, suggesting that it may also be $\mathcal{K} < 1$. On assuming that $\theta_{eq}(r = 0) = \theta_0 \neq \frac{\pi}{2}$, Eq. (23) gives

$$\mathcal{K} = -\cos 2\theta_0 \,. \tag{24}$$

This suggests that the best way to evaluate the parameter $\mathcal{K}$ is to perform careful pure uniaxial experiments and, if $\theta_{eq} \neq 0$, determine it by using Eq. (24). For instance, Wang et al. [41] present results for $r = 0$ stating that the preferential orientation is perpendicular to the main strain direction, but from their Fig. 3a and b cells seem to have an alignment between 70° and 80°, which corresponds to a value of $\mathcal{K} \in [0.766, 0.936]$. Similarly, the interval of $\theta_{eq} \in [80°, 85°]$ found in the experiments in [39, 41] for $r = 0$ would lead to an evaluation of $\mathcal{K} \in [0.936, 0.985]$. On the other hand, when cells are oriented perpendicularly for $r = 0$, as for instance in [25], we can only argue that $\mathcal{K} > 1$.

Furthermore, in another work, Wang et al. [40] observed that, in the presence of the drug N-acetylcysteine, cells reoriented to an angle close to 70°, while for the control case the preferential angle was close to 85°. This result might indicate that N- acetylcysteine interferes with the mechanical machinery of the cell: in model terms, such an effect can be translated into a change in the coefficients of the elastic energy. Specifically, accounting for the fact that $r = 0$, the change in angle corresponds to a decrease from $\mathcal{K} \approx 0.98$ to $\mathcal{K} \approx 0.77$, and therefore to a decrease (in absolute value) in the slope of the straight line in Fig. 3.

Another observation is that, if we invert Eq. (20) to write

$$r = \frac{\mathcal{K} + \cos 2\theta_{eq}}{\mathcal{K} - \cos 2\theta_{eq}} ,$$

and we take for instance the slope indicated by the experiments in [22], then an equilibrium $\theta_{eq} \in [80°, 100°]$ is achieved for $r < 0.146$. So, a wide range of values for r would lead to an almost perpendicular orientation. This might explain why, in many papers, one has a description that only refers to a perpendicular arrangement, without specifying the precise angle of orientation.

Finally, we conclude this Section with a remark about the minimum number of parameters necessary to model the reorientation behaviour. In particular, we can observe that having only $K_\parallel \neq 0$ in Eq. (21) would give $\mathcal{K} = 1$. Then, adding only a non-vanishing $K_\perp$ would decrease $\mathcal{K}$ to $\mathcal{K} < 1$. In order to have $\mathcal{K} > 1$, as observed for instance in [22], it is sufficient to include a non-vanishing K_ϕ, i.e., a response to shear, or alternatively a positive coupling term $K'_\parallel$ between the invariants I_1 and I_4. In any case, taking as an example the value suggested by the experiments in [22], one can evaluate

$$\frac{K_\phi}{K_\parallel} \approx 0.21 ,$$

in the first case, or

$$\frac{K'_\parallel}{K_\parallel} \approx 0.065 ,$$

in the second case. In particular, we can argue that a small contribution of the mixed term $K'_\parallel$ is enough to recover the experimental values.

5.2 Nonlinear Elasticity

The fact that in some experiments the applied strain can be higher than 30% [13, 22] suggests to drop the linear elasticity assumption and to see why also in this nonlinear regime the equilibrium orientation appears to be the same. Actually, a result very similar to Eq. (20) is reached in [24] starting from the following quite general nonlinear constitutive orthotropic model

$$\hat{U} = \frac{1}{2}\mathbf{I} \cdot \mathbb{K}\mathbf{I} + \mathcal{V}, \tag{25}$$

where $\mathbf{I} := (\mathrm{I}_4 - 1, \mathrm{I}_5 - 1, \mathrm{I}_6 - 1, \mathrm{I}_7 - 1, \mathrm{I}_8)^{\mathrm{T}}$ and $\mathbb{K}$ is the symmetric matrix of coefficients (that might possibly depend on the isotropic invariants), while $\mathcal{V}$ is the purely isotropic contribution that depends on $(\mathrm{I}_1, \mathrm{I}_2, \mathrm{I}_3)$. It is also observed that the same results are achieved with a Fung-type energy of the form

$$U_F = C\left(\exp \hat{U} - 1\right), \tag{26}$$

that is often used in biomechanical applications.

In spite of the nonlinear framework and of the generality of the elastic energy, in [24] equilibria are explicitly determined together with their stability and a bifurcation analysis is fully discussed. In summary, it is proved that:

- The symmetry requirements in Eq. (14) imply that the coefficients coupling I_8 with the other invariants must vanish;
- If the energy is independent of the invariants involving $\mathbb{C}^2$, namely I_5 and I_7, equilibria are given by a formula akin to Eq. (20), with $\dfrac{1}{1+r}$ replaced by its nonlinear generalization $\Lambda := \dfrac{\lambda_x^2 - 1}{\lambda_x^2 - \lambda_y^2}$;
- The inclusion of the dependence on I_5 induces a modification from the linear dependence that is acceptable only for very small values of the related elastic coefficient (see Fig. 4), so its contribution does not seem really relevant;

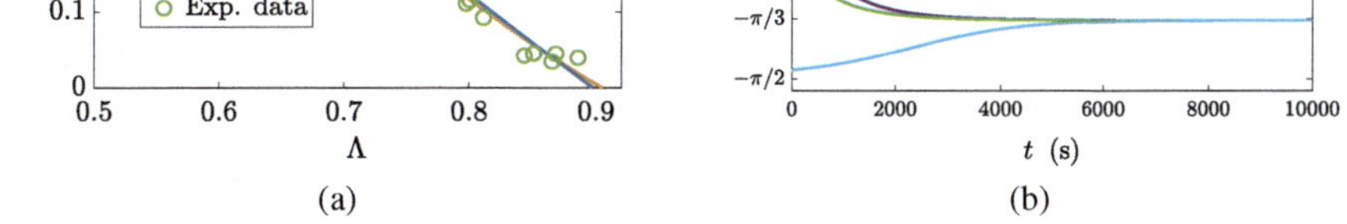

Fig. 4 (**a**): Comparison of experimental data from [13] and [22] for $\lambda_y = 0.889$ taking $k_{55} = 0$ and the best fitting value $k_{55} \approx 0.02$. (**b**): Evolution of the angle in time for $\Lambda = 0.71$ and $k_{55} = 0.02$, for different initial conditions

- The inclusion of the dependence on I_7 or on other coupling terms involving I_5 and I_7 leads to unobserved consequences, mainly for deformations close to the pure uniaxial stretching (i.e., with $\lambda_y = 1$).

Therefore, according to the analysis in [24], the best model appears to be the one that does not include a dependence of the energy on $\mathbb{C}^2$ terms, yielding a nonlinear generalization of Eq. (20).

Also Lazopoulos and Pirentis [19], working in a finite elasticity framework, proposed a model for the cell as a pre-stressed elastic material, whose energy depends only on I_4 but in a non-convex way. Taking into account a misprint, the chosen energy is

$$U = K_1(I_4 - 1)^2 - K_2(I_4 - 1)^4 + K_3(I_4 - 1)^6 ,$$

which gives rise to a co-existence of phases and a saddle-node bifurcation. Thanks to this choice, the model is able to obtain the two symmetric oblique orientations for cells. Instead, in [20] an isotropic Mooney-Rivlin model is used and in [36] a linear elastic model including pre-stresses is deduced focusing on the behaviour of single stress fibres at the microscopic scale. Finally, in [31] the effect of cytoskeletal fluidisation and resolidification is also taken into account. At variance with most experiments, these models yield an equilibrium angle that depends on the applied strain and on the initial orientation. An energetic approach with a wiggly energy to model the attachment of focal adhesions is proposed in [1].

5.3 Temporal Evolution

The models discussed so far mainly focused on the equilibrium orientation of cells, without describing the dynamics of the angle in time. The evolution of cell orientation can be obtained resorting for instance to Lagrangian mechanics arguments, relating the time rate of the orientation angle θ with variations in the virtual work done by the stress acting on the cell due to stress fibre alignment. Considering an overdamped regime, which corresponds to neglecting inertial effects, gives

$$\eta \frac{d\theta}{dt} = -\frac{\partial U}{\partial \theta}(\theta, t) , \tag{27}$$

where η is a viscous-like coefficient measuring cell resistance to internal remodelling. Alternatively, Eq. (27) can be thought of as a simple dissipative dynamics, through which the cell evolves to target the minimum of the elastic energy, as done

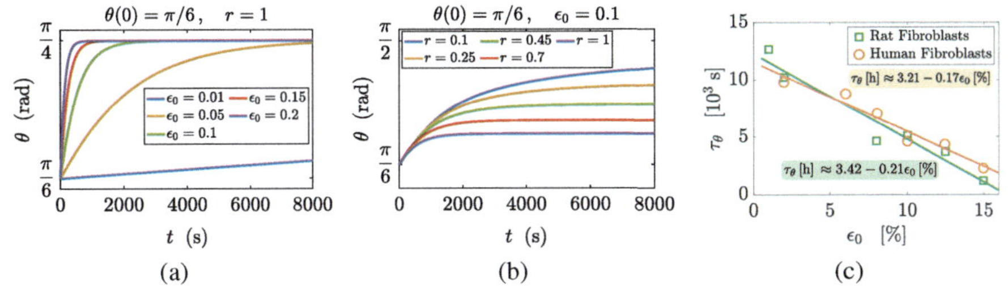

Fig. 5 Evolution of θ according to Eq. (27) (**a**) changing ϵ_0 for fixed $r = 1$ and (**b**) changing r for fixed ϵ_0. (**c**) Reorientation time as a function of imposed strain. Experimental values are taken from [17]

also in [1]. By using the energy in Eq. (19), we can specialize the evolution equation to

$$\eta \frac{d\theta}{dt} = \epsilon^2 \Bigg\{ 4(1-r^2)(K'_{\parallel} - K'_{\perp}) \\ +2(1+r)\big[K_{\parallel}[(r+1)\cos^2\theta - r] + K_{\perp}[(r+1)\cos^2\theta - 1]\big] \\ - \left(\frac{K_\phi}{2} + K_{\parallel\perp}\right)[2(r+1)\cos^2\theta - r - 1]\Bigg\} \sin\theta\cos\theta \, . \tag{28}$$

Such an equation is for instance used in [1, 22, 42] to get plots like those in Fig. 5, representing the trend of the orientation angle to evolve towards its stable equilibrium θ_{eq} obtained as discussed in Sect. 5. In the articles treating the cell as a force dipole like those by De and coworkers [11, 12, 34] a similar equation is employed to determine the evolution of the dipole magnitude.

Although the form of Eq. (27) has been postulated on a phenomenological basis in [22], it can also be recovered within a very general remodelling framework, as done in [7]. In this work, the orientations of different types of stress fibres are considered as additional state variables, for which appropriate thermodynamically consistent evolution equations are derived.

Specifically, with reference to Eq. (28), Fig. 5a shows the influence of the stretch amplitude ϵ_0, for a fixed value of the biaxiality ratio. It can be observed that the amplitude modifies the velocity with which reorientation towards the preferential angle happens, and small amplitudes are associated with very slow processes. Instead, the theoretical equilibrium orientation is not altered by variations in ϵ_0 and remains equal to $\pi/4$ due to the fixed value of r. The influence of variations in the biaxiality ratio is reported in Fig. 5b, showing the tendency to evolve towards the stable equilibrium angle predicted by the bifurcation diagram of Fig. 3.

The fact that reorientation happens faster if the stretching amplitude is increased can be readily argued from Eq. (27). In fact, first of all we notice that, in the linear elastic framework, the elastic strain energy can be written in non-dimensional terms as $U = K_{\parallel}\epsilon^2\bar{U}$, where $\bar{U}$ is dimensionless and independent of ϵ. Hence, one can write the dimensionless evolution equation as

$$\frac{d\theta}{dt} = -\frac{1}{\tau_\theta}\frac{\partial\bar{U}}{\partial\theta}(\theta, t)\,, \tag{29}$$

where $\tau_\theta = \dfrac{\eta}{K_{\parallel}\epsilon^2}$ is a characteristic time that decreases with the stiffness of the material and with the strain magnitude. A similar trend is also obtained in the nonlinear case. Actually, this reasoning predicts that the reorganization time behaves like the inverse of the square of the imposed strain. This might be an explanation of why, for very small amplitudes, reorientation does not happen. Indeed, for $\epsilon \to 0$ the characteristic time would blow up, so that the phenomenon becomes too slow to be observed on the time scale of the experiment and the process is actually hindered by random fluctuations.

However, as shown in Fig. 5c, data in [17] would suggest an almost linear decrease of τ_θ, while the (few) data in [33] seem to suggest that $\tau_\theta \approx \epsilon^{-1.5}$. Though this is not ϵ^{-2}, it still presents a blowing up when $\epsilon \to 0$.

6 Linear Viscoelasticity

As mentioned in the Introduction section, in order to clearly identify a reorientation of cells, the stretching frequency must be sufficiently high. This behaviour cannot be explained by a purely elastic model, which by definition implies an immediate response. Instead, it calls for the introduction of a characteristic response time of the system that needs to be compared with the periodic deformation time scale. The existence of such a characteristic time might be related to the reorganization of the acto-myosin cytoskeleton and of the ensemble of focal adhesions. In this respect, it is known that the characteristic turnover times of both phenomena are of the order of tens of seconds, or even few minutes (see, for instance, [6]).

At the microscopic level, there are many mathematical models aimed at describing the evolution of stress fibres and focal adhesions, as well as their viscoelastic behaviour. Restricting to models applied to the understanding of cell response under stretch, we mention here the one proposed by Qian et al. [32], who started from the dynamics of nucleation and development of both focal adhesions and stress fibres to describe the formation/disassembly of these sub-cellular structures. Their model monitors changes in the areal density of ligand-receptor bonds in the adhesion cluster and in the density of contracting filaments in the stress fibre. Since disassembly is enhanced by the stress acting on the fibres, they tend to form in the direction which is most favourable energetically.

With the same ideas, Chen et al. [5] proposed a Maxwell-like force-deformation relation, also including the increase of size and strength of focal adhesions with stress to mimic catch-bond behaviours [44].

Working instead in a Continuum Mechanics framework at a macroscopic scale, a viscoelastic model is proposed in [23] to explain why, on the time scale of experiments, the reorientation phenomenon does not occur for small frequencies. The proposed model for the stress $\mathbb{T}(t|\theta)$ at time t, given the history of orientations θ, reads

$$\mathbb{T}(t|\theta) = \int_{-\infty}^{t} \frac{1}{\lambda} e^{-(t-\tau)/\lambda} \mathbb{C}_0(\theta(\tau))[\mathbb{E}(t) - \mathbb{E}(\tau)]\, d\tau\,, \tag{30}$$

where $\mathbb{C}_0(\theta(\tau))$ is the fourth-order elasticity tensor depending on the orientation angle θ at a past time τ and the exponential is the memory kernel with relaxation time λ (for more details on this type of models the reader can for instance refer to [2] for the isotropic case and to [30] for the anisotropic case). As usual in rheology, for memory kernels of exponential type it is useful to differentiate Eq. (30) and to rewrite the constitutive equation in the following Maxwell-like differential form:

$$\lambda \frac{d\mathbb{T}}{dt}(t|\theta) + \mathbb{T}(t|\theta) = \mathbb{C}\!\!\!/_0(t|\theta) \frac{d\mathbb{E}}{dt}(t)\,, \tag{31}$$

where

$$\mathbb{C}\!\!\!/_0(t|\theta) := \int_{-\infty}^{t} e^{-(t-\tau)/\lambda} \mathbb{C}_0(\theta(\tau))\, d\tau = \int_{0}^{+\infty} \lambda e^{-s} \mathbb{C}_0(\theta(t-\lambda s))\, ds \tag{32}$$

is a functional on the exponentially weighted history of past orientations. We observe that, in the isotropic case, $\mathbb{C}\!\!\!/_0$ is twice the so-called elastic viscosity, i.e., the area under the relaxation kernel (times the identity tensor).

Equation (31) is then solved together with an equation for the evolution of θ, governed by the variation of the work done by $\mathbb{T}$ over possible orientations. Assuming again to work in an overdamped regime, one then has

$$\frac{d\theta}{dt}(t) = -\frac{1}{K\lambda_\theta} \frac{\partial \mathbb{T}}{\partial \theta}(t|\theta) : \mathbb{E}(t)\,, \tag{33}$$

where, as in the experiments, the periodic deformation is externally imposed to the specimen, and therefore $\mathbb{E}$ can be taken as independent of θ. In Eq. (33) a characteristic Young modulus K, e.g., $K_\parallel$ encountered in Sect. 5, is introduced to put in evidence the existence of a parameter $\lambda_\theta := \eta/K$ related to the time the cell takes to reorient itself. Moreover, Eq. (33) implies that, for a given strain $\mathbb{E}$, θ tends to assume a value such that the variation of $\mathbb{T}$ with respect to θ either vanishes or becomes orthogonal to $\mathbb{E}$.

In Eqs. (31) and (33) there are two intrinsic characteristic times: λ refers to the viscous behaviour of cells, arising, for instance, from the renewal of focal adhesions with the substratum, while λ_θ is related to the characteristic time of reorganization of stress fibres and consequently to the change in cell orientation. Now, it is useful to discuss how the model behaves when the imposed oscillation period is much shorter or longer than the characteristic remodelling times, which as already mentioned is of the order of tens of seconds/minutes [6]. More in detail, in [23] it is proved that, in the high frequency regime, that corresponds to an imposed cyclic strain faster that the reorganization process, i.e., $\lambda\omega$, $\lambda_\theta\omega \gg 1$, as characteristic of most successful experiments, one can approximate

$$\mathbb{T}(t|\theta) \approx \mathbb{C}_0(\theta(t))\mathbb{E}_0 e^{i\omega t}\,. \tag{34}$$

Hence, the stress in such a regime resembles an elastic-like response, as it is classically obtained in viscoelasticity. In this case, therefore, Eq. (33) can be simplified to

$$\frac{d\theta}{dt} = -\frac{1}{\eta}\left[\frac{\partial \mathbb{C}_0}{\partial\theta}\mathbb{E}\right] : \mathbb{E} = -\frac{2}{K\lambda_\theta}\frac{\partial U}{\partial\theta} \tag{35}$$

where

$$U(\theta, t) := \frac{1}{2}\mathbb{E}(t) : \mathbb{C}_0(\theta)\mathbb{E}(t) \tag{36}$$

is the elastic energy. Thus, in the high frequency regime, the change in cell orientation is driven by the minimization of an elastic energy with respect to the orientation angle, as discussed in Sect. 5.

Conversely, if we consider a low frequency regime in which the period of the cyclic strain imposed to the specimen is much longer than the characteristic time λ of cell relaxation, i.e., $\lambda\omega \ll 1$, the reorientation process is faster than the external oscillations and the cell is able to accommodate the cyclic elongations. In this case, in [23] it is proved that

$$\mathbb{T}(t|\theta) \approx \lambda\overline{\mathbb{C}}_0(t|\theta)\frac{d\mathbb{E}}{dt}(t)\,, \tag{37}$$

corresponding to a viscous-like response, being

$$\overline{\mathbb{C}}_0(t|\theta) := \int_0^{+\infty} s e^{-s}\mathbb{C}_0(\theta(t-\lambda s))\,ds\,.$$

Results of some numerical simulations for this model are shown in Fig. 6, highlighting the role of the frequency in reorientation. As evidenced in Fig. 6a, angular frequencies below a minimum threshold, approximately about 0.01 Hz or 0.06 rad/s, lead to very slow reorientation processes and are unable to induce a significantly fast response in cells. Indeed, the reorientation time becomes too slow to be meaningful in comparison with the cell cycle time, and therefore the reorientation process is hindered by random fluctuations. Conversely, for high frequencies the realignment process is initiated, with a characteristic time which is a decreasing function of the frequency, as highlighted by experiments [17].

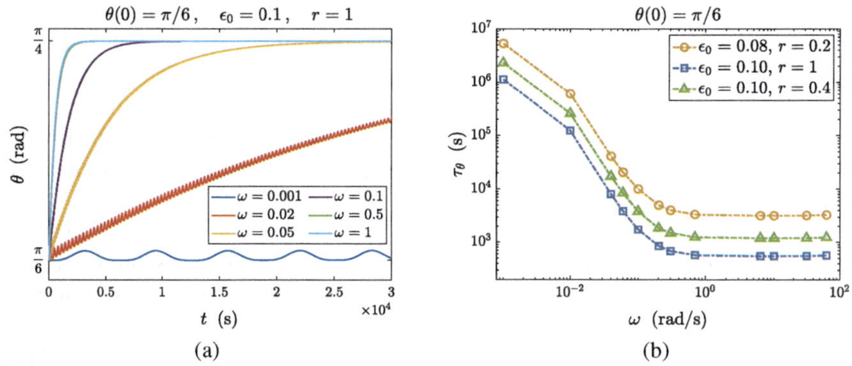

Fig. 6 (**a**) Evolution of the orientation angle according to Eq. (33) for a fixed biaxiality ratio $r = 1$ and different angular frequencies. When low frequencies are applied, the reorientation process is slow, while for increasing frequencies it rapidly reaches the expected steady state. (**b**) Plot of the model characteristic time τ_θ as a function of the angular frequency in logarithmic scale, for different values of ϵ_0 and r. Coherently with experiments [17] and with other models [11], a saturation is observed for high frequencies

Such an effect can be observed clearly in Fig. 6b, where the characteristic time τ_θ of the model is plotted as a function of the angular frequency ω in logarithmic scale. However, as the frequency becomes higher, a saturation appears and further increases do not lead to a faster reorientation. This behaviour appears coherent with experimental results [17] and other model outcomes [11].

So, viscoelasticity plays a relevant role when the period of imposed oscillation is much smaller that the relaxation times related to remodelling phenomena in the cell, because it has enough time to adapt. On the other hand, we have shown the robustness of the equilibrium orientation with respect to elastic constitutive relationships.

7 Concluding Remarks

In this Chapter, we have reviewed some mechanics-based approaches that have been proposed in the literature to describe the phenomenon of cell reorientation when the substrate on which they are laid undergoes a periodic stretch.

In Sect. 5 it was shown that, for a large class of elastic constitutive models, the minimization of the elastic energy best allows to capture the equilibrium orientation angle as a function of the biaxiality ratio of the imposed deformation and of a dimensionless parameter containing all the mechanical characteristics of the cell and of the substratum.

In order to describe the fact that cell orientation is not observed for too low frequencies of oscillation, in Sect. 6 a viscoelastic constitutive model was used. This corresponds to taking into account of the remodelling of focal adhesions and of the viscoelastic behaviour of the cytoskeleton. However, this aspect should be further developed in order to include the dynamics of slip bonds and the development of catch bonds in response to the imposed traction. In addition, in all the models reviewed, the active response of the acto-myosin machinery to the imposed deformation is not considered. This might be an important aspect to include, once experiments are available to clarify the related mechanotransduction dynamics.

Acknowledgments This work was partially supported by the Italian National Group of Mathematical Physics (GNFM–INdAM), the Italian Ministry of Education, Universities and Research through the MIUR grant Dipartimenti di Eccellenza 2018–2022, Project No. E11G18000350001, and the Scientific Research Programmes of Relevant National Interest Project No. 2017KL4EF3. GL acknowledges the support of GNFM through the grant *Progetto Giovani 2023* n. CUPE53C22001930001.

Competing Interests The authors have no conflicts of interest to declare that are relevant to the content of this chapter.

References

1. Abeyaratne, R., Puntel, E., Tomassetti, G.: An elementary model of focal adhesion detachment and reattachment during cell reorientation using ideas from the kinetics of wiggly energies. J. Elast. (2022)
2. Astarita, G., Marrucci, G.: Principles of Non-Newtonian Fluid Mechanics. McGraw-Hill, New York (1974)
3. Barron, V., Brougham, C., Coghlan, K., McLucas, E., O'Mahoney, D., Stenson-Cox, C., McHugh, P.E.: The effect of physiological cyclic stretch on the cell morphology, cell orientation and protein expression of endothelial cells. J. Mater. Sci. Mater. Med. **18**, 1973–1981 (2007)
4. Buck, R.C.: Reorientation response of cells to repeated stretch and recoil of the substratum. Exper. Cell Res. **127**, 470–474 (1980)
5. Chen, B., Kemkemer, R., Deibler, M., Spatz, J., Gao, H.: Cyclic stretch induces cell reorientation on substrates by destabilizing catch bonds in focal adhesions. PLoS ONE **7**, e48346 (2012)
6. Chen, Y., Pasapera, A.M., Koretsky, A.P., Waterman, C.M.: Orientation-specific responses to sustained uniaxial stretching in focal adhesion growth and turnover. Proc. Natl. Acad. Sci. USA **110**, E2352–E2361 (2013)
7. Ciambella, J., Lucci, G., Nardinocchi, P., Preziosi, L.: Passive and active fiber reorientation in anisotropic materials. Int. J. Eng. Sci. **176**, 103688 (2022)
8. Dartsch, P.C., Hammerle, H., Betz, E.: Orientation of cultured arterial smooth muscle cells growing on cyclically stretched substrates. Acta Anatomica **125**, 108–113 (1986)
9. De, R.: A general model of focal adhesion orientation dynamics in response to static and cyclic stretch. Commun. Biol. **1**, 81 (2018)
10. De, R., Safran, S.: Dynamical theory of active cellular response to external stress. Phys. Rev. E **78**, 031923–031940 (2008)
11. De, R., Zemel, A., Safran, S.: Do cells sense stress or strain? Measurement of cellular orientation can provide a clue. Biophys. J. **94**, L29–L31 (2008)
12. De, R., Zemel, S., Safran, A.: Dynamics of cell orientation. Nat. Phys. **3**, 655–659 (2007)

13. Faust, U., Hampe, N., Rubner, W., Kirchgessner, N., Safran, S., Hoffmann, B., Merkel, R.: Cyclic stress at mHz frequencies aligns fibroblasts in direction of zero strain. PLoS ONE **6**, e28963 (2011)
14. Giverso, C., Loy, N., Lucci, G., Preziosi, L.: Cell orientation under stretch: A review of experimental findings and mathematical modelling. J. Theor. Biol. **572**, 111564 (2023)
15. Gérémie, L., Ilker, E., Bernheim-Dennery, M., Cavaniol, C., Viovy, J.-L., Vignjevic, D.M., Joanny, J.-F., Descroix, S.: Evolution of a confluent gut epithelium under on-chip cyclic stretching. Phys. Rev. Res. **4**, 023032 (2022)
16. Holzapfel, G.A., Ogden, R.W.: Constitutive modelling of passive myocardium: a structurally based framework for material characterization. Philos. Trans. R. Soc. A **367**, 3445–3475 (2009)
17. Jungbauer, S., Gao, H., Spatz, J.P., Kemkemer, R.: Two characteristic regimes in frequency-dependent dynamic reorientation of fibroblasts on cyclically stretched substrates. Biophys. J. **95**, 3470–3478 (2008)
18. Kim, B.-S., Nikolovski, J., Bonadio, J., Mooney, D.J.: Cyclic mechanical strain regulates the development of engineered smooth muscle tissue. Nat. Biotechnol. **17**, 979–983 (1999)
19. Lazopoulos, K.A., Pirentis, A.: Substrate stretching and reorganization of stress fibers as a finite elasticity problem. Int. J. Solids Struct. **44**, 8285–8296 (2007)
20. Lazopoulos, K.A., Stamenovic, D.: A mathematical model of cell reorientation in response to substrate stretching. Mol. Cell. Biol. **3**, 43–48 (2006)
21. Liu, B., Qu, M.-J., Qin, K.-R., Li, H., Li, Z.-K., Shen, B.-R., Jiang, Z.-L.: Role of cyclic strain frequency in regulating the alignment of vascular smooth muscle cells in vitro. Biophys. J. **94**, 1497–1507 (2008)
22. Livne, A., Bouchbinder, E., Geiger, B.: Cell reorientation under cyclic stretching. Nat. Commun. **5**, 3938 (2014)
23. Lucci, G., Giverso, C., Preziosi, L.: Cell orientation under stretch: Stability of a linear viscoelastic model. Math. Biosci. **337**, 108630 (2021)
24. Lucci, G., Preziosi, L.: A nonlinear elastic description of cell preferential orientations over a stretched substrate. Biomech. Model. Mechanobiol. **20**, 631–649 (2021)
25. Mao, T., He, Y., Gu, Y., Yang, Y., Yu, Y., Wang, X., Ding, J.: Critical frequency and critical stretching rate for reorientation of cells on a cyclically stretched polymer in a microfluidic chip. AMS Appl. Mater. Interf. **13**, 13934–13948 (2021)
26. Matsumoto, T., Delafontaine, P., Schnetzer, K.J., Tong, B.C., Nerem, R.M.: Effect of uniaxial cyclic stretch on the morphology of monocytes/macrophages in culture. J. Biomech. Eng. **118**, 420–422 (1996)
27. Morioka, M., Parameswaran, H., Naruse, K., Kondo, M., Sokabe, M., Hasegawa, Y., Suki, B., Ito, S.: Microtubule dynamics regulate cyclic stretch-induced cell alignment in human airway smooth muscle cells. PLoS ONE **6**, e26384 (2011)
28. Ogden, R.W.: Nonlinear elasticity, anisotropy, material stability and residual stresses in soft tissue. In: Holzapfel, G.A., Ogden, R.W. (eds.) Biomechanics of Soft Tissue in Cardiovascular Systems, pp. 65–108. Springer, Berlin (2003)
29. Palumbo, S., Carotenuto, A.R., Cutolo, A., Deseri, L., Pugno, N., Fraldi, M.: Mechanotropism of single cells adhering to elastic substrates subject to exogenous forces. J. Mech. Phys. Solids **153**, 104475 (2021)
30. Pettermann, H.E., Cheyrou, C., DeSimone, A.: Modeling and simulation of anisotropic linear viscoelasticity. Mech. Time-Dependent Mater. **25**, 679–689 (2020)
31. Pirentis, A., Peruski, E., Iordan, A.L., Stamenovic, D.: A model for stress fiber realignment caused by cytoskeletal fluidization during cyclic stretching. Cell. Mol. Bioeng. **4**, 67–80 (2011)
32. Qian, J., Liu, H., Lin, Y., Chen, W., Gao, H.: A mechanochemical model of cell reorientation on substrates under cyclic stretch. PLoS ONE **8**, e65864 (2013)
33. Roshanzadeh, A., Nguyen, T.T., Nguyen, K.D., Kim, D.-S., Lee, B.-K., Lee, D.-W., Kim, E.-S.: Mechanoadaptive organization of stress fiber subtypes in epithelial cells under cyclic stretches and stretch release. Sci. Rep. **10**, 18684 (2020)
34. Safran, S., De, R.: Nonlinear dynamics of cell orientation. Phys. Rev. E **80**, 060901 (R) (2009)

35. Spencer, A.J.M.: Constitutive theory for strongly anisotropic solids. In: Spencer, A.J.M. (ed.) Continuum Theory of the Mechanics of Fibre-Reinforced Composites, pp. 1–32. Springer, Wien (1984)
36. Stamenovic, D., Lazopoulos, K.A., Pirentis, A., Suki, B.: Mechanical stability determines stress fiber and focal adhesion orientation. Cell. Mol. Bioeng. **2**, 475–485 (2009)
37. Wang, H., Ip, W., Boissy, R., Grood, E.S.: Cell orientation response to cyclically deformed substrates: experimental validation of a cell model. J. Biomech. **28**(12), 1543–1552 (1995)
38. Wang, J.H.-C.: Substrate deformation determines actin cytoskeleton reorganization: a mathematical modeling and experimental study. J. Theor. Biol. **202**, 33–41 (2000)
39. Wang, J.H.-C., Goldschmidt-Clermont, P., Wille, J., Yin, F.C.-P.: Specificity of endothelial cell reorientation in response to cyclic mechanical stretching. J. Biomech. **34**, 1563–1572 (2001)
40. Wang, J.H.-C., Goldschmidt-Clermont, P., Yin, F.C.-P.: Contractility affects stress fiber remodeling and reorientation of endothelial cells subjected to cyclic mechanical stretching. Ann. Biomed. Eng. **28**, 1165–1171 (2000)
41. Wang, S., Lu, D., Zhang, Z., Jia, X., Yang, L.: Effects of mechanical stretching on the morphology of extracellular polymers and the mRNA expression of collagens and small leucine-rich repeat proteoglycans in vaginal fibroblasts from women with pelvic organ prolapse. PLoS ONE **13**, e0193456 (2018)
42. Xu, G., Feng, X., Gao, H.: Orientations of cells on compliant substrates under biaxial stretches: a theoretical study. Biophys. J. **114**, 701–710 (2018)
43. Zemel, A., Bischofs, I.G., Safran, S.: Active elasticity of gels with contractile cells. Phys. Rev. Lett. **97**, 128103 (2006)
44. Zhu, C.: Mechanochemistry: a molecular biomechanics view of mechanosensing. Ann. Biomed. Eng. **42**, 388–404 (2014)

Modelling of Cancer Stem Cell Driven Solid Tumors

Thomas Hillen and Alexandra Shyntar

Abstract Cancer stem cells (CSCs) are pluripotent cancer cells, which are less sensitive to treatments, and which can generate new tumors once transplanted into a healthy tissue environment. They have been identified in many cancers and they are the driving force of cancer growth and metastasis. Mathematical modelling of CSCs has contributed to our increased understanding of CSC interactions. Here we will consider CSC feedback mechanisms, treatments with radiation therapy, differentiation promoters, and dedifferentiation inhibitors. In addition, we will consider spatially explicit models. The dominating effect in these models is the *tumor growth paradox*. It says that a tumor under treatment might grow larger than a similar tumor without treatment. Using geometric singular perturbation methods, we prove mathematically that such a tumor growth paradox can arise. In the spatial context, it leads to a *tumor invasion paradox*, as we will explain. Many of the models presented here lead to interesting mathematical questions and to the development of new mathematical techniques. We end our chapter with a collection of open problems, which we think, are amenable to further mathematical analysis.

1 Introduction

The term "cancer" is a notion for a large collection of diseases, which share certain common hallmarks of cancer [23–25]. Medical research in oncology over the last decades has revealed a plenitude of cancer types, cancer dynamics, and cancer treatments. Cancer is no longer seen as an isolated disease; it rather arises as a complex interaction of cells of different genotypes, phenotypes, epigenetics, in a biochemical network of chemokines, growth factors, and antigens, in interaction with immune response, microenvironment, and treatments [23–25].

T. Hillen (✉) · A. Shyntar
University of Alberta, Edmonton, AB, Canada
e-mail: thillen@ualberta.ca; shyntar@ualberta.ca

A. d'Onofrio et al. (eds.), *Problems in Mathematical Biophysics*, SEMA SIMAI Springer Series 38, https://doi.org/10.1007/978-3-031-60773-8_7

Just over 20 years ago, Hanahan and Weinberg published their landmark paper *The Hallmarks of Cancer* [23] where they outlined 6 key differences ("hallmarks") between cancer tissue and normal tissue. These original six hallmarks were updated 11 years later in a second paper—*Hallmarks of Cancer: The Next Generation* [24], followed in 2022 by a third paper—*Hallmarks of Cancer: New Dimensions* [25]. As things stand now, the hallmarks of cancer consist of eight hallmark capabilities: (H1) sustained proliferative signalling, (H2) evading growth suppressors, (H3) resisting cell death, (H4) replicative immortality, (H5) angiogenesis, (H6) invasion and metastasis, (H7) reprogrammed metabolism, (H8) avoidance of immune destruction, plus two enabling characteristics: (H9) genome instability, (H10) tumor promoting inflammation; along with four more emerging hallmarks and enabling characteristics: (H11) phenotypic plasticity, (H12) epigenetics, (H13) the microbiome, and (H14) cell senescence. Reading this list is overwhelming. It shows why it is so difficult to properly understand cancer and to efficiently treat it. Researchers with many different expertise are needed to gain a full picture of this disease, and mathematics is one important partner in this research.

1.1 Cancer Stem Cells

A major player in the growth and spread of cancer are cancer stem cells (CSCs). These are pluripotent cells that are able to generate a full tumor if transplanted into new, healthy environment [8, 14, 52]. Although the classification of CSCs is not uniform among cancers, or among scientists, the common features of CSCs often include the ability to self-renew, having unlimited replicative potential, immortality, and having reduced sensitivity to treatments such as radiation and chemotherapy [14, 34, 58]. CSCs are involved in many of the hallmarks mentioned above. They provide sustained growth signals (H1), evade growth suppression (H2), resist cell death (H3), have replicative immortality (H4), are involved in metastasis formation (H6), and can become senescent (H14). They are likely involved in the other hallmarks too. Hence, a good understanding of CSCs and their role inside the tumor environment is essential. CSCs have been identified in most cancers [14], and their relevance in cancer progression has been understood.

One of the first modelling papers of CSCs is the article by Dingli and Michor [13] titled *Successful therapy must eradicate cancer stem cells*. This paper from 2006 was published at a time where the existence of CSCs was still debated. In [13], they developed a simple mathematical model that can explain treatment failures based on CSC treatment resistance. Mathematical modelling is a powerful method for gaining an understanding of complex processes. Since the early models for cancer stem cells [1, 13, 20, 61], mathematical modelling of cancer stem cells is at full swing [3, 5, 6, 21, 28, 33, 36, 41, 49, 53, 57, 60, 62, 64, 66, 67].

CSCs are similar to healthy stem cells, however, healthy stem cells for example in the hematopoietic system, give rise to various cell lineages of progenitor cells, transient amplifying cells and differentiated cells. Such a distinction is less clear for CSCs, hence in our modelling we only distinguish between stem and non-stem cancer cells. As our focus here is on solid tumors, we denote non-stem cancer cells as tumor cells (TCs).

1.2 Outline

For this book chapter, we begin with a very basic cancer stem cell model, and progressively consider model extensions to include many of the effects mentioned above. Writing down the basic CSC model already leads to confusion, since there are several ways to formulate the equations. For example, some authors use a probability of CSC self renewal as a model parameter, while others use a fraction of CSC offspring. Further, some scientists consider asymmetric divisions, while others consider symmetric divisions only. In the end, all these formulations are mathematically equivalent, and we will illustrate this equivalence in Sect. 1.3.

Having established the equivalence of the different model formulations, we add the defining feature of a solid tumor in tissue, which is having limited access to space and nutrients. We express this by a competition function F (see Sect. 1.4). The competition function leads directly to one of the key properties of the CSC model, which is the *tumor growth paradox*. It shows that increased treatment can lead to a larger tumor. This is a surprising effect that is related to the self-renewal properties of CSCs. In Sect. 2, we recall the arguments of [28] and use geometric singular perturbation theory to mathematically understand this phenomenon.

The model in [28] shows that treatments and immune responses can lead to CSC enriched cancers. Hence, it is a natural question to see if differentiation of CSCs into TCs can be promoted, or if dedifferentiation of TCs into CSCs plays a role. Section 3.2 considers treatment of cancer with differentiation promoters such as TGF-β or all-trans-retonic (ATRA) therapy. We combine this treatment with radiation and compare various model outcomes. We find that this combination of treatments shows great improvements for head and neck cancer and for metastatic brain cancer, but not so much for breast cancer. These results were published in [3].

In Sect. 3.3, we turn the differentiation process around and consider dedifferentiation. Dedifferentiation describes the process where differentiated cells, TCs, revert back and show stem cell like behavior. This process has been identified in many cancers [7, 35, 37]. In Sect. 3.3, we review a model from [53], which was developed to include the dedifferentiation promoter survivn and a survivin inhibitor YM155. Based on mice experiments of [31], we parametrize the model and again show that a combination therapy is beneficial. We also consider feedback mechanisms as modelled by [36, 55, 74] in Sect. 3.4. In that case, we account for chemokines that are released from CSCs and from TCs feedback on the self-renewal probability of CSCs. We show that some of these feedbacks can lead to an Allee effect.

In Sect. 4, we consider spatial versions of the CSC model. The first such model is developed closely to the experimental observations and takes the form of a birth-jump process (17). We show some general results on existence, uniqueness and

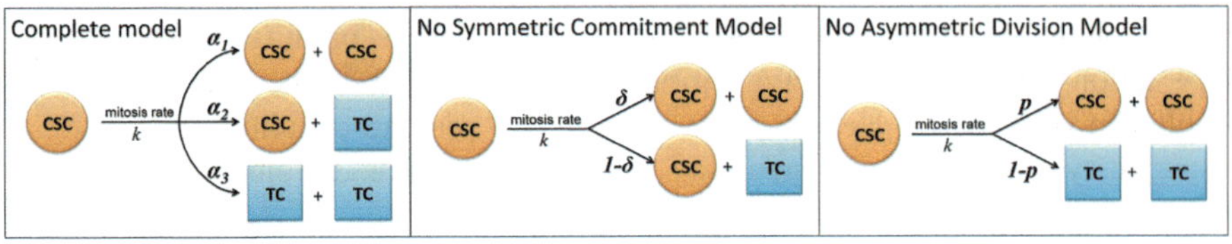

Fig. 1 Schematics for the three ways a CSC model can be derived. In all model diagrams k is the mitosis rate. In the complete model, α_1, α_2, α_3 represent the probability of offspring comprised of two CSCs, probability of offspring comprised of one CSC and one TC, and probability of offspring comprised of two TCs, respectively. In the No Symmetric Commitment Model, δ represents the fraction of CSC offspring. In the No Asymmetric Division Model, p represents the probability of obtaining two CSCs as offspring

positivity from [6, 17, 40]. The integral terms of the birth-jump model can be expanded as second order derivatives, which then leads to a reaction-diffusion model for CSCs and TCs. For this reduced model, we explain the *tumor invasion paradox*, which was proven in [60]. It describes the effect that a tumor with an increased death rate of TC invades faster.

We finish with a conclusion section, where the significance of these results is discussed and interesting open questions are presented.

1.3 Various CSC Models

To model the reproduction of cancer stem cells (CSCs) and their differentiation into non-stem cell tumor cells (TCs), we can use various approaches. We can use the probability of symmetric cell division into two CSCs, or use the frequency of CSC offspring through division, or use transition rates for the various possible mitosis events. We show a schematic in Fig. 1. The left image in Fig. 1 shows the full pathway, where CSCs divide with a rate $k > 0$ and then produce two CSCs with probability α_1 (symmetric division), two TCs with probability α_3, or one CSC and one TC with probability α_2 (asymmetric division). In the middle image of Fig. 1, it is assumed that one offspring replaces the mother cell and the second offspring is either a CSC with probability δ or a TC with probability $1 - \delta$. In the third image of Fig. 1, we only consider symmetric divisions into two CSCs with probability p and into two TCs with probability $1 - p$. These various formulations of symmetric versus asymmetric cell division have led to some confusion in the literature. Here, we quickly show that these approaches are all mathematically equivalent where a simple redefinition of the parameters leads from one form to the other.

We begin with the complete model that is illustrated on the left of Fig. 1. We denote the CSC population by $u(t)$ and the TC population by $v(t)$, and we assume that $k > 0$ is the (constant) mitosis rate. Upon mitosis, a CSC can divide symmetrically into two CSCs, asymmetrically into one CSC and one TC, or symmetrically into two TCs. Each of these transitions happen with a probability of α_1, α_2, and α_3, respectively. Note that $\alpha_1 + \alpha_2 + \alpha_3 = 1$ and if we multiply such a fraction by k, then we obtain the effective rates for these transitions. For example $k\alpha_2$ is the rate of asymmetric division. We ignore TC self-renewal for this argument.

The complete model can be expressed as

$$\begin{aligned} \dot{u} &= 2\alpha_1 ku - \alpha_1 ku - \alpha_3 ku, \\ \dot{v} &= \alpha_2 ku + 2\alpha_3 ku. \end{aligned} \tag{1}$$

Since $\alpha_1 = 1 - \alpha_2 - \alpha_3$, we can already eliminate one parameter:

$$\begin{aligned} \dot{u} &= (1 - \alpha_2 - 2\alpha_3)ku, \\ \dot{v} &= (\alpha_2 + 2\alpha_3)ku. \end{aligned}$$

Note that if $\alpha_2 + 2\alpha_3 > 1$, then the CSC population declines. Now, we set $\delta := 1 - (\alpha_2 + 2\alpha_3)$ to obtain the model

$$\begin{aligned} \dot{u} &= \delta ku, \\ \dot{v} &= (1 - \delta)ku. \end{aligned} \tag{2}$$

These are the equations for the asymmetric division model that is shown in the middle of Fig. 1. Note that in general δ can be negative, which means that CSCs can decline. If $\delta > 0$, then δ can be seen as the fraction (or probability) of offspring that are CSCs.

Finally, in the symmetric division model on the right of Fig. 1, we denote by p the probability that division leads to two CSCs and by $1 - p$ the probability that division leads to two TCs. The equations for this model are

$$\begin{aligned} \dot{u} &= 2pku - ku = (2p - 1)ku, \\ \dot{v} &= 2(1 - p)ku. \end{aligned} \tag{3}$$

If $\delta = 2p - 1$, then $1 - \delta = 2(1 - p)$ and we obtain model (2) again. In reverse, if we define $p = \frac{1}{2}(\delta + 1)$, then model (2) leads to model (3). Hence, (2) and (3) are fully equivalent.

Model (1) has two free parameters α_2 and α_3 and we cannot uniquely recover those from the knowledge of δ or p. However, once α_2 and α_3 are given, the dynamics of (1) are fully described by any of the models (2) or (3). In the following sections, we will mostly use (2) and in cases where the probability p is important, we will also use model (3).

1.4 The Base Model

To obtain the base model for CSC driven solid tumors, we add three more effects into the model (2). First of all, we introduce a competition function $F(n)$, where $n = u + v$ denotes the total cell population. As n increases, the availability of space and nutrients decreases until a certain carrying capacity K is reached. For $n > K$ no reproduction is possible due to limited space and limited nutrients and the cells become quiescent. To be specific, we assume

(F) $F(n)$ is Lipschitz continuous, $F(0) = 1$, F is non-increasing, and $F(n) = 0$ for all $n \geq K$.

Our standard example is $F(n) = \max\{1 - n, 0\}$ and $K = 1$.

We also include TC self renewal with rate k_2 and TC cell death with rate a. We do not (yet) include CSC death, since we assume that CSC are essentially immortal if unperturbed. Then our base model becomes

$$\begin{aligned} \dot{u} &= \delta k F(n) u, \\ \dot{v} &= (1-\delta) k F(n) u + k_2 F(n) v - a v. \end{aligned} \tag{4}$$

2 The Tumor Growth Paradox

To explain how the tumor growth paradox arises from a mathematical point of view, we simplify a bit and assume that the growth rates for CSCs and TCs are the same and equal to one: $k = k_2 = 1$. The same arguments also hold for more general values, but the calculations would include additional growth rate ratios. With this, model (4) simplifies to a two-parameter model for the fraction of CSC offspring δ and the TC death rate a

$$\begin{aligned} \dot{u} &= \delta F(n) u, \\ \dot{v} &= (1-\delta) F(n) u + F(n) v - a v. \end{aligned} \tag{5}$$

Definition 1 System (5) shows a tumor growth paradox, if there exist parameter values $a_1 > a_2$ and an open time interval (t_1, t_2), with $0 < t_1 < t_2$ such that $n^{a_1}(0) = n^{a_2}(0)$ and

$$n^{a_1}(t) > n^{a_2}(t), \qquad \text{for all} \qquad t \in (t_1, t_2),$$

where $n^{a_i} = u^{a_i} + v^{a_i}$ denotes the solution of (5) for death rate $a = a_i$, $i = 1, 2$.

In other words, if we consider two versions of (5) with different death rates $a_1 > a_2$ and the same initial condition, then there is a time where the tumor with the larger death rate of TCs becomes larger than the tumor with the lower death rate of TCs. A paradox.

In [28], we prove:

Theorem 1 *Assume the conditions **(F)** and assume $\delta > 0$ is small. Then* (5) *shows a tumor growth paradox.*

The proof uses a multiscale argument, which we like to present here in some detail. It gives a lot of insight into the dynamics of the model, and it prepares the arguments for later sections. The parameter δ is assumed to be small enough such that the conditions of the Fenichel theorems of geometric singular perturbation theory are satisfied [27]. Please see [28] for those details on the scaling. Here, we simply assume δ is small and as an example we use $\delta = 0.01$ in our simulations. We use this δ in an asymptotic expansion of model (5). If we set $\delta = 0$ in (5), then we obtain the leading order term, also called the *fast system:*

$$\begin{aligned} \dot{u} &= 0 \\ \dot{v} &= F(n)u + F(n)v - av. \end{aligned} \tag{6}$$

We rescale time by setting $\tau = \delta t$, which describes a macroscopic time, also called the slow time scale. If we perform this transformation in (5) and then look at the leading order term, we obtain the *slow system*

$$\begin{aligned} \tfrac{d}{d\tau}U &= F(N)U \\ 0 &= F(N)(U+V) - aV. \end{aligned} \tag{7}$$

Here we use capital letters $U, V, N = U + V$, to distinguish the slow system (7) from the fast system (6). The second equation from (7) is an algebraic equation for U and V. In the (U, V) phase space it defines a manifold, called the *slow manifold*

$$M := \{(U, V) \in \mathbb{R}^2 : \; F(U+V)(U+V) = aV\}.$$

We plot this manifold for three different values of $a = 0.1, 0.5, 1$ in Fig. 2a. The first equation of (7) describes the dynamics on M. As $F(N) > 0$ for $N < 1$, the dynamics on M is a continued growth until $N = 1$.

The fast system (6) describes the fast approach onto the manifold M. In our case $u(t) =$ constant and the v-nullcline of the fast system is the slow manifold M. The Fenichel theory of geometric singular perturbation, which is nicely presented in [27], now ensures that the orbits of the full model (5) follow the fast system until they reach M, and then follow the slow system on M. In Fig. 2b, we show three orbits of the model (5) for different values of $a = 0.1, 0.5, 1$, which all start at the same initial value $(u(0), v(0)) = (0.01, 0.05)$, and all end at time $t = 1000$. In all three cases we see a quick vertical growth until the corresponding slow manifold is reached. Then the orbits follow their corresponding M. Notice that the blue curve corresponds to the lowest death rate $a = 0.1$, while the red curve corresponds to the larger rate $a = 1$. The blue curve raises much higher in the v-component, while the red curve advances further in u-direction. To see the total population $n = u + v$ at

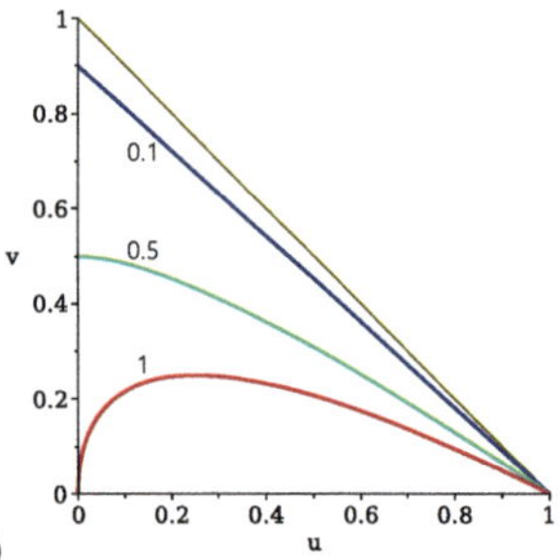

(a)

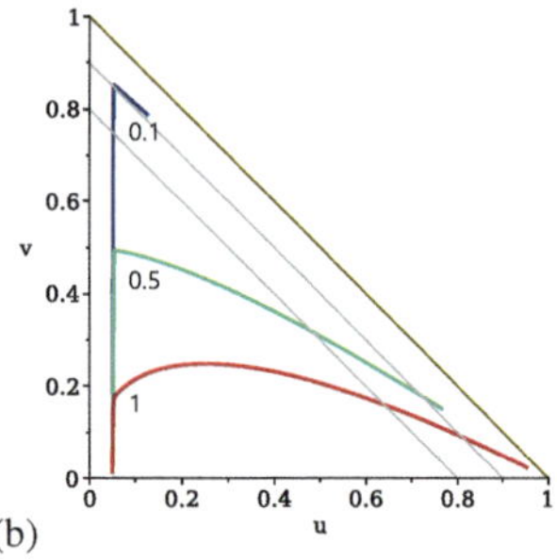

(b)

Fig. 2 (**a**) The slow manifolds M in the phase space for $a = 0.1$ in blue, $a = 0.5$ in green and $a = 1$ in red (the values are also indicated on top of the corresponding line). (**b**) Three orbits with a common initial condition but with different a values illustrating the tumor growth paradox. The additional diagonal lines denote the levels of $u + v = 0.8, 0.9, 1.0$

$t = 1000$, we include three helper lines for the levels of $n = 0.8, 0.9$, and 1. Using this information, we see clearly that the red curve (large a) has advanced further beyond the 0.9 level than the green curve (medium a) and the blue curve (small a), showing the tumor growth paradox in action. We also note that larger a selects for a CSC enriched tumor (larger u-values).

3 Inclusion of Treatments and Feedbacks

To discuss the inclusion of treatments, we will present three scenarios. In the first case, we use a hypothetical treatment on-off situation, which enables us to explain the dynamics using the slow manifold technique from above. In that case, no attempt was made to fit to specific cancer data or cancer treatments. In Sect. 3.2 however, we review the results from [3] and parameterize the model for differentiation therapy of head and neck cancer, breast cancer, and metastatic brain cancer. We will see that combination of differentiation promoters and radiation treatment can have large beneficial effects for the patients. In Sect. 3.3, we consider a model given in [53], where we included an anti-survivin treatment for non small cell lung cancer (NSCLC). Survivin promotes TC dedifferentiation into CSCs, and a treatment with YM155 (*sepantronium bromide*) inhibits survivin, i.e. inhibiting dedifferentiation [31, 45]. A fit to mouse data of NSCLC shows a significant effect of YM155 treatment. In Sect. 3.4 we include feedback mechanisms that arise through cytokines that are released from the growing tumor. We find growth inhibition and Allee effects related to those feedbacks.

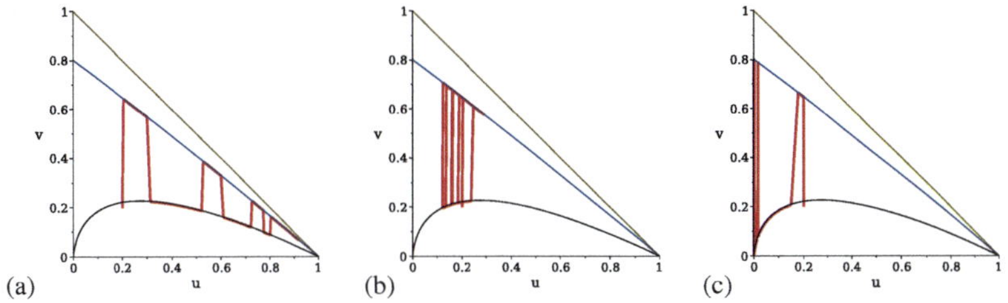

Fig. 3 Typical treatment scenarios with initial condition $(u_0, v_0) = (0.2, 0.2)$. (**a**) The CSCs are minimally affected and $\kappa = 0.001$. The disease progresses along the slow manifolds of on-treatment and off-treatment. (**b**) Here $\kappa = 0.008$ and CSCs are sensitive to the treatment. We obtain a trade off between tumor shrinkage during treatment and tumor regrowth in between treatments. If treatment was repeated indefinitely, we would end up in a periodic orbit. In (**c**), we use $\kappa = 0.02$, which leads to full elimination of the cancer. The slow manifold for no-treatment is shown as a thin blue line and the on-treatment manifold is shown in black. Note that the slow manifold for off-treatment manifold is above the on-treatment manifold

3.1 General Treatments

The slow manifold analysis is quite useful when including treatments. Mathematically, treatment changes the death rate a of TCs and it also affects CSCs. During treatment a is large and between treatments it is low. As we have seen, the slow manifold M depends on the value of a, where the manifolds with high values of a are lower in the phase plane than the manifolds for low a (see Fig. 2). For a periodic treatment (such as fractionated radiation), we will have two manifolds to consider, a treatment-on manifold and a treatment-off manifold. The solution will jump between these. We illustrate this behavior in Fig. 3a, where we clearly see jumps between a higher arc (no treatment) and a lower arc (treatment).

The model that we use for Fig. 3 is slightly modified, since if treatment is applied the CSCs will be affected, but to a lower degree. CSCs have various resistance mechanism [14]. Hence, we consider

$$\begin{aligned} \dot{u} &= \delta F(n)u - a\kappa R(t)u \\ \dot{v} &= (1-\delta)F(n)u + F(n)v - av - aR(t)v, \end{aligned} \tag{8}$$

where $R(t)$ denotes the radiation treatment schedule. We multiply the radiation term by the natural death rate a such that the effect of R can be better compared to the death rate of v. As a simple example, we assume radiation is on for 200 time units and off for 200 time units. During treatments, $R(t) = 10$ and in between treatments $R(t) = 1$. The simulations for initial condition $(u_0, v_0) = (0.2, 0.2)$ are shown in Fig. 3. Now the behavior depends on the size of the CSC resistance factor κ. If κ is small ($\kappa = 0.001$ in Fig. 3a), then the disease progresses. If κ is intermediate ($\kappa = 0.008$ in Fig. 3b) then we reach a periodic solution. This corresponds to cancer control but not to cancer removal. And finally, for large enough sensitivity of CSCs ($\kappa = 0.02$ in Fig. 3c), we can observe treatment success.

3.2 *Differentiation Therapy*

Since CSCs are less sensitive to treatments [14, 34, 47], it might be possible to use drugs that force CSCs to differentiate, thereby reducing the number of CSCs and increasing the sensitivity for treatments. As discussed in [38, 43, 69, 74], members of the TGF-β family are known to increase differentiation. TGF-β also affects other cancer characteristics such as immune evasion, but here we focus on the role of CSCs differentiation. In case of leukemia, a differentiation promoting therapy, called ARTA therapy, was discussed in [58], and in case of melanoma, interferon β and mezerein were considered as differentiation promoters [39].

Youssefpour et al. [74] developed a very detailed CSC model, which not only includes CSCs and TCs, but also transient amplifying cells, chemokine induced feedback mechanisms, spatial dependence and mechanical forces. They show that a differentiation promoter in combination with radiation treatment can have beneficial effects, allowing reduced radiation doses to be used to achieve treatment success. In [3], they use our simple model from above, to specifically consider parameter values for three cancers: head and neck, breast, and metastatic brain tumors.

To include a differentiation promoter, we follow the choice of [74] and modify the probability of symmetric CSC division as function of a differentiation promoter concentration $C(t)$ as

$$p(t) = p_{\min} + (p_{\max} - p_{\min})\frac{1}{1 + \psi C(t)}, \tag{9}$$

where $p_{\max}$ is the self-renewal probability without any treatment and $p_{\min}$ is the minimum self renewal probability at maximal treatment. The parameter ψ represents the sensitivity of the CSC to the chosen drug. Since the effect of the differentiation promoter affects the probability p, we use the model formulation (3) as our base model. We recall that the p relates to δ as $\delta = 2p - 1$. In our simulations, we chose $p_{\max} = 0.505$, which corresponds to $\delta = 0.01$, as chosen before in Sect. 2. We chose $p_{\min} = 0.2$ as in [74], which gives a negative $\delta = -0.6$, i.e. CSCs die out at large doses of $C(t)$.

To describe radiation damage, we use the standard linear-quadratic model (LQ-model). It describes the surviving fraction of cells that are exposed to a dose d as

$$S(d) = e^{-\alpha d - \beta d^2}. \tag{10}$$

The parameters α and β denote the radio sensitivities of the tissue at hand. These values are well known for cancer and healthy tissues, and we will consider typical values for three cancer types: head and neck, metastatic brain cancer, and breast

Table 1 Model parameters. The parameters for head and neck cancer were taken from [19], the values for metastatic brain cancer from [75], and those for breast cancer from [50, 51]

Parameter	Head and neck	Brain	Breast
α/β	10	12	2.88
α (Gy^{-1})	0.35	0.3	0.08
β (Gy^{-2})	0.035	0.025	0.027
d (Gy)	2.53	3.8	2.26
k (1/h)	0.027	0.021	0.007

cancer (see Table 1). The α term corresponds to a single hit action, i.e. DNA damage that results from direct interaction of radiation with DNA [18, 72]. The β-term represents double-hit action, where two non-lethal single hit events interact to form lethal damage. In our model, we assume that CSCs have an increased ability of damage repair, hence we chose $\beta = 0$ for CSC.

There are (at least) two methods to include fractionated radiotherapy into an ODE model of cancer. One method is to stop the ODE model at the treatment time, and apply the surviving fraction (10) to the solution. The new values will then be used as new initial conditions to solve the ODE for the next non-treatment period. Alternatively, we can include the surviving fraction directly into the ODE formulation via the hazard function approach. As shown in detail in [22], the hazard function for fractionated treatment of dose d per fraction is given as $h(t) = (\alpha + \beta d)\dot{D}$, where $\dot{D}$ is the dose rate during treatment (for example 2 Gy per hour). In our case we have two hazard functions, one for CSC and one for TC as

$$h_{\mathrm{CSC}}(t) = \alpha \dot{D}, \qquad h_{\mathrm{TC}}(t) = (\alpha + \beta d)\dot{D}.$$

Hence our modified model for combination treatment with differentiation promoter and radiation treatment becomes

$$\begin{aligned} \dot{u} &= (2p(t) - 1)kF(n)u - h_{\mathrm{CSC}}(t)u \\ \dot{v} &= 2(1 - p(t))kF(n)u + kF(n)v - av - h_{\mathrm{TC}}(t)v, \end{aligned} \tag{11}$$

where we also include the growth rate k.

In [3], we present a full analysis of this model for different cancers and with different schedules for $C(t)$. For these cases, we also computed the tumor control probability and analysed how it depends on the combination treatment. Here, we take a simpler approach, which conveys the same message. We consider radiation treatment for the first week (5 treatment days), and vary the probability of CSC self renewal between its minimum $p_{\min}$ and maximum $p_{\max}$ values. The other model parameters are chosen from the literature shown in Table 1.

Figure 4 shows the corresponding solutions of model (11). We show the CSC population in red and the TC population in blue. The case of maximal CSC self renewal $p = p_{\max}$ as shown in bold lines while the minimal self renewal $p = p_{\min}$ cases are shown as thin line. The value $p = p_{\max}$ corresponds to the absence of differentiation promoter, while the case $p = p_{\min}$ assumes maximal differentiation

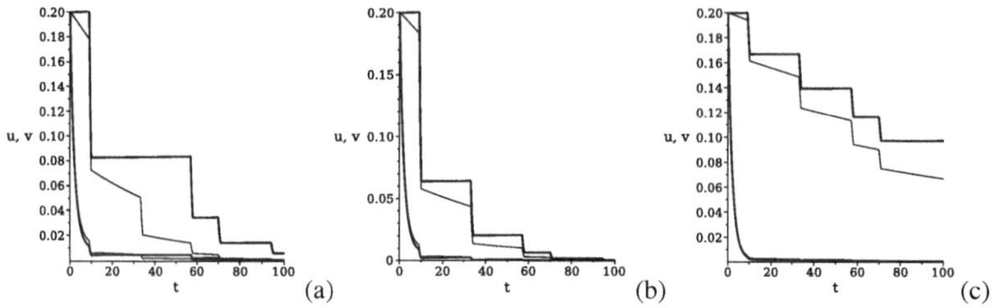

Fig. 4 In each figure, CSCs are represented by the red curves and TCs are represented by the blue curves. Bold curves represent the absence of a differentiation promoter in treatment and the thin curves represent the presence of a differentiation promoter in treatment. (For a black and white print: the curves corresponding to CSCs are above the curves corresponding to TCs in each figure.) The time is shown in hours (120 h = 5 days), and radiation treatment of $2Gy$ is applied on each day at 9 AM for one hour. Figures (**a**), (**b**), (**c**) represent the cases for head and neck cancer, metastatic brain cancer, and breast cancer, respectively. In (**a**), the treatment with a differentiation promoter is significantly better at reducing the CSCs, whereas there is less difference between treatments in (**c**), but the treatment with a differentiation promoter is still more effective

promoter effect. Figure 4a and b, and c show the corresponding plots for head and neck cancer, metastatic brain cancer, and breast cancer, respectively.

We observe the largest effect of differentiation promoter for the head and neck case (A). The thin lines are clearly below the bold lines, and the tumor declines much quicker under differentiation treatment. In the case of metastatic brain cancer (B), we see that treatments with and without differentiation promoter are efficient, while for breast cancer both treatments are less efficient. In each case we get slight improvement by using a differentiation promoter. For a more detailed analysis of these cases see [3].

3.3 *Survivin and Dedifferentiation*

Since the work of Takahashi [68], it is known that non-stem cancer cells can, potentially, reverse differentiation and express stem-cell like behavior. This process of *dedifferentiation* has been reported in many cancers [7, 35, 37]. As non stem tumor cells (TCs) are more sensitive to radiation, dedifferentiation can lead to radioresistance. Dahan [10] found that the inhibitor of apoptosis protein *survivin* supports dedifferentiation. Survivin has a low expression in healthy tissue [2], while it is is over expressed in embryonic tissue and in cancer tissue. Furthermore, survivin expression increases upon radiation treatment, thereby reducing apoptosis and increasing radioresistance [10]. Nakahara [45] found that *sepratonium bromide* (YM155) reduces the expression of survivin, and can be used as survivin inhibitor in treatments. Iwasa [32] used these observations to design a detailed mouse model. They consider immuno suppressed mice that carry a human NSCLC and perform four types of experiments: (i) a control case of no treatment, (ii) a radiation treatment of 2 Gy per day for 5 days, (iii) a YM155 treatment, and (iv) a combination of radiation and YM155 treatments. In [53] a CSC-based mathematical model was developed to fit to the data of Iwasa [32]. The model fits the data very well and supports the conclusion that survivin plays an important role in cancer radio resistance.

Here, we review the model of [53] and explain the effects of survivin inhibition on a simplified treatment model. The above base model (4) is extended by a third compartment $s(t)$ for the survivin concentration, plus dynamics that result from survivin:

$$\begin{aligned} \dot{u} &= \delta k F(n)u + \mu(s)v - \tau_s(s)u, \\ \dot{v} &= (1-\delta)kF(n)u + k_2 F(n)v - \mu(s)v - \tau_d(s)v, \\ \dot{s} &= \omega_d \tau_d(s)v + \omega_s \tau_s(s)u - \sigma s, \end{aligned} \tag{12}$$

where $\mu(s)$ describes the survivin-induced dedifferentiation and $\tau_d(s)$, $\tau_s(s)$ are the survivin dependent death rates of CSCs and TCs, respectively. The parameters k and k_2 denote the base rates of mitosis for CSCs and TCs, respectively. The factors ω_d, ω_s denote the survivin production rates based on cell apoptosis events and the rate σ denotes the rate of survivin clearance. We chose to describe the availability of space with the nonlinear function

$$F(n) = \max\{1 - n^4, 0\}$$

since cells are deformable and they can squeeze into open spaces [46]. The functional forms of the survivin dependent rates were chosen according to [53, 74]. The survivin dependent death rates are decreasing functions of s, since survivin is an anti-apoptotic protein:

$$\tau_s(s) = \frac{\tau_{\text{dmax}}}{1+\theta_s s}, \qquad \tau_d(s) = \frac{\tau_{\text{smax}}}{1+\theta_d s}. \tag{13}$$

The survivin-induced dedifferentiation rate $\mu(s)$ is chosen to be a sigmoid function

$$\mu(s) = \frac{\mu_{\max}}{1 + \left(\frac{\mu_{\max}-\mu_{\min}}{\mu_{\min}}\right)^{1-\frac{s}{s_2}}}. \tag{14}$$

And finally, radiation was modelled by the linear quadratic model as done before in (10). In [53], a detailed model fitting of the model (12), (13), (14) to the experimental data of Iwasa [32] was performed. In Table 2, we list the parameters that best fit those data.

In Fig. 5, we show some model simulations with the parameter values from Table 2. In Fig. 5a, the dashed lines show the tumor evolution as a function of time for the untreated (control) case. The dark blue (dashed) line shows the CSC compartment, while the light blue line shows the TCs. Survivin concentration is shown in pink. We clearly see the growth of a stem-cell driven tumor. The thin

Table 2 Parameter values used for the survivin model (12)

Parameter	Meaning	Value from [53]
Initial conditions		
$u(0)$	Initial CSC density	0.0025
$v(0)$	Initial TC density	0.05
$s(0)$	Initial *survivin* concentration	0.0004
Model parameters		
δ	Fraction of CSC after mitosis of CSC	0.01
$k(time^{-1})$	CSC mitosis rate	0.0659
$k_2(time^{-1})$	TC mitosis rate	0.6256
$\tau_{\text{smax}}(time^{-1})$	Max CSC death rate	0.002
$\tau_{\text{dmax}}(time^{-1})$	Max TC death rate	0.5
$\mu_{\text{max}}(time^{-1})$	Max rate of dedifferentiation	1.0997
$\omega_s(time^{-1})$	CSC *survivin* release	77
$\omega_d(time^{-1})$	Max TC *survivin* release	55
$\sigma(time^{-1})$	*Survivin* decay rate	0.475
$\theta_s(concentration^{-1})$	CSC *survivin* sensitivity	250
$\theta_d(concentration^{-1})$	TC *survivin* sensitivity	125
$\mu_{min}(time^{-1})$	Min rate of dedifferentiation	0.000001
$s_2(concentration)$	$\frac{\mu_{max}}{2}$ *survivin* concentration	0.1187
Radiation parameters		
$\alpha/\beta\ (Gy)$	NSCLC LQ ratio	10
$\alpha_s(Gy^{-1})$	CSC DNA damage single tract	0.02465
$\alpha_d(Gy^{-1})$	TC DNA damage single tract	0.2465
$\beta_s(Gy^{-2})$	CSC DNA damage double tract	0.002465
$\beta_d(Gy^{-2})$	TC DNA damage double tract	0.02465

solid lines show the case where survivin production is inhibited (for example by YM155). In that case, the tumor is TC dominated (golden curve) with a very small CSC compartment (black line) and zero survivin (purple). The tumor grows slower and the composition is very different. We expect that the increased TC fraction in the case of inhibited survivin will make this tumor much more radiosensitive.

To show this effect, we apply radiation of 5 Gy on day 3 and day 7 in Fig. 5b. Here, the dashed lines show the growth of the total tumor $u(t) + v(t)$ as functions of time without radiation. The base case is in black and the survivin-inhibited case in red. The corresponding thin lines include radiation, which can clearly be seen by the jumps on days 3 and 7. Radiation in the base case (thin black line) has a positive effect on delaying the cancer growth for a few days. But it does not control the cancer. However, radiation applied to the survivin-inhibited case (thin red line), has a drastic effect. On day 7, the cancer gets close to zero, which can be sufficient for treatment success if random perturbations are included. Even if it is not eradicated, the growth has been significantly delayed through the combination therapy of survivin inhibition and radiation treatment.

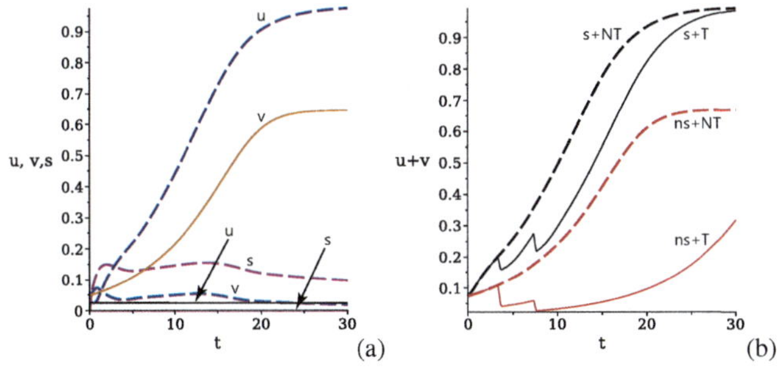

Fig. 5 (**a**) Dashed curves show the solutions of the survivin model (12) without treatment, CSCs dark blue labelled with u, TCs blue labelled with v, survivin pink labelled with s; the solid lines are the model with survivin inhibition, CSCs in black labelled with u, TCs in gold labelled with v and survivin in purple labelled with s. (**b**) Total cancer population $u(t) + v(t)$ as function of time for the four cases: survivin + no treatment (s + NT) in dashed black, survivin + treatment (s + T) in solid black, no survivin + no treatment (ns + NT) in dashed red and no survivin + treatment (ns + T) in solid red

These simulations, and the more detailed analysis in [53], confirm Iwasa's [32] conclusion that survivin inhibition might be a suitable strategy to reduce radio-resistance in cancer cells.

3.4 Inclusion of Feedback Mechanisms

Rodriguez-Berens et al. [55, 73] used a simple variation of (4) to study feedback mechanisms that control CSC self renewal. In the experiments of Lander and others [38, 73], two effects were identified. The first effect is that a growing TC compartment down-regulates the proliferation rate k. Secondly, the growing TC compartment reduces the probability of CSC self renewal p. Hence in [55], they assume that both $k(v)$ and $p(v)$ depend on the TC compartment as strictly decreasing functions, which converge to 0 as $v \to \infty$. Their model reads

$$\begin{aligned} \dot{u} &= (2p(v) - 1)k(v)u, \\ \dot{v} &= 2(1 - p(v))k(v)u - av. \end{aligned} \tag{15}$$

Compared to (4), this model has no TC self renewal (i.e. $k_2 = 0$) and also no volume mechanism (i.e. $F(n) \equiv 1$). The authors chose $p(0) > 0.5$ as otherwise the cancer would not grow in the first place. Using the above model (15), they find in [55] that growth control depends significantly on the self renewal probability $p(v)$ and feedback on p is essential. Roughly speaking, we need to achieve $p(v) < 0.5$ for large values of v such that CSCs can decay. Feedback in the division rate $k(v)$ will only change the speed of growth, but will not control tumor growth [55].

In the context of colorectal cancers, it has been found that healthy stem cells and also cancer stem cells release the molecule Wnt as a self renewal promoter [9], thereby creating a positive feedback between CSC growth and CSC self renewal. Hence in Konstorum et al. [36], they included such a positive feedback into the CSC model. A self renewal promoter $w(t)$ is included, and, as before, TGF-β is modelled as differentiation promoter $C(t)$. The CSC model then becomes

$$\begin{aligned} \dot{u} &= (2p(C,w)-1)ku, \\ \dot{v} &= 2(1-p(C,w))ku - av, \\ \dot{C} &= \nu v - \mu C, \\ \dot{w} &= w\left(\frac{\beta u w}{1+\lambda w} - 1\right), \end{aligned} \tag{16}$$

where the probability of CSC self renewal is increasing in w and decreasing in C with a functional form as in (9)

$$p(C,w) = p_{\min} + (p_{\max} - p_{\min})\frac{\xi w}{1+\xi w}\frac{1}{1+\psi C},$$

and $p_{\min}, p_{\max}, \xi, \psi > 0$ are constants. The differentiation promoter C is produced by the TC at rate $\nu > 0$ and decays at rate $\mu > 0$. The growth activator w shows saturated growth with a finite carrying capacity and the coefficients β and λ are constant. In [36], the model (16) is simplified by assuming that C and v are in quasi steady state. In other words this means that we look at the dynamics of (16) on the slow manifold. For this reduced system, the existence of an Allee effect is shown in [36]. If the CSC and self promoter w are too small, then the CSC population dies off. The tumor can only grow if there is a sufficiently large CSC and w compartment. This observation has consequences for treatment since cancer does not need to be fully extinct via treatment. Rather, it is sufficient to reduce its size below an Allee threshold, such that the tumor can no longer self sustain and dies out. It is really interesting to see that such a positive feedback loop between CSC and w can produce an Allee effect. In the above two models (15) and (16), we did not include TC self renewal (i.e. $k_2 = 0$) and neither the volume filling constraints $F(n)$. It is an interesting question to include TC self renewal $k_2 \neq 0$ and volume filling terms $F(n)$ in these models to see how the results change with the inclusion of these effects.

4 Spatially Explicit Models for CSCs

As we now understand the cancer stem cell model and its modifications very well, it is time to put it into the spatial context. We could simply, as many authors do, add diffusion terms to the equations of (4). This however, appears too naive for our purpose. Rather, we take guidance from the work of Enderling et al. [16, 70], where a spatial cancer stem cell model was developed as an individual based cellular automaton. They consider a square grid where each grid cell can be either occupied by a CSC or a TC, or be empty. Cells can divide if nearby empty cells are available. Otherwise, they stay quiescent or die. In addition, cells are able to randomly move to nearby grid cells, if space is available. The CSCs have unlimited replicative potential, while the TCs can only divide a limited amount of time. In simulations, it is seen that CSCs surround themselves with TCs, which in turn occupy space and inhibit further CSC divisions. CSCs become trapped and tumor growth stops. Once TCs are removed, say via treatment, space for CSC division and movement becomes available which results in more CSCs, eventually leading to a larger tumor. This effect is known as the *tumor growth paradox*. In [16], the ability of cells to move randomly seemed important, as otherwise the tumor growth paradox would not arise.

4.1 Birth-Jump Models

Here we take the rules of Enderling's cellular automaton model [16] and formulate them as a spatial continuous model, using the framework of birth-jump models [29]. A birth-jump process is a process where population growth and spread are not decoupled but rather interdependent. The standard reaction-diffusion type models assume that the movement process is independent from population kinetics such as mitosis or death. This is expressed mathematically by using summation of the corresponding terms in the equations. Birth-jump models, however, were developed to describe movement that is directly coupled to a birth event [29]. One application of this scenario is in metastatic cancer. Once a cancer reaches a metastatic stage, a large fraction of newly generated cancer cells are released immediately into the blood stream, and are transported away to distant sites. Another application of birth-jump models are forest fires, where burning fire branches are launched into the air, carried by the wind, landing ahead of the fire to create new fires (*fire spotting*) [42]. Here of course, we are interested in cancer spread. In Enderlings model [15], newly generated cancer cells need to find an open grid cell in their neighborhood to settle. Hence, spread and birth is directly coupled and a birth-jump approach seems natural. Here, we assume that upon division one daughter cell replaces the mother, while the other daughter cell is redistributed locally to some empty space. Mathematically, we introduce a spatial relocation kernel $K(x, y)$ and as before a function for the volume effect $F(n)$. We denote the spacial domain by Ω and for K, F, Ω we assume:

$$(A.1)\quad K \geq 0,\, K \in C(\bar{\Omega}, \bar{\Omega}),\, K \in L^2(\Omega \times \Omega),$$
$$\int_\Omega K(x, y)dx = 1,\, \int_\Omega K(x, y)dy \leq 1.$$
$$(A.2)\quad F : \mathbb{R}^+ \to [0, 1],\, F(0) = 1,\, F(n) = 0 \quad \text{for} \quad n \geq 1$$
$$F \quad \text{is non-increasing and Lipschitz continuous.}$$
$$(A.3)\quad \Omega \subset \mathbb{R}^m \text{ is either a smooth bounded domain, or it is } \Omega = \mathbb{R}^m.$$

Then the spatially dependent cancer stem cell birth-jump model, introduced in [28], reads

$$u_t(x, t) = d_u \Delta u(x, t) + \delta k \int_\Omega K(x, y) F(n(x, t)) u(y, t) dy \tag{17}$$
$$v_t(x, t) = d_v \Delta v(x, t) + (1 - \delta) k \int_\Omega K(x, y) F(n(x, t)) u(y, t) dy$$
$$+ k_2 \int_\Omega K(x, y) F(n(x, t)) v(y, t) dy - av,$$

where again $n(x, t) = u(x, t) + v(x, t)$ and d_u, d_v are diffusion coefficients. The above system needs to be equipped with appropriate boundary conditions, for example with the no-flux boundary conditions

$$\frac{\partial u}{\partial n} = \frac{\partial v}{\partial n} = 0, \quad \text{where } n \text{ denotes an outer normal on} \quad \partial\Omega. \tag{18}$$

Maddalena et al. [40] provided a full solution theory for (17) and (18). The diffusion terms are the leading order differential operators. As they define strongly continuous semigroups of L^2 and H^1 spaces, we can consider the integral terms as compact perturbations. Then perturbed semigroup theory can be applied [48] to obtain unique local solutions. Furthermore, Maddalena et al. [40] show the following

Theorem 2 *Assume $n \leq 3$ and $u_0, v_0 \in H^2(\Omega)$. Then there exists a unique global solution of (17) with $u(., t), v(., t) \in H^2(\Omega)$ and*

$$u(x, t) \geq 0,\; v(x, t) \geq 0,\; 0 \leq u(x, t) + v(x, t) \leq 1, \qquad \textit{for all } (x, t).$$

Moreover, Maddalena et al. [40] show with an energy principle, that all stationary states are spatially homogeneous. Hence, the model (17) excludes formation of small and stationary tumor masses. This is a serious short coming, since not all tumors continue to grow. The reason is the included diffusion terms, which always allow CSC and TC to escape and invade further.

In Borsi et al. [6] they consider the above model (17) without diffusion terms, i.e. $d_u = d_v = 0$. The leading order terms are now the non-linear integral operators and a semigroup method is not applicable. Instead, the authors use a fixed point argument. They assume for the initial conditions $u_0(x)$, $v_0(x)$ that

$$(B) \qquad u_0,\ v_0 \in C(\bar{\Omega}),\ u_0 \geq 0,\ v_0 \geq 0,\ 0 \leq u_0 + v_0 \leq 1.$$

Using a-priori estimates and a fixed point argument, it was shown in [6]:

Theorem 3 *Assume (A), (B) and* $d_u = d_v = 0$. *Then (17) has a unique global solution*

$$u, v \in C(\bar{\Omega} \times [0, \infty)),$$

with

$$u(x,t) \geq 0,\ v(x,t) \geq 0,\ 0 \leq u(x,t) + v(x,t) \leq 1, \qquad \text{for all } (x,t).$$

Also, in Borsi et al. [6] numerical simulations were performed which again confirm the tumor growth paradox. We expect that the model (17) without diffusion would support finite size stationary tumors, but to prove this mathematically is an open question.

4.2 Reaction Diffusion Models

Spatial integral operators are in some sense close to differential operators. This applies for symmetric kernels $K(|x-y|)$, which have small variance and even smaller higher moments. For this case, we can expand the integral operator as

$$\begin{aligned} &\int_\Omega K(|x-y|)F(n(x,t))u(y,t)dy \qquad (19)\\ &\approx F(n(x,t))\Big[Au(x,t) + B_{ij}(x)\partial_i\partial_j u(x,t) + C_{ijkl}(x)\partial_i\partial_j\partial_k\partial_l u(x,t) + \text{h.o.t.}\Big] \end{aligned}$$

where the partial symbols ∂_j stand for the partial spatial derivatives in direction x_j. The coefficients A, B, C denote the first even moments of K, where A is a scalar, B a 2-tensor and C a 4-tensor:

$$A := \int_\Omega K(|y|)dy, \qquad B_{ij}(x) = \frac{1}{2}\int_\Omega (x-y)_i(x-y)_j K(|y|)dy$$

$$C_{ijkl}(x) = \frac{1}{24}\int_\Omega (x-y)_i(x-y)_j(x-y)_k(x-y)_l K(|y|)dy$$

where the indices denote vector components and we use summation convention for repeated indices. Note that our earlier assumption (A) implies that $A = 1$.

Fasano et al. [17] looked at this case in one dimension up to order two. In this case, the diffusion cancer stem cell model becomes

$$\begin{aligned} u_t &= d_u u_{xx} + \delta k F(n)(u + B u_{xx}), \\ v_t &= d_v v_{xx} + (1-\delta) k F(n)(u + B u_{xx}) + k_2 F(n)(v + B v_{xx}) - a v. \end{aligned} \tag{20}$$

The authors in [17] note that this problem is not well defined as a biological model, since the Bu_{xx}-term in the second equation could lead to negative solutions for v, which is biologically unrealistic. They argue, however, that such a case is artificial and not be relevant in applications. Nevertheless, they show for model (20):

Theorem 4 *Let $u_0, v_0 \in C^{2+\gamma}(\mathbb{R})$ such that*

$$u_0(x) \geq 0, \;\; v_0(x) \geq 0, \;\; 0 \leq u_0(x) + v_0(x) \leq 1 - M$$

for some small $M > 0$. Then (20) has a unique local classical solution $u(x,t), v(x,t)$ which satisfies $u(x,t) + v(x,t) \leq 1$ for all times $t < T^*$.

In [60], we go one step further and also assume that B is small such that the B-terms can be ignored. This means the model simplifies further and becomes

$$\begin{aligned} u_t &= d_u u_{xx} + \delta k F(n) u, \\ v_t &= d_v v_{xx} + (1-\delta) k F(n) u + k_2 F(n) v - a v. \end{aligned} \tag{21}$$

This model does preserve positivity. It is also the model we would have obtained if we simply added diffusion terms to (4). However, now we know how it was derived and what assumptions were used to get to this model. Existience and uniqueness for model (21) is now covered by standard arguments for parabolic equations. The diffusion terms define a semigroup, and the other terms are globally Lipschitz-continuous perturbations, implying global classical solutions of (21) for standard boundary conditions [48].

A further interesting open problem arises if fourth order terms from the above expansion (19) were included. The effect on spatial invasion and regularity of solutions has not been studied yet.

4.3 The Tumor Invasion Paradox

With the above model (21), we come back full circle to the tumor growth paradox. In the spatial context, the tumor growth paradox leads to an increased cancer invasion speed for increased TC death rates. We call this a *tumor invasion paradox*. We can prove the tumor invasion paradox for a special case of (21). We assume the base mitosis rates for CSC and TC are the same and equal to one $k = k_2 = 1$. We also assume that the diffusion coefficients for CSC and TC are the same, since these are cells of similar size, and that diffusion is slow such that $d_u = d_v = \delta d$, for an order one value d. The resulting system becomes

$$\begin{aligned} u_t &= \delta d u_{xx} + \delta F(n) u, \\ v_t &= \delta d v_{xx} + (1-\delta) F(n) u + F(n) v - a v. \end{aligned} \tag{22}$$

As in Sect. 2, we consider δ to be small and use perturbation arguments. To leading order ($\delta \to 0$), we obtain the fast system

$$\begin{aligned} u_t &= 0, \\ v_t &= F(n) u + F(n) v - a v, \end{aligned} \tag{23}$$

which, as before, describes fast convergence towards the slow manifold

$$M := \{(u, v) : a v = F(u+v)(u+v)\}.$$

Rescaling time as $\tau = \delta t$ gives us the slow system, which to leading order becomes

$$\begin{aligned} u_\tau &= d u_{xx} + F(u+v) u, \\ 0 &= F(u+v) u + F(u+v) v - a v. \end{aligned} \tag{24}$$

The second equation ensures that the dynamics are on the slow manifold M, while the first equation of (24) describes the dynamics on M.

In [60], we analyse the properties of the slow manifold in detail. We show that on M, v can be written as a graph on u, i.e. $v = v_a(u)$. We use an index a to indicate that this map depends on the death rate a. Using this representation, we can write the first equation of (24) as an equation for u alone as

$$u_\tau = d u_{xx} + F(u + v_a(u)) u. \tag{25}$$

In [60], we show that this Eq. (25) is a bistable equation of Fisher-KPP type and we can look for travelling wave solutions. Travelling wave solutions are solutions that retain their shape and either propagate to the left or right with a constant speed. For example, a travelling wave solution propagating to the right can be written as $u(x, t) = \varphi(x - ct)$ where c is a constant that defines the speed of the wave and $\varphi(z)$ with $z = x - ct$ gives the wave profile [12]. Additionally, φ satisfies

$$\varphi(-\infty) = 1 \quad \text{and} \quad \varphi(\infty) = 0.$$

This means that on the left the population density has reached its maximum size of 1, but the population has not yet arrived to the space on the right [12]. Since we know that (25) is of Fisher-KPP type, we can directly apply the known results on travelling waves for Fisher-KPP equations [12, 71]. To formulate the result, we define invasion initial conditions.

Definition 2 An *invasion initial condition* for $u_0(x)$ is a non-increasing function $u_0(x)$ that satisfies

$$\lim_{x \to -\infty} u_0(x) = 1, \qquad u_0(x) = 0 \text{ for all } x \geq x^*, \qquad \text{for some} \quad x^* \in \mathbb{R}.$$

We show in [60]:

Theorem 5 *We assume the previous assumptions (A) and (B), $a > 0$ and that $F'(n) < 0$ for $0 \leq n \leq 1$. We consider invasion initial conditions for $u_0(x)$ and denote the solution of (25) with $u(\tau, x)$.*

1. *If $a \geq 1$ then $u(\tau, x)$ converges to a travelling wave with minimum wave speed*

$$c^* = 2\sqrt{d}.$$

2. *If $a < 1$ then $u(\tau, x)$ converges to a travelling wave with minimum wave speed*

$$c^* = 2\sqrt{da}.$$

The cases connect continuously for $a \to 1$.

We see that for large a ($a > 1$) the wave speed is independent of its value. However, for small a we see an invasion paradox. Reducing the death rate a reduces the invasion speed, quite contrary to what is expected.

We simulate (25) by choosing $F = 1 - n$ where $n = u + v$. We choose this simple function F for the ease of computation since in [60] it was shown that any F satisfying the assumptions in Theorem 5 the resulting numerical solutions are very similar. We set $d = 1$ for simplicity and focus on studying the death rates $a = 3, 1, 0.5, 0.1$ as these highlight the key dynamics of the solutions to (25). By using the *pdepe* solver in MATLAB, we numerically obtain the travelling wave solutions of (25) shown in Fig. 6. In these simulations, the initial condition is a step function where $u_0(x) = 1$ for $x < x^*$ and $u_0(x) = 0$, otherwise. We set $x^* = 20$ and stop the simulation at $\tau = 80$ to illustrate the main behaviour of the solutions given our parameter choices.

In Fig. 6, we can clearly see the invasion paradox. For $a \geq 1$, we see that there is not much difference in the travelling wave solutions or invasion speeds as Theorem 5 states. However, for $a < 1$, as a decreases the travelling wave solutions become slower, illustrating the invasion paradox.

In summary, the invasion paradox is an underlying property of (21). This shows that in addition to the tumor growth paradox, the invasion paradox may be at play. This means that cancer treatments may additionally increase tumor spread [60].

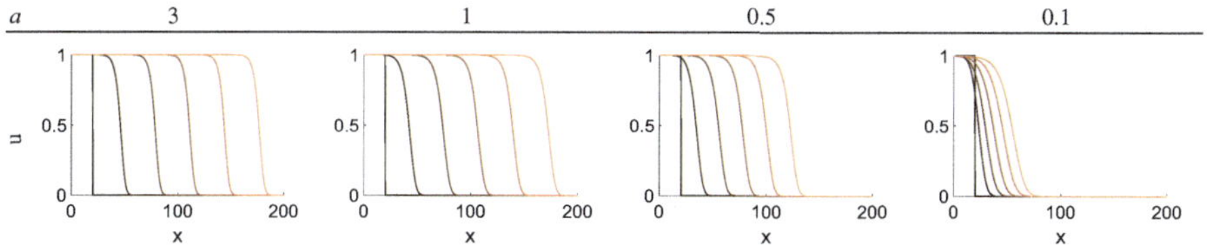

Fig. 6 Travelling wave solutions at $\tau = 0, 16, 32, 48, 64, 80$ of system (25). The value of a is given in column headers. The initial condition is a step function satisfying the invasion initial condition with $x^* = 20$. For death rates $a = 3, 1$ no invasion paradox is present, whereas for $a = 0.5, 0.1$ the invasion paradox can be observed, where the solutions for $a = 0.5$ invade faster into space than the solutions with $a = 0.1$

Fortunately, shorter treatments were shown to not significantly increase spread [60], but the invasion paradox becomes more significant in long term treatments.

5 Conclusions and Open Questions

The inclusion of CSC dynamics into solid tumor modelling adds another dimension to the discussion of tumor growth. On the one hand, CSCs are essential to keep a tumor growing, while on the other hand, CSCs can be blocked from space and nutrients making them quiescent. Once we wake them up, by liberating space or supplying nutrients, they divide again and possibly create a worse situation than before. Mathematically, this was expressed through the *tumor growth paradox* and the *tumor invasion paradox*. Through the spatial modelling, we observe a dichotomy between treatment-related tumor cell death and treatment induced spatial invasion. A fine balance needs to be achieved so that all CSCs are killed before the tumor can spread any further.

We included many cancer related effects into the modelling, such as treatments, combination therapies, feedback mechanisms, and spatial dependence. But this is just scratching the surface. Many interesting questions remain, and many of them are accessible to mathematical modelling. Here we like to summarize some of the questions that we find interesting for further study.

1. **Hallmarks:** We related the CSC modelling to the hallmarks (H1) sustained growth signals, (H2) evasion of growth suppression, (H3) resistance of cell death, (H4) replicative immortality, (H6) metastasis formation, and (H14) senescence. We expect that CSC play important roles in the other hallmarks as well, and a detailed modelling of those seems to be a natural place to continue. For example, the interaction of immune cells and CSCs is relatively poorly understood. There is indication [56] that CSCs evade immune destruction. This implies that an immune response would select for CSCs [28]. CSCs might also be able to re-educate immune cells to become pro-tumor immune cells [54, 59].

2. **Allee effect:** The Allee effect model (16) did not include the volume constraint $F(n)$ and TC self reproduction. It would be interesting to extend the model (16) by including those terms and analyzing how the Allee threshold changes with those terms.
3. **Geometric singular perturbation theory for PDEs:** We saw that the slow manifold analysis is a quite powerful tool to understand the paradoxical behavior in tumor dynamics. For the ODE cases this is justified through the Fenichel theorems [27] for geometric singular perturbation theory. Unfortunately, such a theory is missing for the PDE models. Hence our scaling analysis in the PDE case is just a formal expansion without an abstract backbone behind it. Some early results for PDEs are available in the work of Bates et al. [4].
4. **Compact-support steady states for birth-jump models:** It would be important to understand possible steady states of the non-local spatial model (17) without diffusion, ($d_u = d_v = 0$). Since the steady states represent solid tumors that stopped growing. We saw that if diffusion is added, steady states are constants [40]. However, without diffusion, we expect that non-constant steady states with compact support are possible. This will, of course, depend on the choice of the integral kernel $K(x, y)$.
5. **Tumor spheroids:** It would also be interesting to see the relative distribution of CSCs in a spatially extended tumor. There is extensive material available about CSC distributions in growing spheroids [30] and this material is just waiting for a mathematical analysis.
6. **Fourth order terms:** In the moment expansion of the integral term (19), Fasano et al. [17] stopped at the second order term (20). Many colleagues asked what would happen if the fourth order terms in (19) were included. In that case the model gets a Cahn-Hilliard [44] type structure, and phase separation phenomena might occur, where one phase describes tumor present and the other phase tumor absent. Interesting dynamics on the phase boundaries might occur.
7. **Stochastic effects:** The modelling presented here does not directly include stochastic effects and these differential equation models are mostly useful for large population sizes. This is problematic once the cell populations become small. In particular, if treatment reduces the cell numbers to almost zero. In those cases stochastic effects will be significant, leading to eradication in some cases and re-growth in others. To properly address these stochastic effects, scientists have used birth-death and branching stochastic processes to estimate probabilistic quantities such as the *tumor control probability* [11, 26, 76], or a *normal tissue complication probability* [63, 65]. These methods require a different framework as presented here, hence we did not discuss them in any detail.

References

1. Agur, Z., Daniel, Y., Ginosar, Y.: The universal properties of stem cells as pinpointed by a simple discrete model. J. Math. Biol. **44**(1), 79–86 (2002)
2. Altieri, D.: Survivin, versatile modulation of cell division and apoptosis in cancer. Oncogene **22**, 8581–8589 (2003)
3. Bachman, J., Hillen, T.: Mathematical optimization of the combination of radiation and differentiation therapies of cancer. Frontiers in Molecular and Cellular Oncology (2012). https://doi.org/10.3389/fonc.2013.00052. Free online
4. Bates, P., Lu, K., Zeng, C.: Approximately invariant manifolds and global dynamics of spike states. Invent. Math. **174**(2), 355–433 (2008)
5. Beretta, E., Capasso, V., Morozova, N.: Mathematical modelling of cancer stem cells population behavior. Math. Model. Nat. Pheno.**7**(1), 279–305 (2012)
6. Borsi, I., Fasano, A., Primicerio, M., Hillen, T.: A non-local model for cancer stem cells and the tumour growth paradox. Math. Med. Biol. **34**(1), 59–75 (2017)
7. Chaffer, C., Brueckmann, I., Scheel, C., et al.: Normal and neoplastic nonstem cells can spontaneously convert to a stem-like state. Proc. Natl. Acad. Sci. **108**(19), 7950–7955 (2011)
8. Clarke, M.F., Dick, J.E., Dirks, P.B., et al.: Cancer stem cells–perspectives on current status and future directions: AACR workshop on cancer stem cells. Cancer Res. **66**(19), 9339–9344 (2006)
9. Clevers, A., Loh, K., Nusse, R.: Stem cell signaling. an integral program for tissue renewal and regeneration: Wnt signaling and stem cell control. Science **346**(6205), 1248012 (2014)
10. Dahan, P., Martinez Gala, J., Delmas, C., et al.: Ionizing radiations sustain glioblastoma cell dedifferentiation to a stem-like phenotype through survivin: Possible involvment in radioresistance. Cell Death Disease **5**(11) (2014). https://doi.org/10.1038/cddis.2014.509
11. Dawson, A., Hillen, T.: Derivation of the tumour control probability (TCP) for a cell cycle model. Comput. Math. Meth. Med. **7**, 121–142 (2006)
12. De Vries, G., Hillen, T., Lewis, M., et al.: A Course in Mathematical Biology: Quantitative Modeling with Mathematical and Computational Methods. SIAM, Philadelphia (2006)
13. Dingli, D., Michor, F.: Successful therapy must eradicate cancer stem cells. Stem Cells **24**(12), 2603–2610 (2006)
14. Dittmar, T., Zänker, K.: Role of Cancer Stem Cells in Cancer Biology and Therapy. CRC Press, Boca Raton (2013)
15. Enderling, H., Alexander, N., Clark, E., et al: Dependence of invadopodia function on collagen fiber spacing and crosslinking: computational modeling and experimental evidence. Biophys J. **95**(5), 2203–2218 (2008)
16. Enderling, H., Anderson, A., Chaplain, M., et al.: Paradoxical dependencies of tumor dormancy and progression on basic cell kinetics. Cancer Res. **69**(22), 8814–8821 (2009)
17. Fasano, A., Mancinia, A., Primicerio, M.: Tumours with cancer stem cells: a PDE model. Math. Biosci. **272**, 76–80 (2016)
18. Fowler, J.: The linear quadratic formula and progress in fractionated radiotherapy. Br. J. Radiol. **62**, 679–694 (1989)
19. Fowler, J.F.: 21 years of biologically effective dose. Br. J. Radiol. **83**(991), 554–568 (2010)
20. Ganguli, R., Puri, I.: Mathematical model for the cancer stem cell hypothesis. Cell Proliferation **39**, 3–14 (2006)
21. Gao, X., McDonald, J., Hlatky, L., Enderling, H.: Acute and fractionated irradiation differentially modulate glioma stem cell division kinetics. Cancer Res. **73**(5), 1481–90 (2013)
22. Gong, J., dos Santos, M.M., Finlay, C., Hillen, T.: Are more complicated tumor control probability models better? Math. Med. Biol. **30**(1), 1–19 (2011)
23. Hanahan, D., Weinberg, R.: The hallmarks of cancer. Cell **100**(1), 57–70 (2000)
24. Hanahan, D., Weinberg, R.: Hallmarks of cancer: The next generation. Cell **144**, 646–674 (2011)
25. Hanahan, D.: Hallmarks of cancer: New dimensions. Cancer Discovery **12**, 31–46 (2022)

26. Hanin, L.G.: A stochastic model of tumor response to fractionated radiation: limit theorems and rate of convergence. Math. Biosci. **91**(1), 1–17 (2004)
27. Hek, G.: Geometric singular perturbation theory in biological practice. J. Math. Biol. **60**(3), 347–386 (2010)
28. Hillen, T., Enderling, H., Hahnfeldt, P.: The tumor growth paradox and immune system-mediated selection for cancer stem cells. Bull. Math. Biol. **75**(1), 161–184 (2013)
29. Hillen, T., Greese, B., Martin, J., de Vries, G.: Birth-jump processes. J. Biol. Dyn. **9**, 104–127 (2014). https://doi.org/10.1080/17513758.2014.950184
30. Ishiguro, T., Ohata, H., Sato, A., et al.: Tumor-derived spheroids: relevance to cancer stem cells and clinical applications. Cancer Sci. **108**(3), 283–289 (2017)
31. Iwasa, Y., Nowak, M., Michor, F.: Evolution of resistance during clonal expansion. Genetics **172**, 2557–2566 (2006)
32. Iwasa, T., Okamoto, I., Suzuki, M., et al.: Radiosensitizing effect of YM155, a novel small-molecule survivin suppressant, in non-small cell lung cancer cell lines. Clin. Cancer Res. **14**(20), 6496–6504 (2008)
33. Johnston, M., Maini, P., Chapman, S., et al.: On the proportion of cancer stem cells in a tumour. J. Theor. Biol. **266**, 708–711 (2010)
34. Kim, J., Tannock, I.: Repopulation of cancer cells during therapy: an important cause of treatment failure. Nat. Rev. Cancer **5**, 516–525 (2005)
35. Klevebring, D., Rosin, G., Ma, R., et al.: Sequencing of breast cancer stem cell populations indicates a dynamic conversion between differentiation states in-vivo. Breast Cancer Res. **16**(4) (2014). https://doi.org/10.1186/bcr3687
36. Konstorum, A., Lowengrub, J., Hillen, T.: Feedback regulation in a cancer cell model can cause an Allee effect. Bull. Math. Biol. **78**(4), 754–785 (2016)
37. Lagadec, C., Vlashi, E., Donna, L.D., et al.: Radiation-induced reprogramming of breast cancer cells. Stem Cells **30**(5), 833–844 (2012)
38. Lander, A., Gokoffski, K., Wan, F., et al.: Cell lineages and the logic of proliferation control. PLoS Biol. **7**(1), e15 (2009)
39. Leszczyniecka, M., Roberts, T., Dent, P., et al.: Differentiation therapy of human cancer: basic science and clinical applications. Pharmacol. Ther. **90**, 105–156 (2001)
40. Maddalena, L.: Analysis of an integro-differential system modelling tumor growth. Appl. Math. Comput. **245**, 152–157 (2014)
41. Marciniak-Czochra, A., Stiehl, T., Ho, A., et al.: Modeling of asymmetric cell division in hematopoietic stem cells–regulation of self-renewal is essential for efficient repopulation. Stem Cells Develop. **18**(3), 377–386 (2009)
42. Martin, J., Hillen, T.: The spotting distribution of wildfires. Appl. Sci. **6**(6), 177–211 (2016)
43. Meulmeester, E., Ten Dijke, P.: The dynamic roles of TGF-beta in cancer. J. Pathol. **223**(2), 205–218 (2011)
44. Miranville, A.: The Cahn–Hilliard Equation: Recent Advances and Applications. SIAM, Philadelphia (2019)
45. Nakahara, T., Kita, A., Yamanaka, K., et al.: YM155, a novel small-molecule survivin suppressant, induces regression of established human hormone-refractory prostate tumor xenografts. Cancer Res. **67**(17), 8014–8021 (2007)
46. Painter, K., Hillen, T.: Volume-filling and quorum sensing in models for chemosensitive movement. Can. Appl. Math. Q. **10**(4), 501–543 (2002)
47. Pajonk, E., Vlashi, E., McBride, W.: Radiation resistance of cancer stem cells: The 4 R's of radiobiology revisited. Stem Cells **24**(4), 639–648 (2010)
48. Pazy, A.: Semigroups of Linear Operators and Applications to Partial Differential Equations. Springer, New York (1983)
49. Poleszczuk, J., Hahnfeldt, P., Enderling, H.: Evolution and phenotype selection of cancer stem cells. PLoS Comput. Biol. **11**(3), e1004,025 (2015)
50. Qi, X.S., White, J., Li, X.A.: Is α/β for breast cancer really low? Radiot. Oncol**100**(2), 282–288 (2011)

51. Rew, D.A., Campbell, I.D., Taylor, I., Wilson, G.D.: Proliferation indices of invasive breast carcinomas after in vivo 5-bromo-2′-deoxyuridine labelling: A flow cytometric study of 75 tumours. Br. J. Surg. **79**(4), 335–339 (1992)
52. Reya, R., Morrison, S., Clarke, M., Weissman, I.: Stem cells, cancer, and cancer stem cells. Nature **414**(6859), 105–111 (2001)
53. Rhodes, A., Hillen, T.: Mathematical modeling of the role of survivin on dedifferentiation and radioresistance in cancer. Bull. Math. Biol. **78**(6), 1162–1188 (2016)
54. Rhodes, A., Hillen, T.: A mathematical model of the immune-mediated theory of metastasis. J. Theor. Biol. **482**, 109999 (2019)
55. Rodriguez-Berens, I., Komarova, N., Wodarz, D.: Evolutionary dynamics of feedback escape and the development of stem-cell driven cancers. Proc. R. Acad. Sci. **108**(47), 18983–18911 (2011)
56. Schatton, T., Frank, M.: Antitumor immunity and cancer stem cells. Ann. NY Acad. Sci. **1176**, 154–169 (2009)
57. Scott, J., Hjelmeland, A., Chinnaiyan, P., et al.: Microenvironmental variables must influence intrinsic phenotypic parameters of cancer stem cells to affect tumourigenicity. PLoS Comput. Biol. **10**(1), e1003433 (2014)
58. Sell, S.: Stem cell origin of cancer and differentiation therapy. Crit. Rev. Oncol. **51**, 1–28 (2004)
59. Shahriyari, L.: A new hypothesis: some metastases are the result of inflammatory processes by adapted cells, especially adapted immune cells at sites of inflammation. F1000 Res. **5**(175) (2016). https://doi.org/10.12388/f1000research.8055.1
60. Shyntar, A., Patel, A., Rhodes, M., et al.: The tumor invasion paradox in cancer stem cell driven tumors. Bull. Math. Biol. **84**(139) (2022)
61. Sole, R., Rodriguez-Caso, C., Diesboeck, T., Saldance, J.: Cancer stem cells as the engine of unstable tumor progression. J. Theor. Biol. **253**, 629–637 (2008)
62. Sottoriva, A., Verhoeff, J., Borovski, T., et al.: Cancer stem cell tumor model reveals invasive morphology and increased phenotypical heterogeneity. Cancer Res. **70**(1), 46–56 (2010)
63. Stavreva, N., Stavrev, P., Warkentin, M.W.B., Fallone, B.G.: Derivation of the expressions for γ_{50} and d_{50} for different individual TCP and NTCP models. Phys. Med. Biol. **47**(6) (2002)
64. Stiehl, T., Marciniak-Czochra, A.: Mathematical modeling of leukemogenesis and cancer stem cell dynamics. Math. Model. Nat. Pheno. **7**, 166–202 (2012)
65. Stocks, T., Hillen, T., Gong, J., Burger, M.: A stochastic model for the normal tissue complication probability (ntcp) in radiation treatment of cancer. Math. Med. Biol. 1–24 (2016)
66. Sun, Z., Komarova, N.: Stochastic control of proliferation and differentiation in stem cell dynamics. Math. Biosci. **240**, 231–240 (2012)
67. Swierniak, A., Kimmel, M., Smieja, J.: Mathematical modelling as a tool for planning anticancer therapy. Eur. J. Pharmacol. **625**, 108–121 (2009)
68. Takahashi, K., Yamanaka, S.: Induction of pluripotent stem cells from mouse embryonic and adult fibroblast cultures by defined factors. Cell **126**(4), 663–676 (2006)
69. Watabe, T., Miyazono, J.: Roles of TGF-beta family signaling in stem cell renewal and differentiation. Cell Res. **19**(1), 103–115 (2009)
70. Weekes, S., Barker, B., Bober, S., et al.: A multicompartment mathematical model of cancer stem cell-driven tumor growth dynamics. Bull. Math. Biol. **76**(7), 1762–1782 (2014)
71. Weinberger, H., Lewis, M., Li, B.: Analysis of linear determinacy for spread in cooperative models. J. Math. Biol. **45**, 183–218 (2002)
72. Wheldon, T.: Mathematical Models in Cancer Research. Adam Hilger, London (1988)
73. Wodarz, D., Komarova, N.: Dynamics of Cancer, Mathematical Foundations of Oncology. World Scientific, Singapore (2014)
74. Youssefpour, H., Li, X., Lander, A., Lowengrub, J.: Multispecies model of cell lineages and feedback control in solid tumors. J. Theor. Biol. **304**(0), 39–59 (2012)
75. Yuan, J., Wang, J.Z., Lo, S., et al.: Hypofractionation regimens for stereotactic radiotherapy for large brain tumors. Int. J. Radiat. Oncol. Biol. Phys. **72**(2), 390–397 (2008)
76. Zaider, M., Minerbo, G.: Tumor control probability: a formulation applicable to any temporal protocol of dose delivery. Phys. Med. Biol. **45**, 279–293 (2000)

Optimal Strategies for Pathological Epidermis: The Optimal Control of Basal Cell Dynamics

Mimmo Iannelli and Gabriela Marinoschi

Abstract In normal epidermis, the basal layer has the goal of sustaining in a controlled way the epidermis renewal, producing non proliferating differentiated keratinocytes that migrate into the suprabasal zone. However, hyperproliferation and migration of proliferating cells into the supra-basal zone may occur in correspondence of epidermal disorders such as actinic keratosis and psoriasis. In Gandolfi et al. (J. Math. Biol. 73:1595–1626, 2016; SIAM J. Appl. Math. 80:543–571, 2020), we developed and studied an age-structured model of the cell populations dynamics in the basal layer, describing the normal epidermis state together with some pathological states related to hyper-proliferation. Here we approach the problem of controlling such situations through agents that, inducing cell's death, reduce the effect of proliferation. An optimal control problem is formulated in connection with a cost function that aims to reduce the distance from the steady state of the normal epidermis. Existence of an optimal control is proved and necessary conditions of optimality are derived.

This study was intended to continue the work on epidermis modeling done since so many years in collaboration with Alberto Gandolfi [3, 4] (Gandolfi et al., J. Math. Biol. 73:1595–1626, 2016; SIAM J. Appl. Math. 80:543–571, 2020). Indeed it started when Alberto was still with us, building the project. We planned to meet after the summer 2019 to continue working: we miss his contribution and his friendship.

M. Iannelli (✉)
Department of Mathematics, University of Trento, Trento, Italy
e-mail: mimmo.iannelli@unitn.it

G. Marinoschi
Gheorghe Mihoc-Caius Iacob Institute of Mathematical Statistics and Applied Mathematics, Romanian Academy, Bucharest, Romania
e-mail: gabriela.marinoschi@acad.ro

A. d'Onofrio et al. (eds.), *Problems in Mathematical Biophysics*, SEMA SIMAI Springer Series 38, https://doi.org/10.1007/978-3-031-60773-8_8

1 Introduction

The epidermis is a multi-layered epithelium composed by cells named keratinocytes that undergoes a continuous renewal process [7, 8]. In normal epidermis, cell proliferation is mostly confined in the lowest cell layer, the basal layer, and has the goal of sustaining in a controlled way the epidermis renewal, producing non proliferating differentiated keratinocytes that migrate into the suprabasal zone. Differentiated cells undergoes a process of maturation during which keratin is accumulated in the cytoplasm and, at the end of this process, these cells die forming the corneous cells of the stratum corneum. The corneous cells are eventually shed from the epidermis surface. Some epidermal disorders such as actinic keratosis and psoriasis, however, are characterized by hyperproliferation, and proliferating cells can be found also in the suprabasal layers [1, 10].

In some previous papers [3–6], we developed and studied an age-structured model of the cell populations dynamics in the basal layer, aimed at describing the normal epidermis state and some hyperproliferative (but non tumoral) pathological states. This model is briefly described in the following, with some modifications introduced in view of the control problem we intend to study. Indeed, here we intend to study the optimal treatment with agents that induce cell death, focusing on their effects on the basal cell compartment. We will search for the optimal time administration of a given agent, knowing the sensitivities to the agent of the proliferating and differentiated cells. We will seek a trade-off among the following objectives:

(a) reducing the flow of proliferating cells towards the suprabasal region;
(b) keeping the flow of differentiated cells as close as possible to a reference flow, namely the flow of differentiated cells in normal epidermis;
(c) keeping low the cost of the therapy.

Our model considers three cell types residing in the basal layer: proliferating cells, differentiated cells, and apoptotic (dead) cells. Proliferating cells divide both symmetrically, producing two proliferating or two differentiated cells, and asymmetrically, producing one proliferating cell and one differentiated cell; differentiated cells detach and migrate into the suprabasal region after spending some time in the basal layer; apoptotic cells produced by apoptosis of both proliferating and differentiated cells normally after some time dissolve into liquids. In the occurrence of pathological states related to hyper-proliferation, proliferating cells may migrate into the super-basal layer.

Since basal cells are linked to an underlying membrane (basement membrane) their total surface density $N(t)$ at time t (number of cells per unit surface) is required to satisfy the physical constraint

$$0 \leq N(t) < N_{max}, \tag{1}$$

where N_{max} is the cell density corresponding to a complete occupancy of the available space.

The populations of proliferating, differentiated, and dead cells are described by age densities of the number of cells per unit surface denoted by $\nu_P(a,t)$, $\nu_D(a,t)$, $\nu_A(a,t)$, respectively. The equations with their respective boundary conditions at $a = 0$ read

$$\begin{cases} \dfrac{\partial \nu_P}{\partial t} + \dfrac{\partial \nu_P}{\partial a} = -(\beta_P(a) + \eta_P(a) + \sigma_P(a)\mu(t))\nu_P(a,t), \\ \nu_P(0,t) = r\theta(N(t)) \displaystyle\int_0^{a_P^+} \beta_P(a)\nu_P(a,t)da, \end{cases} \tag{2}$$

$$\begin{cases} \dfrac{\partial \nu_D}{\partial t} + \dfrac{\partial \nu_D}{\partial a} = -(\eta_D(a, N(t)) + \sigma_D(a)\mu(t))\nu_D(a,t), \\ \nu_D(0,t) = r(1-\theta(N(t))) \displaystyle\int_0^{a_P^+} \beta_P(a)\nu_P(a,t)da, \end{cases} \tag{3}$$

$$\begin{cases} \dfrac{\partial \nu_A}{\partial t} + \dfrac{\partial \nu_A}{\partial a} = -(\beta_A(a, N(t)) + \eta_A(a))\nu_A(a,t), \\ \nu_A(0,t) = \left(\displaystyle\int_0^{a_P^+} \sigma_P(a)\nu_P(a,t)da + \int_0^{a_D^*} \sigma_D(a)\nu_D(a,t)da \right)\mu(t) \\ \qquad\qquad\qquad +(2-r) \displaystyle\int_0^{a_P^+} \beta_P(a)\nu_P(a,t)da \end{cases} \tag{4}$$

where $N(t)$ is given by

$$N(t) = \int_0^{a_P^+} \nu_P(a,t)da + \int_0^{a_D^*} \nu_D(a,t)da + \int_0^{a_A^+} \nu_A(a,t)da. \tag{5}$$

These equations must be completed with initial conditions at $t = 0$

$$\nu_P(a,0) = \nu_P^0(a) \geq 0, \quad \nu_D(a,0) = \nu_D^0(a) \geq 0, \quad \nu_A(a,0) = \nu_A^0(a) \geq 0\,, \tag{6}$$

satisfying the constraint

$$\int_0^{a_P^+} \nu_P^0(a)da + \int_0^{a_D^*} \nu_D^0(a)da + \int_0^{a_A^+} \nu_A^0(a)da = N^0 < N_{max}\,. \tag{7}$$

Concerning Eq. (2) describing the dynamics of proliferating cells, the parameters have the following meaning

- $\beta_P(a) \geq 0$ $(a \in (0, a_P^+))$ denotes the per cell division rate of proliferating cells.

The upper bound of the cell cycle duration a_P^+ is supposed to be finite, and so $\beta_P(a)$ will blow up at $a = a_P^+$, that is

$$\int_0^{a_P^+} \beta_P(a)da = +\infty. \tag{8}$$

- $r \in (0, 2]$ gives the mean number of viable daughter cells after a cell division.

  Under normal, unperturbed conditions $r = 2$, whereas damaging factors may induce abortive mitoses and in this case it will be $r < 2$.
- $\theta(N) \in [0, 1]$ gives the average fraction of viable daughter cells that are proliferating cells.

  As mentioned above, the one cell division can originate either two proliferating cells or two differentiated cells, or one proliferating cell and one differentiated cell [2, 9], thus the production of the two types is asymmetric and the function $\theta(N)$ takes into account this asymmetry. Actually, $\theta(N)$ depends on N because cells may indeed sense cell crowding by responding to the mechanical stress due to cell-cell contact (see [6] and references therein) and/or to the concentration of chemical factors that may accumulate in the cell microenvironment; as a consequence the output of cell proliferation is forced to shift toward differentiated cells, as N approaches its physical limit N_{max}. Thus we assume that the function $\theta(\cdot)$ satisfies

$$\theta(\cdot) \in C^1([0, N_{max}]), \quad \theta(0) = 1, \quad \theta'(x) < 0, \quad \theta(N_{max}) < \frac{1}{2} \tag{9}$$

  Indeed, the last condition forces the production toward differentiated cells. In [6] we adopted the following form

$$\theta(N) = \frac{\chi^p}{\chi^p + \left(\frac{N}{N_{max}}\right)^p},$$

  with $\chi = 0.95$ and $p = 3$. This form allowed simulations predicting realistic proportions of the two types of cells in the steady state (see [6] for more detailed information).
- $\eta_P(a)$ represents the per cell rate of migration into the suprabasal region of proliferating cells.

  This rate is actually identically equal to zero in normal epidermis, but we will consider the pathological situation when also proliferating cells migrate into the supra-basal layer.
- The per cell mortality rate $\sigma_P(a)\mu(t)$ describes the action of possible cell killing agents. Specifically $\sigma_P(a)$, satisfying

$$0 \leq \sigma_P(a) \leq \sigma_P^+,$$

denotes the age-sensitivity (age-specific mortality per dose unit) of proliferating cells to the chosen agent, while $\mu(t)$ is the dose rate of the treatment. In hyperproliferative disorders, the therapy is in fact often based on the use of cytotoxic agents, such as radiations or chemicals.

Differentiated cells are described by Eq. (3). They originate from the (asymetric) division of the proliferating population and the terms r, $\beta_P(\cdot)$, $\theta(\cdot)$ are the same as described above. Furthermore

- $\eta_D(a, N)$ represent the per cell rate of migration into the suprabasal region of differentiated cells.

  In particular we assume

$$\eta_D(a, N) = \eta_D^0(a) + \eta_D^1(a)H(N), \tag{10}$$

  where $\eta_D^0(a) \geq 0$ satisfies

$$\int_0^{a_D^*} \eta_D^0(a)da = +\infty, \tag{11}$$

  and $\eta_D^1(a) \geq \bar{\eta}_D > 0$ is bounded.

  Condition (11) impose that this rate blows up at some age a_D^* much shorter than the maximum age of the differentiated cells since they leave fastly the basal layer to continue their life in the supra-basal layer.

  On the other hand, the rate $\eta_D(a, N)$ depends on the total number of cells through the factor $H(N)$, where the function $H(\cdot)$ is supposed to satisfy

$$\begin{aligned} &H(\cdot) \in C^1\left([0, N_{max})\right), \quad H'(x) \geq 0, \\ &H(0) = 0, \quad \lim_{x \to N_{max}} H(x) = +\infty\,. \end{aligned} \tag{12}$$

  The last condition means that cell crowding forces cell detachment.
- The per cell mortality rate $\sigma_D(a)\mu(t)$ describes the action of possible cell killing agents on differentiated cells.

  The role of this term is the same of $\sigma_P(a)\mu(t)$ appearing in Eq. (2). Here we also assume the age specific sensivity $\sigma_D(a)$ to satisfy

$$0 \leq \sigma_D(a) \leq \sigma_D^+,$$

  We note that the sensitivities may be quite different for proliferating and differentiated cells, we recall for instance that some chemicals act preferably on proliferating cells in the cell cycle S-phase (thus in a specific age range) and may be substantially ineffective on non proliferating cells.

Finally, Eq. (4) concerns apoptotic cells that are important because, though dead, they still occupy space in the basal layer and in some pathological situation can also migrate into the supra-basal zone. Their compartment gathers all the cells coming from the killing action and those from the failing of the division process of proliferating cells when $r < 2$. Their dynamics is regulated by dissolution and detachment through the following parameters

- $\beta_A(a, N(t))$, $a \in (0, a_A^+)$, denotes the per cell rate of dissolution of dead cells into liquids. It is defined similarly to (10) by

$$\beta_A(a, N) = \beta_A^0(a) + \beta_A^1(a)H(N), \tag{13}$$

  where $\beta_A^0(a) \geq 0$ with

$$\int_0^{a_A^+} \beta_A^0(a)da = +\infty, \tag{14}$$

  and $\beta_A^1(a) \geq \bar{\beta}_A > 0$ is bounded. Indeed we are assuming that apoptotic cells have a limited age (duration) a_A^+ and that crowding forces their dissolution.
- $\eta_A(a)$ represents the per cell rate of migration into the suprabasal region of apoptotic cells. As already said for proliferating cells, this rate is identically equal to zero in the case of non pathological epidermis.

The age densities of the flows of proliferating, differentiated, and possibly apoptotic cells into the suprabasal region are given by

$$\begin{aligned} S_P(a, t) &= \eta_P(a)\nu_P(a, t), \\ S_D(a, t) &= \eta_D(a, N(t))\nu_D(a, t), \\ S_A(a, t) &= \eta_A(a)\nu_A(a, t). \end{aligned} \tag{15}$$

In the case of normal epidermis, $\eta_P(a)$ can be assumed identically zero. This is not true in a pathological condition of hyperproliferation, when the the flow $S_P(a, t)$ will be present. Other possible alterations of the normal conditions such as accelerated cell division and changes in the balance between the basal production of proliferating cells and differentiated cells, can be accounted for in the model by changing the rate $\beta_P(a)$ and the function $\theta(N)$, respectively. However, property (9) is always assumed to hold. This assumption implies that in all the pathological alterations here considered, a stationary state of the basal cell populations can be established (see [6]).

With the aim of analysing the optimal treatment described above we consider the following control problem on an assigned finite time horizon.

Problem 1 *Find the function* $\mu(t) \in \mathcal{U}$, *where*

$$\mathcal{U} = \{\mu \in L^\infty(0, T), \ \ 0 \leq \mu(t) \leq \mu_{max}\}, \tag{16}$$

that minimizes the functional

$$
\begin{aligned}
\mathcal{J}(\mu) = & \int_0^T \int_0^{a_P^+} \eta_P(a) v_P(a,t) dadt \\
& + \frac{\gamma_1}{2} \int_0^T \int_0^{a_D^*} \left[(\eta_D^0(a) + \eta_D^1(a) H(N(t))) v_D(a,t) - S_D^*(a) \right]^2 dadt \\
& + \frac{\gamma_2}{2} \int_0^T \mu(t) dt,
\end{aligned} \tag{17}
$$

where $v_P(a,t)$ *and* $v_D(a,t)$ *are the solutions to the differential system* (2)–(5), *with initial conditions* (6), (7).

In (17) the known function $S_D^*(a)$ is the age density of the steady state flow of differentiated cells in the normal epidermis (see [6]), while $\gamma_1, \gamma_2 \geq 0$ are relative weights.

We end this section by summarizing the mathematical assumptions that we will consider fulfilled throughout the paper

$$
\begin{aligned}
& \beta_P \in L^1_{loc}([0, a_P^+)), \quad \beta_P(a) \geq 0 \text{ a.e.}, \quad \int_0^{a_P^+} \beta_P(a) da = +\infty, \\
& \beta_A^0 \in L^1_{loc}([0, a_A^+)), \quad \beta_A^0(a) \geq 0 \text{ a.e.}, \quad \int_0^{a_A^+} \beta_A^0(a) da = +\infty, \\
& \beta_A^1 \in L^\infty(0, a_A^+), \quad \beta_A^1(a) \geq \bar{\beta}_A > 0 \text{ a.e.}, \\
& \eta_P \in L^\infty(0, a_P^+), \quad \eta_P(a) \geq 0 \text{ a.e.}, \\
& \eta_D^0 \in L^1_{loc}([0, a_D^*)), \quad \eta_D^0(a) \geq 0 \text{ a.e.}, \quad \int_0^{a_D^*} \eta_D^0(a) da = +\infty, \\
& \eta_D^1 \in L^\infty(0, a_D^*), \quad \eta_D^1(a) \geq \bar{\eta}_D > 0 \text{ a.e.}, \\
& \sigma_P \in L^\infty(0, a_P^+), \quad 0 \leq \sigma_P(a) \leq \sigma_P^+ \text{ a.e.}, \\
& \sigma_D \in L^\infty(0, a_D^*), \quad 0 \leq \sigma_D(a) \leq \sigma_D^+ \text{ a.e.}, \\
& S_D^* \in L^2(0, a_D^*)
\end{aligned} \tag{18}
$$

and the functions $\theta(N)$ and $H(N)$ satisfy (9) and (12), respectively.

2 The State System

In order to simplify the equations, we will transform the problem considering the new variables $\tilde{\nu}_P(a,t)$, $\tilde{\nu}_D(a,t)$, $\tilde{\nu}_A(a,t)$ through the following definitions

$$\tilde{\nu}_P(a,t) = \frac{\nu_P(a,t)}{M_P(a)E_P(a)},\ \tilde{\nu}_D(a,t) = \frac{\nu_D(a,t)}{E_D^0(a)},\ \tilde{\nu}_A(a,t) = \frac{\nu_A(a,t)}{M_A^0(a)E_A(a)} \tag{19}$$

where

$$M_P(a) = \exp\left(-\int_0^a \beta_P(\sigma)d\sigma\right), \quad M_A^0 = \exp\left(-\int_0^a \beta_A^0(\sigma)d\sigma\right)$$

$$E_P(a) = \exp\left(-\int_0^a \eta_P(\sigma)d\sigma\right), \quad E_D^0(a) = \exp\left(-\int_0^a \eta_D^0(\sigma)d\sigma\right),$$

$$E_A(a) = \exp\left(-\int_0^a \eta_A(\sigma)d\sigma\right)$$

In these new variables problem (2)–(5) becomes (dropping the decoration $\tilde{}$ for simplicity)

$$\begin{cases} \dfrac{\partial \nu_P}{\partial t} + \dfrac{\partial \nu_P}{\partial a} + \sigma_P(a)\mu(t)\nu_P(a,t) = 0, \\ \nu_P(0,t) = r\theta(N(t)) \displaystyle\int_0^{a_P^+} K_P(a)\nu_P(a,t)da, \end{cases} \tag{20}$$

$$\begin{cases} \dfrac{\partial \nu_D}{\partial t} + \dfrac{\partial \nu_D}{\partial a} + (\eta_D^1(a)H(N(t)) + \sigma_D(a)\mu(t)\nu_D(a,t) = 0, \\ \nu_D(0,t) = r(1-\theta(N(t))) \displaystyle\int_0^{a_P^+} K_P(a)\nu_P(a,t)da, \end{cases} \tag{21}$$

$$\begin{cases} \dfrac{\partial \nu_A}{\partial t} + \dfrac{\partial \nu_A}{\partial a} + \beta_A^1(a)H(N(t))\nu_A(a,t) = 0, \\ \nu_A(0,t) = \left(\displaystyle\int_0^{a_P^+} \sigma_P(a)k_P(a)\nu_P(a,t)da + \int_0^{a_D^*} \sigma_D(a)\mu(t)k_D(a)\nu_D(a,t)da\right)\mu(t) \\ \qquad\qquad + (2-r)\displaystyle\int_0^{a_P^+} K_P(a)\nu_P(a,t)da \end{cases} \tag{22}$$

$$N(t) = \int_0^{a_P^+} k_P(a)\nu_P(a,t)da + \int_0^{a_D^*} k_D(a)\nu_D(a,t)da + \int_0^{a_A^+} k_A(a)\nu_A(a,t)da, \tag{23}$$

where

$$K_P(a) = \beta_P(a)M_P(a)E_P(a), \; k_P(a) = M_P(a)E_P(a),$$
$$k_D(a) = E_D^0(a), \; k_A(a) = M_A^0(a)E_A(a). \tag{24}$$

Of course, problem (20)–(23) is endowed with initial conditions $(\nu_P^0, \nu_D^0, \nu_A^0)$ coming from (6) through the normalization (19) and satisfying the constraint

$$\int_0^{a_P^+} k_P(a)\nu_P^0(a)da + \int_0^{a_D^*} k_D(a)\nu_D^0(a)da + \int_0^{a_A^+} k_A(a)\nu_A^0(a)da = N^0 < N_{max}. \tag{25}$$

In the new variables we also have

$$\mathcal{J}(\mu) = \int_0^T \int_0^{a_P^+} G_P(a)\nu_P(a,t)dadt$$
$$+\frac{\gamma_1}{2}\int_0^T \int_0^{a_D^*} \left[\left(G_D^0(a) + G_D^1(a)H(N(t))\right)\nu_D(a,t) - S_D^*(a)\right]^2 dadt$$
$$+\frac{\gamma_2}{2}\int_0^T \mu(t)dt, \tag{26}$$

where

$$G_P(a) = \eta_P(a)M_P(a)E_P(a), \; G_D^0(a) = \eta_D^0(a)E_D^0(a), \; G_D^1(a) = \eta_D^1(a)E_D^0(a).$$

Concerning problem (20)–(23) we have an existence and uniqueness result that can be proven adapting the proof from [6], where the same theorem is proven under slightly different assumptions. Namely we have

Theorem 1 *Let the assumptions* (18) *be fulfilled and consider the initial conditions*

$$\nu_P^0 \in L^\infty(0, a_P^+), \quad \nu_D^0 \in L^\infty(0, a_D^*), \quad \nu_A^0 \in L^\infty(0, a_A^+), \tag{27}$$

satisfying the constraint (25). *Then problem* (2)–(5) *has a unique solution*

$$\boldsymbol{\nu}(a,t;\boldsymbol{\nu}^0(\cdot)) \equiv (\nu_P(a,t), \; \nu_D(a,t), \; \nu_A(a,t)),$$

where the components

$$\nu_P \in L^\infty\big((0, a_P^+) \times (0,T)\big),$$
$$\nu_D \in L^\infty\big((0, a_D^*) \times (0,T)\big),$$
$$\nu_A \in L^\infty\big((0, a_A^+) \times (0,T)\big),$$

are non-negative and satisfy the problem in the sense that the directional derivatives $\left(\frac{\partial}{\partial t}+\frac{\partial}{\partial a}\right)$ *exist and satisfy* (20)–(23). *Moreover the total population* $N(t)$ *satisfies*

$$N(t) < N_{max}.$$

Actually, from the proof of this theorem we know that the solution of (20)–(23) can be represented in integral form as (by $a \wedge t$ we denote the minimum between a and t)

$$\begin{aligned}
\nu_P(a,t) &= B_P(t-a)e^{-\int_0^{a\wedge t}\sigma_P(a-\sigma)\mu(t-\sigma)d\sigma}\\
\nu_D(a,t) &= B_D(t-a)e^{-\int_0^{a\wedge t}\left[H(N(t-\sigma))\eta_D^1(a-\sigma)+\sigma_D(a-\sigma)\mu(t-\sigma)\right]d\sigma}\\
\nu_A(a,t) &= B_A(t-a)e^{-\int_0^{a\wedge t}H(N(t-\sigma))\beta_A^1(a-\sigma)d\sigma}
\end{aligned} \tag{28}$$

where the functions $B_P(t)$, $B_D(t)$, $N(t)$ are continuous, $B_A(t) \in L^\infty(0,T)$, and solve the following system of integral equations

$$B_P(t) = r\theta(N(t))\int_0^\infty \mathcal{K}_P(a,t;\mu)B_P(t-a)da, \tag{29}$$

$$B_D(t) = r(1-\theta(N(t)))\int_0^\infty \mathcal{K}_P(a,t;\mu)B_P(t-a)da, \tag{30}$$

$$\begin{aligned}
B_A(t) = \mu(t)&\int_0^\infty \mathcal{H}_P(a,t;\mu)B_P(t-a)da\\
&+\mu(t)\int_0^\infty \mathcal{K}_D(a,t;N,\mu)B_D(t-a)da\\
&+(2-r)\int_0^\infty \mathcal{K}_P(a,t;\mu)B_P(t-a)da,
\end{aligned} \tag{31}$$

$$\begin{aligned}
N(t) = &\int_0^\infty \mathcal{L}_P(a,t\,;\mu)B_P(t-a)da\\
&+\int_0^\infty \mathcal{L}_D(a,t\,;N,\mu)B_D(t-a)da\\
&+\int_0^\infty \mathcal{L}_A(a,t\,;N)B_A(t-a)da,
\end{aligned} \tag{32}$$

with

$$
\begin{aligned}
&\mathcal{K}_P(a,t;\mu) = K_P(a)e^{-\int_0^{a\wedge t}\sigma_P(a-\sigma)\mu(t-\sigma)d\sigma},\\
&\mathcal{H}_P(a,t;\mu) = \sigma_P(a)k_P(a)e^{-\int_0^{a\wedge t}\sigma_P(a-\sigma)\mu(t-\sigma)d\sigma},\\
&\mathcal{K}_D(a,t;N,\mu) = \sigma_D(a)k_D(a)e^{-\int_0^{a\wedge t}\left[H(N(t-\sigma))\eta_D^1(a-\sigma)+\sigma_D(a-\sigma)\mu(t-\sigma)\right]},\\
&\mathcal{L}_P(a,t\,;\mu) = k_P(a)e^{-\int_0^{a\wedge t}\sigma_P(a-\sigma)\mu(t-\sigma)d\sigma},\\
&\mathcal{L}_D(a,t\,;N,\mu) = k_D(a)e^{-\int_0^{a\wedge t}\left[H(N(t-\sigma))\eta_D^1(a-\sigma)+\sigma_D(a-\sigma)\mu(t-\sigma)\right]d\sigma},\\
&\mathcal{L}_A(a,t\,;N) = k_A(a)e^{-\int_0^{a\wedge t}H(N(t-\sigma))\beta_A^1(a-\sigma)d\sigma},
\end{aligned}
\tag{33}
$$

where all the functions initially defined in $[0,a_P^+]$, $[0,a_D^*]$, $[0,a_A^+]$ are extended as zero outside their interval of definition and the following initial heredities are assigned

$$
\begin{aligned}
B_P(t) &= \nu_P^0(-t),\ t\in[-a_P^+,0],\\
B_D(t) &= \nu_D^0(-t),\ t\in[-a_D^*,0],\\
B_A(t) &= \nu_A^0(-t),\ t\in[-a_A^+,0].
\end{aligned}
$$

We also draw the following

Corollary 1 *Under the assumptions of Theorem 1, there exist constants α, C_{max} and N^{**}, independent of $\mu(t)$ such that*

$$
B_P(t)+B_D(t)+B_A(t)\le e^{\alpha t}C_{max},\tag{34}
$$

$$
N(t)\le N^{**}.\tag{35}
$$

Proof Indeed, from the proof of the theorem where the estimates are stated (see [6]), we can see that they are independent of $\mu(t)$. In fact the first inequality is fulfilled where α is any constant large enough to have

$$
\int_0^{a_P^+} e^{-\alpha s}K_P(s)ds < \frac{1}{4}
$$

and C_{max} is a constant depending on $K_P(a)$, ν_P^0, ν_D^0 and on the bound μ_{max} (see (16)). To get the second inequality, we first need to chose N_P^* such that

$$
\int_0^{a_P^+} k_P(a)\nu_P^0(a)da < N_P^* < N_{max},\quad \theta(N_P^*) < \frac{1}{2},
$$

then we chose $N^{**} > N_P^*$ large enough to satisfy

$$H(N^{**})(\bar{\eta}_D \wedge \bar{\beta}_A)(N^{**} - N_P^*) > 8e^{\alpha T} \left\| \nu_P^0 \right\|_\infty .$$

All these conditions are independent of $\mu(t)$. □

We note that problem (29)–(32) in the variables B_P, B_D, B_A, $N(t)$ is equivalent to the original problem because the solution to (29)–(32) produces a solution to (20)–(23) via the representation (28) and viceversa. Then in the following we adopt (29)–(32) as the state system for the optimization problem.

The representation above allows to state some estimates which are independent of $\mu(t)$ and that we will need later, namely we can estimate the continuity modulus of $N(t)$, $B_P(t)$, $B_D(t)$.

Lemma 1 *Let assumptions of Theorem 1 be satisfied. Then the solution of* (2)–(5) *satisfies*

$$|N(t) - N(t_0)| \le C_N \left(|t - t_0| + \int_0^\infty |k_P(t-a) - k_P(t_0 - a)| da \right), \tag{36}$$

$$|B_P(t) - B_P(t_0)| \le C_{PD}\Big(|t - t_0| + \int_0^\infty |k_P(t-a) - k_P(t_0 - a)| da + \int_0^\infty |K_P(t-a) - K_P(t_0 - a)| da \Big), \tag{37}$$

$$|B_D(t) - B_D(t_0)| \le C_{PD}\Big(|t - t_0| + \int_0^\infty |k_P(t-a) - k_P(t_0 - a)| da + \int_0^\infty |K_P(t-a) - K_P(t_0 - a)| da \Big), \tag{38}$$

where C_N and C_{PD} are constants independent of $\mu(t)$.

Proof Considering the first term in (32) we have

$$
\begin{aligned}
&\int_0^{a_P^+} \mathcal{L}_P(a,t\,;\mu)B_P(t-a)da \\
&\quad = \int_0^t k_P(a)B_P(t-a)e^{-\int_0^a \sigma_P(a-\sigma)\mu(t-\sigma)d\sigma}da \\
&\qquad\qquad + \int_t^\infty k_P(a)\nu_P^0(a-t)e^{-\int_0^t \sigma_P(a-\sigma)\mu(t-\sigma)d\sigma}da \\
&\quad = \int_0^t k_P(t-a)B_P(a)e^{-\int_0^{t-a} \sigma_P(t-a-\sigma)\mu(t-\sigma)d\sigma}da \\
&\qquad\qquad + \int_0^\infty k_P(a+t)\nu_P^0(a)e^{-\int_0^t \sigma_P(a+t-\sigma)\mu(t-\sigma)d\sigma}da \\
&\quad = \int_0^t k_P(t-a)B_P(a)e^{-\int_0^{t-a} \sigma_P(\sigma)\mu(a+\sigma)d\sigma}da \\
&\qquad\qquad + \int_0^\infty k_P(a+t)\nu_P^0(a)e^{-\int_0^t \sigma_P(a+\sigma)\mu(\sigma)d\sigma}da,
\end{aligned}
$$

where all the function initially defined in $[0,a_P^+]$, $[0,a_D^*]$, $[0,a_A^+]$ are extended as zero outside their interval of definition. Then, using (24), (33) and (34), we have

$$
\begin{aligned}
&\left|\int_0^{a_P^+} \mathcal{L}_P(a,t\,;\mu)B_P(t-a)da - \int_0^{a_P^+} \mathcal{L}_P(a,t_0\,;\mu)B_P(t_0-a)da\right| \\
&\quad \le \left(e^{\alpha T}C_{max} + \|\nu_P^0\|_\infty\right)\Big[\left(1+t_0\sigma_P^+\mu_{max}\right)|t-t_0| \\
&\qquad\qquad + \int_0^\infty |k_P(t-a)-k_P(t_0-a)|da\Big].
\end{aligned}
$$

Similar calculations give

$$
\begin{aligned}
&\left|\int_0^{a_D^*} \mathcal{L}_D(a,t\,;N,\mu)B_D(t-a)da - \int_0^{a_D^*} \mathcal{L}_D(a,t_0\,;N,\mu)B_D(t_0-a)da\right| \\
&\quad \le \left(e^{\alpha T}C_{max} + \|\nu_D^0\|_\infty\right)\Big[\left(1+t_0\sigma_D^+\mu_{max}+t_0H(N^{**})\|\eta_D^1\|_\infty\right)|t-t_0| \\
&\qquad\qquad + \int_0^\infty |k_D(t-a)-k_D(t_0-a)|da\Big],
\end{aligned}
$$

and

$$\left| \int_0^{a_A^+} \mathcal{L}_A(a, t\,; N) B_A(t-a) da - \int_0^{a_A^+} \mathcal{L}_A(a, t_0\,; N) B_A(t_0 - a) da \right|$$
$$\leq \left(e^{\alpha T} C_{max} + \|\nu_A^0\|_\infty \right) \Big[\left(1 + t_0 H(N^{**}) \|\eta_A^1\|_\infty \right) |t - t_0|$$
$$+ \int_0^\infty |k_A(t-a) - k_A(t_0 - a)| da \Big],$$

where we have used (35) to estimate terms including $H(N(t))$ such as, for instance,

$$\left| e^{\int_0^{t-a} H(N(a+\sigma))\eta_D^1(\sigma)d\sigma} - e^{\int_0^{t_0-a} H(N(a+\sigma))\eta_D^1(\sigma)d\sigma} \right|$$
$$\leq \left| \int_{t_0-a}^{t-a} H(N(a+\sigma))\eta_D^1(\sigma)d\sigma \right| \leq H(N^{**}) \|\eta_D^1\|_\infty |t - t_0|.$$

As a consequence of all these estimates we have (36). Finally we use (36) in (29) and (30), and by similar estimates we prove (37) and (38).

□

3 Existence of the Optimal Control

We are now ready to consider the existence of an optimal control. Indeed, we have

Theorem 2 *Let assumptions* (18) *be satisfied, then there exist* $\mu^* \in L^\infty(0, T)$ *such that*

$$\mathcal{J}(\mu^*) = \min_{\mu \in \mathcal{U}} \mathcal{J}(\mu).$$

Proof Since all the terms are non negative, the functional (26) is bounded from below and an infimum exists

$$d = \inf_{\mu \in \mathcal{U}} \mathcal{J}(\mu) \geq 0.$$

Let us take a minimizing sequence $\mu^n(t) \in \mathcal{U}$ such that

$$d \leq \mathcal{J}(\mu^n) \leq d + \frac{1}{n}.$$

Let $B_P^n(t)$, $B_D^n(t)$, $B_A^n(t)$, $N^n(t)$ be the solution to system (29)–(32) corresponding to μ^n. Namely

$$B_P^n(t) = r\theta(N^n(t)) \int_0^{a_P^+} \mathcal{K}_P(a, t; \mu^n) B_P^n(t-a)da, \tag{39}$$

$$B_D^n(t) = r(1-\theta(N^n(t))) \int_0^{a_P^+} \mathcal{K}_P(a, t; \mu^n) B_P^n(t-a)da, \tag{40}$$

$$\begin{aligned} B_A^n(t) = \mu^n(t) &\int_0^{a_P^+} \mathcal{H}_P(a, t; \mu^n) B_P^n(t-a)da \\ &+\mu^n(t) \int_0^{a_D^*} \mathcal{K}_D(a, t; N^n, \mu) B_D^n(t-a)da \\ &\qquad +(2-r) \int_0^{a_P^+} \mathcal{K}_P(a, t; \mu^n) B_P^n(t-a)da, \end{aligned} \tag{41}$$

$$\begin{aligned} N^n(t) = &\int_0^{a_P^+} \mathcal{L}_P(a, t\,; \mu^n) B_P^n(t-a)da \\ &+ \int_0^{a_D^*} \mathcal{L}_D(a, t\,; N^n, \mu^n) B_D^n(t-a)da \\ &\qquad + \int_0^{a_A^+} \mathcal{L}_A(a, t\,; N^n) B_A^n(t-a)da. \end{aligned} \tag{42}$$

Note that the densities corresponding to μ^n are determined by the solution to (39)–(42) through the formulas

$$\begin{aligned} \nu_P^n(a,t) &= B_P^n(t-a) e^{-\int_0^{a\wedge t} \sigma_P(a-\sigma)\mu^n(t-\sigma)d\sigma}, \\ \nu_D^n(a,t) &= B_D^n(t-a) e^{-\int_0^{a\wedge t} [H(N^n(t-\sigma))\eta_D^1(a-\sigma)+\sigma_D(a-\sigma)\mu^n(t-\sigma)]d\sigma}, \\ \nu_A^n(a,t) &= B_A^n(t-a) e^{-\int_0^{a\wedge t} H(N^n(t-\sigma))\beta_A^1(a-\sigma)d\sigma}. \end{aligned} \tag{43}$$

Now, since μ^n is a sequence uniformly bounded in $L^\infty(0,T)$ we can take a weak*- convergent subsequence

$$\mu^n \to \mu^* \quad \text{weak* in } L^\infty(0,T). \tag{44}$$

On the other hand, thanks to (36), (37), (38), the sequences $B_P^n(t)$, $B_D^n(t)$, $N^n(t)$ are equi-continuous so that we can take a further subsequence such that

$$N^n \to N^*, \quad B_P^n \to B_P^*, \quad B_D^n \to B_D^* \quad \text{uniformly in } [0, T]. \tag{45}$$

Furthermore, by (34), the sequence $B_A^n(t)$ is uniformly bounded so that the subsequence can be chosen such that

$$B_A^n \to B_A^* \quad \text{weak* in } L^\infty(0, T).$$

Then we can pass to the limit in (39)–(42) and show that B_P^*, B_D^*, B_A^* and N^* satisfy (29)–(32) corresponding to μ^*. Indeed, the right hand sides in (39) and (40) converge because of (45) and because, thanks to (44), we have for instance (similar terms converge analogously)

$$\begin{aligned}&\int_0^a \sigma_P(a-\sigma)\mu^n(t-\sigma)d\sigma \\ &\qquad = \int_{t-a}^t \sigma_P(a-t+\sigma)\mu^n(\sigma)d\sigma \to \int_{t-a}^t \sigma_P(a-t+\sigma)\mu^*(\sigma)d\sigma\end{aligned}$$

for a and t fixed.

Note that on the same subsequence we can go to the limit in (43) obtaining the densities $\nu_P^*(a,t)$, $\nu_D^*(a,t)$, $\nu_A^*(a,t)$ that solve the problem (20)–(23) corresponding to μ^*. Thus we can go to the limit in

$$\begin{aligned}\mathcal{J}(\mu^n) &= \int_0^T \int_0^{a_P^+} G_P(a)\nu_P^n(a,t)dadt \\ &+ \frac{\gamma_1}{2}\int_0^T \int_0^{a_D^*} \Big[\Big(G_D^0(a) + G_D^1(a)H(N^n(t))\Big)\nu_D^n(a,t) - S_D^*(a)\Big]^2 dadt \\ &+ \frac{\gamma_2}{2}\int_0^T \mu^n(t)dt\end{aligned}$$

and get

$$\mathcal{J}(\mu^*) = d,$$

i.e., μ^* is an optimal control.

□

4 Optimality Conditions

Let μ^* be an optimal controller and B_P^*, B_D^*, B_A^* and N^*, the corresponding optimal state . Let $\lambda \in (0, 1)$ and introduce

$$\mu^\lambda = \mu^* + \lambda\omega,$$

where

$$\omega = \overline{\mu} - \mu^*, \text{ with } \overline{\mu} \in \mathcal{U}.$$

We denote by B_P^λ, B_D^λ, B_A^λ and N^λ the solution to the state system (29)–(32) corresponding to μ^λ and consider the following system in variation obtained differentiating (29)–(32)

$$\begin{aligned}
D_P(t) &= \int_0^t \mathcal{A}_P^*(a,t)D_P(t-a)da + \mathcal{D}_P^*(t)D_N(t) - \int_0^t \mathcal{E}_P^*(a,t)(t-a)da,\\
D_D(t) &= \int_0^t \mathcal{A}_D^*(a,t)D_P(t-a)da - \mathcal{D}_P^*(t)D_N(t) - \int_0^t \mathcal{E}_D^*(a,t)(t-a)da,\\
D_A(t) &= \int_0^t \mathcal{A}_A^*(a,t)D_P(t-a)da + \int_0^t \mathcal{B}_A^*(a,t)D_D(t-a)da\\
&\quad - \int_0^t \mathcal{D}_A^*(a,t)D_N(t-a)da + \mathcal{E}^*(t)(t) - \int_0^t \mathcal{E}_A^*(a,t)(t-a)da,\\
D_N(t) &= \int_0^t \mathcal{A}_N^*(a,t)D_P(t-a)da + \int_0^t \mathcal{B}_N^*(a,t)D_D(t-a)da\\
&\quad + \int_0^t \mathcal{C}_N^*(a,t)D_A(t-a)da - \int_0^t \mathcal{D}_N^*(a,t)D_N(t-a)da\\
&\quad - \int_0^t \mathcal{E}_N^*(a,t)(t-a)da.
\end{aligned} \tag{46}$$

where

$$\mathcal{A}_P^*(a,t) = r\theta(N^*(t))\mathcal{K}_P(a,t;\mu^*),$$

$$\mathcal{A}_D^*(a,t) = r[1-\theta(N^*(t))]\mathcal{K}_P(a,t;\mu^*),$$

$$\mathcal{D}_P^*(t) = r\theta'(N^*(t))\int_0^\infty \mathcal{K}_P(a,t;\mu^*)B_P^*(t-a)da,$$

$$\mathcal{E}_P^*(a,t) = \int_0^\infty \mathcal{A}_P^*(\tau+a,t)\sigma(\tau)d\tau,$$

$$\mathcal{E}_D^*(a,t) = \int_0^\infty \mathcal{A}_D(\tau+a,t)\sigma(\tau)d\tau,$$

$$\mathcal{A}_A^*(a,t) = \mu^*(t)\mathcal{H}_P(a,t;\mu^*) + (2-r)\mathcal{K}_P(a,t;\mu^*),$$

$$\mathcal{B}_A^*(a,t) = \mu^*(t)\mathcal{K}_D(a,t;N^*,\mu^*),$$

$$\mathcal{D}_A^*(a,t) = \mu^*(t)H'(N^*(t-a))\int_0^\infty \mathcal{K}_D(a+\tau,t;N^*,\mu^*)\eta_D^1(\tau)d\tau,$$

$$\mathcal{E}^*(t) = \int_0^\infty \big[\mathcal{H}_P(a,t;\mu^*)B_P^*(t-a) + \mathcal{K}_D(a,t;N^*,\mu^*)B_D^*(t-a)\big]da,$$

$$\begin{aligned}\mathcal{E}_A^*(a,t) =& \mu(t)\int_0^\infty \big[\mathcal{H}_P(a+\tau,t;\mu^*)\sigma_P(\tau) + \mathcal{K}_D^*(a+\tau,t;N^*,\mu^*)\sigma_D(\tau)\big]d\tau \\ &+(2-r)\int_0^\infty \mathcal{K}_P(a+\tau,t,;N^*,\mu^*)B_P^*(t-a-\tau)\sigma_P(\tau)d\tau,\end{aligned}$$

$$\mathcal{A}_N^*(a,t) = \mathcal{L}_P(a,t;\mu^*),$$

$$\mathcal{B}_N^*(a,t) = \mathcal{L}_D(a,t;N^*,\mu^*),$$

$$\mathcal{C}_N^*(a,t) = \mathcal{L}_A(a,t;N^*),$$

$$\begin{aligned}\mathcal{D}_N^*(a,t) = H'(N^*(t-a))\int_0^\infty \Big[&\mathcal{L}_D(a+\tau,t,N^*,\mu^*)B_D^*(t-a-\tau)\eta_D^1(\tau) \\ &+\mathcal{L}_A(a+\tau,t,N^*)B_A^*(t-a-\tau)\beta_D^1(\tau)\Big]d\tau,\end{aligned}$$

$$\begin{aligned}\mathcal{E}_N^*(a,t) = \int_0^\infty \big[&\mathcal{L}_P(a+\tau,t,\mu^*)B_P^*(t-a-\tau)\sigma_P(\tau) \\ &+\mathcal{L}_D(a+\tau,t,N^*,\mu^*)B_D^*(t-a-\tau)\sigma_D(\tau)\big]d\tau.\end{aligned}$$

Then, using standard argument of integral equations theory, we can prove

Proposition 1 *Let the assumptions* (18) *hold, then problem* (46) *has one and only one solution*

$$(D_P, D_D, D_A, D_N) \in C([0,T]) \times C([0,T]) \times L^\infty(0,T) \times C([0,T])$$

Moreover, we have

$$\lim_{\lambda\to 0} \frac{B_P^\lambda - B_P^*}{\lambda} = D_P \quad in\ C([0,T]),$$
$$\lim_{\lambda\to 0} \frac{B_D^\lambda - B_D^*}{\lambda} = D_D \quad in\ C([0,T]),$$
$$\lim_{\lambda\to 0} \frac{N^\lambda - N^*}{\lambda} = D_N \quad in\ C([0,T])$$
$$\lim_{\lambda\to 0} \frac{B_A^\lambda - B_A^*}{\lambda} = D_A \quad in\ L^\infty(0,T).$$

Actually, in view of the variation of (26), we need the following limits

$$y_P = \lim_{\lambda\to 0} \frac{\nu_P^\lambda - \nu_P^*}{\lambda}, \qquad y_D = \lim_{\lambda\to 0} \frac{\nu_D^\lambda - \nu_D^*}{\lambda}$$

and, indeed, using Proposition 1 and (28) we have that the two limits exist and

$$y_P(a,t) = D_P(t-a)e^{-\int_0^{a\wedge t} \sigma_P(a-\sigma)\mu^*(t-\sigma)d\sigma} - \nu_P^*(a,t)\int_0^{a\wedge t} \sigma_P(a-\sigma)\omega(t-\sigma)d\sigma$$

$$\begin{aligned} y_D(a,t) &= D_D(t-a)e^{-\int_0^{a\wedge t}[H(N^*(t-\sigma))\eta_D^1(a-\sigma)+\sigma_D(a-\sigma)\mu^*(t-\sigma)]d\sigma} \\ &\quad -\nu_D^*(a,t)\int_0^{a\wedge t}\Big[H'(N^*(t-\sigma))\eta_D^1(a-\sigma)D_N(t-\sigma) \\ &\qquad +\sigma_D(a-\sigma)\omega(t-\sigma)\Big]\,d\sigma \end{aligned} \tag{47}$$

Thus, since from (26) we have

$$\lim_{\lambda\to 0}\frac{\mathcal{J}(\mu^\lambda)-\mathcal{J}(\mu^*)}{\lambda} = \int_0^T\int_0^\infty G_P(a)y_P(a,t)dadt$$

$$+\gamma_1\int_0^T\int_0^\infty \mathcal{G}_1^*(a,t)v_D^*(a,t)D_N(t)dadt$$

$$+\gamma_1\int_0^T\int_0^\infty \mathcal{G}_2^*(a,t)y_D(a,t)dadt$$

$$+\gamma_2\int_0^T (t)dt \geq 0$$

where

$$\mathcal{G}_1^*(a,t) = \left[\left(G_D^0(a)+G_D^1(a)H(N^*(t))\right)v_D^*(a,t)-S_D^*(a)\right] \times G_D^1(a)H'(N^*(t)),$$

$$\mathcal{G}_2^*(a,t) = \left[\left(G_D^0(a)+G_D^1(a)H(N^*(t))\right)v_D^*(a,t)-S_D^*(a)\right] \times\left(G_D^0(a)+G_D^1(a)H(N^*(t))\right),$$

using (47) we have

$$\begin{aligned}&\int_0^T\int_0^t \mathcal{S}_1^*(a,t)D_P(t-a)dadt+\gamma_1\int_0^T\int_0^t \mathcal{S}_2^*(a,t)D_D(t-a)dadt\\&-\gamma_1\int_0^T\int_0^t \mathcal{S}_3^*(a,t)D_N(t-a)dadt+\gamma_1\int_0^T \mathcal{S}_4^*(t)D_N(t)dt\\&-\int_0^T\int_0^t \mathcal{S}_5^*(a,t)(t-a)dadt+\gamma_2\int_0^T (t)dt \geq 0,\end{aligned} \tag{48}$$

where

$$\begin{aligned}
\mathcal{S}_1^*(a,t) &= G_P(a) e^{-\int_0^{a\wedge t} \sigma_P(a-\tau)\mu^*(t-\tau)d\tau}, \\
\mathcal{S}_2^*(a,t) &= \mathcal{G}_2^*(a,t) e^{-\int_0^{a\wedge t}\left[H(N^*(t-\tau))\eta_D^1(a-\tau)+\sigma_D(a-\tau)\mu^*(t-\tau)\right]d\tau}, \\
\mathcal{S}_3^*(a,t) &= H'(N^*(t-a)) \int_0^\infty \mathcal{G}_2^*(a+\tau,t) v_D^*(a+\tau,t)\eta_D^1(\tau)\,d\tau, \\
\mathcal{S}_4^*(t) &= \int_0^\infty \mathcal{G}_1^*(a,t) v_D^*(a,t) da, \\
\mathcal{S}_5^*(a,t) &= \int_0^\infty G_P(a+\tau) v_P^*(a+\tau,t)\sigma_P(\tau)d\tau \\
&\quad + \gamma_1 \int_0^\infty \mathcal{G}_2^*(a+\tau,t) v_D^*(a+\tau,t)\sigma_D(\tau)\,d\tau.
\end{aligned} \tag{49}$$

Actually, in view of subsequent developments, (48) is conveniently written as

$$\begin{aligned}
&\int_0^T \mathcal{F}_1^*(t) D_P(t)dt + \int_0^T \mathcal{F}_2^*(t) D_D(t)dt \\
&\qquad + \int_0^T \mathcal{F}_3^*(t) D_N(t)dt + \int_0^T \mathcal{F}_4^*(t)(t)dt \ \geq\ 0,
\end{aligned} \tag{50}$$

where

$$\begin{aligned}
\mathcal{F}_1^*(t) &= \int_t^T \mathcal{S}_1^*(\tau-t,\tau)d\tau, \quad \mathcal{F}_2^*(t) = \gamma_1 \int_t^T \mathcal{S}_i^*(\tau-t,\tau)d\tau, \\
\mathcal{F}_3^*(t) &= \gamma_1 \left(\mathcal{S}_4^*(t) - \int_t^T \mathcal{S}_3^*(\tau-t,\tau)d\tau \right), \\
\mathcal{F}_4^*(t) &= \gamma_2 - \int_t^T \mathcal{S}_5^*(\tau-t,\tau)d\tau.
\end{aligned}$$

Now, on the basis of (50), we consider the dual variables $Q_P(t)$, $Q_D(t)$, $Q_A(t)$, $Q_N(t)$ satisfying the dual problem

$$
\begin{aligned}
Q_P(t) &= \int_t^T \mathcal{A}_P^*(a-t,a)Q_P(a)da + \int_t^T \mathcal{A}_D^*(a-t,a)Q_D(a)da \\
&\qquad + \int_t^T \mathcal{A}_A^*(a-t,a)Q_A(a)da \\
&\qquad\qquad + \int_t^T \mathcal{A}_N^*(a-t,a)Q_N(a)da + \mathcal{F}_1^*(t), \\
Q_D(t) &= \int_t^T \mathcal{B}_A^*(a-t,a)Q_A(a)da \\
&\qquad + \int_t^T \mathcal{B}_N^*(a-t,a)Q_N(a)da + \mathcal{F}_2^*(t), \\
Q_A(t) &= \int_t^T \mathcal{C}_N^*(a-t,a)Q_N(a)da, \\
Q_N(t) &= \mathcal{D}_P^*(t)Q_P(t) - \mathcal{D}_P^*(t)Q_D(t) \\
&\qquad - \int_t^T \mathcal{D}_A^*(a-t,a)Q_A(a)da \\
&\qquad\qquad - \int_t^T \mathcal{D}_N^*(a-t,a)Q_N(a)da + \mathcal{F}_3^*(t).
\end{aligned}
\tag{51}
$$

Concerning this problem we have an existence result similar to that stated for problem (46)

Proposition 2 *Let the assumptions* (18) *hold, then problem* (51) *has one and only one solution*

$$(Q_P, Q_D, Q_A, Q_N) \in C([0,T]) \times C([0,T]) \times L^\infty(0,T) \times C([0,T]).$$

We use (51) to elaborate (50). Indeed, multiplying the four equation of (51) respectively by $D_P(t)$, $D_D(t)$, $D_A(t)$, $D_N(t)$, integrating over the interval $[0, T]$, exchanging integrals and summing all the equations, we finally get

$$\int_0^T \mathcal{F}_1^*(t) D_P(t)dt + \int_0^T \mathcal{F}_2^*(t) D_D(t)dt + \int_0^T \mathcal{F}_3^*(t) D_N(t)dt + \int_0^T \mathcal{F}_4^*(t)(t)dt$$

$$= \int_0^T \left\{\gamma_2 - E(t;\mu^*)\right\}(t)dt \geq 0, \tag{52}$$

where

$$\begin{aligned} E(t;\mu^*) = &\int_t^T \mathcal{S}_5^*(\tau - t, \tau)d\tau + \int_t^T \mathcal{E}_P^*(\tau - t, \tau)Q_P(\tau)d\tau \\ &+ \int_t^T \mathcal{E}_D^*(\tau - t, \tau)Q_D(\tau)d\tau + \mathcal{E}^*(t)Q_A(t) \\ &+ \int_t^T \mathcal{E}_A^*(\tau - t, \tau)Q_A(\tau)d\tau + \int_t^T \mathcal{E}_N^*(\tau - t, \tau)Q_N(\tau)d\tau. \end{aligned} \tag{53}$$

We now recall that, considering the multivalued function (actually the maximal monotone graph representing the normal cone of the interval $[a, b]$), $N_{[a,b]} : [a, b] \to 2^{\mathbf{R}}$ defined as

$$N_{[a,b]}(x) = \begin{cases} (-\infty, 0] & \text{for } x = a \\ 0 & \text{for } x \in (a, b) \\ [0, \infty) & \text{for } x = b, \end{cases}$$

we have

$$\left(I + N_{[a,b]}\right)^{-1}(x) = \begin{cases} a \text{ if } & x < a, \\ x \text{ if } & x \in [a, b], \\ b \text{ if } & x > b. \end{cases}$$

Then we can state our main theorem in this section

Theorem 3 *Let μ^* be an optimal controller with ν_P^*, ν_D^*, ν_A^* the corresponding optimal states and Q_P, Q_D, Q_A, Q_N the corresponding solution to problem* (51).*Then, μ^* is implicitly defined through*

$$\mu^*(t) = \left(I + N_{[0,\mu_{max}]}\right)^{-1}\left(\mu^*(t) + E(t;\mu^*) - \gamma_2\right) \tag{54}$$

where $N_{[0,\mu_{max}]}$ is the normal cone for the interval $[0, \mu_{max}]$.

Proof By recalling that $w = \overline{\mu} - \mu^*$, (52) can be still written

$$\int_0^T \left\{E(t;\mu^*) - \gamma_2\right\}(\mu^* - \overline{\mu})(t)dt \geq 0, \tag{55}$$

taking place for all $\overline{\mu} \in \mathcal{U}$. Relation (55) implies that

$$E(t;\mu^*) - \gamma_2 \in N_{[0,\mu_{max}]}(\mu^*(t)), \text{ a.e. } t \in (0,T), \tag{56}$$

Then (54) holds as claimed. □

The previous theorem provides a basic Eq. (54), necessarily satisfied by the optimal control whose existence has been stated in Theorem 2. Note that (54) implies the following conditions that provide informations on μ^*

$$\begin{aligned}
&\mu^*(t) = 0 && \text{on the set} \quad \{t : E(t;\mu^*) < \gamma_2\}, \\
&E(t;\mu^*) - \gamma_2 = 0 && \text{on the set} \quad \{t : \gamma_2 < \mu^*(t) + E(t;\mu^*) < \gamma_2 + \mu_{max}\}, \\
&\mu^*(t) = \mu_{\max} && \text{on the set} \quad \{t : E(t;\mu^*) > \gamma_2\}.
\end{aligned} \tag{57}$$

5 Conclusions

The theoretical framework we have developed in this paper is a starting point for further investigation in the context of realistic applications of the model, using, for instance, realistic values of the parameters from [6]. In particular, Eq. (54) can be exploited for numerical computations. Indeed, $E(t;\mu^*)$ defined in (53) depends on $\mu^*(t)$ through the kernels (49) and through the solution of the system (51) thus (54) represent an equation to be coupled with an iterative method to compute the optimal control.

The procedure we propose in Sects. 2 and 4 is somewhat new and is based on the system of integral Eqs. (29)–(32) and on the dual system (51). The kernels in the integral system are rather involved but still the method we followed is somewhat natural, in spite of the expression of the kernels. An alternative approach would be based on the use of the partial differential system (2)–(4), with its system in variation and the corresponding dual system, but we opted for the integral formulation because it turns out to be more natural to deal with the retarded nature of the equations.

Acknowledgments Mimmo Iannelli and Gabriela Marinoschi acknowledge the support of the project "Analysis and control of mathematical models for the evolution of epidemics, tumors and phase field processes" developed in the framework of the agreement between the Italian CNR and the Romanian Academy. Gabriela Marinoschi acknowledges the support of a grant of the Ministry of Research, Innovation and Digitization, CNCS—UEFISCDI, project number PN-III-P4-PCE-2021-0921, within PNCDI III.

References

1. Bordbar, A., Dias, D., Cabral, A., Beck, S., Boon, M.E.: Assessment of cell proliferation in benign, premalignant and malignant skin lesions. Appl. Immunohistochem Mol. Morphol. **15**, 229–235 (2007)
2. Clayton, E., Doupe, D.P., Klein, A.M., Winton, D.J., Simons, B.D., Jones, P.H.: A single type of progenitor cell maintains normal epidermis. Nature **446**, 185–189 (2007)
3. Gandolfi, A., Iannelli, M., Marinoschi, G.: An age-structured model of epidermis growth. J. Math. Biol. **62**, 111–141 (2011)
4. Gandolfi, A., Iannelli, M., Marinoschi, G.: Time evolution for a model of epidermis growth. J. Evol. Equ. **13**, 509–533 (2013)
5. Gandolfi, A., Iannelli, M., Marinoschi, G.: The steady state of epidermis: mathematical modeling and numerical simulations. J. Math. Biol. **73**, 1595–1626 (2016)
6. Gandolfi, A., Iannelli, M., Marinoschi,G.: The basal layer of the epidermis: a mathematical model for cell production under a surface density constraint. SIAM J. Appl. Math. **80**, 543–571 (2020)
7. Hadgraft, J.: Skin, the final frontier. Int. J. Pharm. **224**, 1–18 (2001)
8. Houben, E., De Paepe, K., Rogiers, V.: A keratinocyte's course of life. Skin Pharmacol. Physiol. **20**, 122–132 (2007)
9. Jones, P.H., Simons, B.D., Watt, F.M.: Sic transit gloria: farewell to the epidermal transit amplifying cell? Cell Stem Cell **1**, 371–381 (2007)
10. Kawashima, K., Doi, H., Ito, Y., Shibata, M.A., Yoshinaka, R., Otsuki, Y.: Evaluation of cell death and proliferation in psoriatic epidermis. J. Dermatol. Sci. **35**, 207–214 (2004)

On the Qualitative Behaviour of Oscillating Biochemical Systems: The Stochastic Approach

Gabriella Mavelli, Alessandro Borri, Pasquale Palumbo, and Alberto Maria Bersani

Abstract Some crucial enzymatic reaction systems, often showing feedbacks, give rise to oscillatory mechanisms. It occurs in well-known biochemical processes, such as, to give some examples, the PER-TIM circuit related to circadian rhythms, the mitogen-activated protein kinase (MAPK) cascade, the glycolysis and so on. In this work, we address the problem of modeling some complex biological processes by means of a stochastic approach, with a focus on the PER circadian clock model. As it is well-known, in such a scenario, a network of biochemical reactions is well described in terms of a Continuous-Time Markov Chain (CTMC), whose stochastic behaviour is fully characterized by the CTMC initial probability distribution and the probabilities per unit of time of each reaction (propensities). In order to cope with the burden of the system dimensionality, we resort to Quasi-Steady State Approximations (QSSAs), in particular to the standard QSSA and the total QSSA (sQSSA and tQSSA, respectively), which are widely used in enzyme kinetics to simplify the equations modeling the reactions, thus speeding up numerical simulations. It has been shown that, both in a deterministic and in a stochastic setting, the tQSSA is valid in a much wider range of parameters and initial conditions than the sQSSA. With the aim of investigating the asymptotic course of the process, we study the stationary distribution of the CTMC by applying the stochastic approach to the non-approximated original chemical networks (later referred to as full system), as well as to the standard and total QSSAs. Our simulation results fully confirm the superiority of the performances of tQSSA with respect to sQSSA in terms of qualitative and

G. Mavelli · A. Borri (✉)
Istituto di Analisi dei Sistemi e Informatica, National Research Council of Italy, Rome, Italy
e-mail: gabriella.mavelli@iasi.cnr.it; alessandro.borri@iasi.cnr.it

P. Palumbo
Department of Biotechnology and Biosciences, University of Milano-Bicocca, Milano, Italy
e-mail: pasquale.palumbo@unimib.it

A. M. Bersani
Dipartimento di Ingegneria Meccanica e Aerospaziale, University of Rome "La Sapienza", Rome, Italy
e-mail: alberto.bersani@sbai.uniroma1.it

A. d'Onofrio et al. (eds.), *Problems in Mathematical Biophysics*, SEMA SIMAI Springer Series 38, https://doi.org/10.1007/978-3-031-60773-8_9

quantitative agreement of probability distributions with the stochastic modeling of the full system. Furthermore, the agreement (in spite of the large complexity reduction) between full system and tQSSA also in the stochastic framework allows to reveal non-trivial correlations among the species involved in the reactions, which cannot be investigated in a purely deterministic framework.

1 Introduction

One of the main contributions of mathematicians to the biological field is one of the best-known models of enzyme kinetics put forth during the first years of the twentieth century by Henri [1–3], and Michaelis and Menten [4], and later continued by Briggs and Haldane in 1925 [5]. This formulation considers a reaction where a substrate S binds an enzyme E reversibly to form a complex C. The complex can then decay irreversibly to a product P and the enzyme, which is then free to bind another molecule of the substrate. These reactions are summarized in the scheme

$$E + S \underset{k_{-1}}{\overset{k_1}{\rightleftarrows}} C \overset{k_2}{\longrightarrow} E + P, \tag{1}$$

where k_1, k_{-1} and k_2 are kinetic parameters associated with the reaction rates (i.e. rate constants).

A very common approximation in the deterministic setting (standard Quasi-Steady State Approximation, or sQSSA) consists in assuming that, after a short transient phase, the complex concentration can be considered approximately constant with respect to the substrate dynamics [6]. The sQSSA utilizes timescale separation to project models of biochemical networks onto lower-dimensional slow manifolds, thus rapidly fluctuating species are not simulated explicitly (see, among the others, [7, 8], and the review paper [9]). In the last decades, many researchers highlighted limits and malfunctioning of the sQSSA, thereby introducing and exploring a new approximation, called total Quasi-Steady State Approximation (tQSSA). Under suitable and biologically consistent hypotheses, the tQSSA-based methods revealed to be very effective in handling the full system of equations, considerably unburdening the computational effort and providing a good approximation at the same time (see, among the others, [9–14]).

In recent years, several papers (see, for example, [15–20]) discussed the validity of the sQSSA in more complex mechanisms. In particular it can be observed that when a system undergoes oscillations [15, 16, 19, 20], any QSSA risks to fail, because the central hypothesis for the quasi-steady state assumption is a substantial equilibrium for the complexes, which cannot be guaranteed in presence of the substrate oscillations. In [20] the criticisms about the application of the sQSSA were devoted to stochastic models, too.

In this paper we want to take a step forward, investigating, in a stochastic framework, the applicability of any QSSA to models providing oscillatory behaviours, and specifically the molecular mechanism of circadian rhythms.

Nearly all living organisms (bacteria, fungi, plants and mammals) have developed the capability of generating autonomously sustained oscillations at the molecular level, known as circadian rhythms, with a period close to 24 h. Countless molecules synchronize their dynamic behaviour to produce periodic oscillations observable on a macroscopic level. This phenomenon is one of the most remarkable manifestations of self-organization in chemical kinetics, which serve to optimize the functioning of unicellular or multicellular organisms. The core biochemical process from which these endogenous oscillations originate in constant environmental conditions is the negative autoregulation (negative feedback) exerted by a protein (*clock protein*) on the expression of its gene [21–24]. Experimental studies during the last two decades have shed much light on the molecular mechanism of circadian rhythms. In *Drosophila* the proteins PER and TIM repress, through the formation of a heterodimer, the activation of the *per* and *tim* genes (PER-TIM model). In mammals the CRY proteins, instead of TIM, form a regulatory complex with a PER protein to inhibit the expression of the *per* and *cry* genes. Light can affect circadian rhythms by enhancing the expression of the genes (e.g. *per* gene in mammals) and by inducing the degradation of the proteins (e.g the TIM protein in *Drosophila*) [21, 22].

With the aim of exploring the design of circadian clock, on the basis of experimental observations, a number of mathematical models based on genetic regulatory processes have been proposed to probe properties of circadian rhythms in some organisms. They were first proposed for *Drosophila* and *Neurospora* [25–31], and later for mammals [32–35]. The mathematical model for the core biochemical mechanism of circadian rhythms in *Drosophila*, based on the negative autoregulation exerted by PER clock protein alone [25, 26], well predicts the fundamental features of the circadian rhythms, such as the occurrence of sustained oscillations in constant environmental conditions (e.g., constant darkness). It reproduces also the entrainment of circadian rhythms by light-dark cycles (that make living organisms able to adapt their physiological behaviour to the natural alternation of day and night) and the effect induced by light pulses on the phase of the oscillation (phase-shifting or long-term suppression). Since similar results have been obtained by models incorporating additional clock gene products [27–31], the core model proposed in [25, 26], which is described by a set of five kinetic equations, is adopted, for simplicity, in many studies both for *Drosophila* and for *Neurospora*. In the latter, FRQ protein is the clock protein for the circadian oscillations.

Since the regulatory mechanism of the oscillations within the rhythm-producing cells involves a small number of proteins and mRNAs, molecular fluctuations are likely to have a great impact on circadian oscillations [36]. Thus, to evaluate the robustness of circadian oscillations with respect to molecular noise, it is necessary to resort to a stochastic approach. In the stochastic scenario molecular fluctuations are taken into account by describing the biochemical reaction system as a birth-and-death stochastic process governed by Chemical Master Equations (CMEs) [37].

CME provides an accurate stochastic description of complex biochemical networks in terms of probability distribution of the underlying chemical populations, in contrast to deterministic methods which handle biochemical processes in terms of the average value dynamics for each involved species. Indeed, deterministic models can be dealt with as first-order approximations of the average value dynamics coming from the stochastic CME approach [37]. Various studies of circadian rhythms in a stochastic setting are present in literature, and the proposed stochastic models have been used to identify and study the factors that contribute to increase robustness [38–41]. In [36], even if the maximum numbers of mRNA and protein molecules in the model are of the order of 3000 and 10,000, respectively (that is rather larger than the one encountered within cells), stochastic simulations produced highly irregular oscillations. In a subsequent work [38], the authors conversely show that, with a maximum number of mRNA and protein molecules of the order at least of tens and hundreds, respectively, robust circadian oscillations take place. Larger numbers of molecules further reinforce the robustness with respect to noise. Concerning the robustness of the oscillatory dynamics with respect to parameter changes, deterministic models predict that, in a precise region of parameter space, the genetic regulatory network at the core of the clock mechanism can produce sustained oscillations of the limit-cycle type, corresponding to the circadian rhythm, whereas outside this range the gene network operates in a stable steady state [25–31]. Similar results are reached in a stochastic framework. Sustained oscillations of stochastic variables (i.e. the numbers of clock proteins and mRNAs) occur in a large domain of parameter values. If parameters are outside such a domain, the variables fluctuate around steady-state levels [38]. The authors show that, if the values of the bimolecular rate constants characterizing the association of the regulatory protein to the gene promoter on DNA are under a given order, the robustness property of the system is lost. This happens because in this case the systems works close to the bifurcation point, where the disrupting effect of molecular noise becomes more significant [38, 41]

The PER-TIM model for circadian oscillations in *Drosophila* [27] has been considered in [42]. The deterministic mathematical model in this case is given by a system of 10 ODEs describing the time evolution of the mean concentration of the 10 species involved. In this work, we consider a version of the stochastic model, easier to handle than the one adopted for the core model, obtained without decomposing the reactions into elementary steps, and by simply including the non-linear kinetic functions in the probabilities of occurrence of the global reaction steps (*stochastic non-developed model*). In the more general case here considered results very similar to those reached by the core model have been obtained, thus validating the use of such a simpler model to study the molecular mechanism of circadian rhythms. Nevertheless, the advantage of using PER-TIM or more detailed models is that they allow us to investigate further features of the circadian oscillations, such as the property of persistence of circadian rhythmicity. The latter is enhanced by the cooperative nature of repression [38]. In some living organisms, cooperativity in the oscillatory mechanism is supported by complexes made up of several clock proteins [21–24, 43, 44]. Robustness with respect to noise may be also enhanced by the presence of multiple interlocked regulatory feedback loops [36, 45]. Similarly, the robustness of circadian rhythms is reported to be enhanced in organisms characterized by syncitial morphology, such as *Neurospora*. Here the different nuclei may be coupled through a common variable, cytosolic FRQ, and the coupling might enhance the robustness even in the presence of reduced amounts of this protein within each nucleus. Other forms of coupling similarly may enhance the robustness of circadian rhythms in various organisms, such as that occurring in mammals through neurotransmitters in the case of the rhythm-producing neurons in the suprachiasmatic nuclei (SCN) [46, 47], or in plants [45]. By means of the PER-TIM model it is furthermore possible to assess the effect of molecular noise on chaotic oscillations, since it admits chaotic behaviour, for suitable parameter values [42].

In our work we propose a CME-based stochastic study focused on the molecular model describing the regulation carried out by PER alone (the so-called PER circadian clock). The main subject of our investigation is the asymptotic probability distribution of the underlying biochemical populations. Within the framework of enzymatic reaction networks, many authors investigated the QSSAs via the CME-approach, with the aim of providing a good approximation of the full model also in this setting (see [48–58]). Our study investigates the applicability of both standard and total QSSA, showing the preminent role of the tQSSA. The methodology exploited is the one proposed in [59–61] managing to cope with the computational burden, that usually arises for CMEs, by means of a proper organization of the probabilities in the CME entries. This enables us to characterize the CME dynamics according to a recursive block-tridiagonal structure. In this way, explicit solutions of the CME are achieved for both standard and total QSSA, according to a smart application of the Gauss Elimination Method [60]. This allows us not to resort to Monte Carlo simulation, that may be computationally demanding as well as lead to misleading results unless allowing for enough stochastic realizations (whose number is not known a priori).

The paper is organized as follows. In Sect. 2 we briefly recall the most important background concerning the sQSSA and the tQSSA for enzymatic reactions. Section 3 proposes the equations of the PER circadian clock in details, dealing with the CME-based stochastic model. Section 4 reports the standard and total QSSA of the PER circadian clock CME-based stochastic model, by first providing them in the deterministic framework. In Sect. 5 we discuss the numerical results, interpreting in particular the asymptotic distribution of the reactant concentrations. Section 6 contains the concluding remarks and some perspectives.

2 Background on Michaelis-Menten Kinetics, and on sQSSA and tQSSA

The *Michaelis-Menten* (MM) *kinetics* gives a very good description of (1), in terms of Ordinary Differential Equations (ODEs). For notational convenience we will use the same symbol to denote both a chemical species and its concentration (i.e. the variables of the ODEs), omitting its dependence on time. We can mathematically describe reaction (1) using the mass action principle—where the growth rates of each reactant are proportional to the instantaneous concentrations of the reactants themselves—and conservation laws. This approach leads to the following (full) system:

$$\begin{cases} \frac{dS}{dt} = -k_1(E_T - C)S + k_{-1}C \\ \frac{dC}{dt} = k_1(E_T - C)S - (k_{-1} + k_2)C, \end{cases} \tag{2}$$

with the initial conditions

$$S(0) = S_T, \qquad C(0) = 0, \qquad E(0) = E_T, \tag{3}$$

and the conservation laws

$$E + C = E_T, \qquad S + C + P = S_T, \tag{4}$$

where E_T and S_T are the total enzyme and substrate concentrations respectively.

The MM reaction is characterized by two phases: a short transient phase of rapid increase of the complex C and a second, slower, phase, called *quasi-steady state phase*, where the complex is considered substantially in equilibrium.

The hypothesis of quasi-steady state simplifies the reaction, leading to an ODE for the substrate, with initial condition $S(0) = S_T$, while the complex is assumed to be in a quasi-steady state, i.e. $\frac{dC}{dt} \approx 0$. The *standard QSSA* (sQSSA) of system (2) is thus achieved:

$$\begin{cases} C \approx \frac{E_T \cdot S}{K_M + S} \\ \frac{dS}{dt} \approx -k_2 C \approx -\frac{V_{max} S}{K_M + S} \\ S(0) = S_T, \end{cases} \tag{5}$$

where $K_M = \frac{k_2 + k_{-1}}{k_1}$ is called the MM constant or affinity constant, and $V_{max} = k_2 E_T$.

Let us consider again the classical MM kinetics (2). Introducing the total substrate at a generic time instant t, $\overline{S}(t) = S(t) + C(t)$, (2) then becomes:

$$\begin{cases} \dfrac{d\overline{S}}{dt} = -k_2 C \\ \dfrac{dC}{dt} = k_1[C^2 - (E_T + \overline{S} + K_M)C + E_T\overline{S}] \\ \overline{S}(0) = S_T, \quad C(0) = 0, \end{cases} \tag{6}$$

with conservation laws

$$E + C = E_T, \quad \overline{S} + P = S_T \,. \tag{7}$$

Assuming that the complex is in a quasi-steady state $\left(\dfrac{dC}{dt} \approx 0\right)$ yields the *total QSSA* (tQSSA) [11], which is valid for a broader range of parameters, with respect to sQSSA, covering both high and low enzyme concentrations:

$$\frac{d\overline{S}}{dt} \approx -k_2\, C_-(\overline{S}), \quad \overline{S}(0) = S_T, \tag{8}$$

where

$$C_-(\overline{S}) = \frac{(E_T + K_M + \overline{S}) - \sqrt{(E_T + K_M + \overline{S})^2 - 4E_T\overline{S}}}{2} \tag{9}$$

is the only biologically allowed solution of $\dfrac{dC}{dt} = 0$ in the second equation (6).

We must underline that the sQSSA, as every QSSA, represents the system dynamics after a (in general short) transient phase, during which the substrates are partially bound and the complexes begin to form. Consequently, since the QSSA is applied considering the complexes substantially constant, the total concentration of free and activated substrate(s) will be considered constant, but its value, due to the presence of complexes, cannot coincide with the initial substrate value, when all the complex concentrations were equal to zero. Setting S_T as initial value for the total amount of (inactive and activated) substrate concentrations in the sQSSA naturally brings to wrong conclusions, since the system is forced to fulfill a conservation law that implicitly neglects all the complex concentrations. In particular we focus on the dynamics of the intermediate complexes, which in common literature either are ignored or are supposed to rapidly become negligible in the quasi steady-state phase, differently from what really happens. This brings to what we call "complex depletion paradox" [62], according to which complexes disappear in the conservation laws, in contrast with the equations of their dynamics. On the other hand the tQSSA cannot produce this situation, because the substrates and the complexes formed by them are included in the same (total) variables and the system cannot distinguish between free substrates and bound ones. For this reason, differently from the sQSSA, the tQSSA saves the conservation laws and produces the same asymptotic values as the full system. The most dramatic consequence is that, when applied to well-known mechanism, like, e.g., the (double) phosphorylation-dephosphorylation cycle, the enzymatic cascades, the van Slyke-Cullen mechanism etc., the sQSSA predicts phenomena which do not appear when the mechanisms are studied by means of the full system of equations describing the systems or by means of the tQSSA [15–20].

3 The PER Circadian Clock: Deterministic and CME-Based Stochastic Model

The model we are focusing on is the original, described in [25, 26], where the *per* mRNA, whose cytosolic concentration is here indicated by M, is synthetized in the nucleus and transfers to the cytosol, where it synthesizes the protein PER and is degraded. The rate of PER synthesis is proportional to the concentration M. Since PER can be multiply synthesized, the model takes into consideration three states of PER: the unphosphorylated state P_0, the monophosphorylated P_1 and the bisphosphorylated P_2. The phosphorylation and dephosphorylation reactions happen following the usual Michaelis-Menten kinetics, catalyzed by suitable enzymes (kinases E_1, E_2 for the former, and phosphatases F_1, F_2 for the latter) and forming temporary complexes C_i , $i = 1, \ldots 4$. P_2 can either be degraded or enter into the nucleus, where it becomes P_N. This last step is reversible, because P_N can transfer to the cytosol and become P_2. In Goldbeter's model, the degradation of M and P_2, which biologically follow complex mechanisms, are described by means of a Michaelis-Menten saturation term, with rates v_m and v_d and Michaelis constants K_m and K_d, respectively. P_N acts on M with a negative feedback, modeled by a Hill function, whose steepness is modulated by the degree n of cooperativity.

The scheme of this mechanism can be summarized as follows:

$$M \xrightarrow{k_s} P_0$$

$$P_0 + E_1 \underset{d_1}{\overset{a_1}{\rightleftarrows}} C_1 \xrightarrow{b_1} P_1 + E_1$$

$$P_1 + F_1 \underset{d_2}{\overset{a_2}{\rightleftarrows}} C_2 \xrightarrow{b_2} P_0 + F_1$$

$$P_1 + E_2 \underset{d_3}{\overset{a_3}{\rightleftarrows}} C_3 \xrightarrow{b_3} P_2 + E_2$$

$$P_2 + F_2 \underset{d_4}{\overset{a_4}{\rightleftarrows}} C_4 \xrightarrow{b_4} P_1 + F_2$$

$$P_2 \longrightarrow \emptyset$$

$$P_2 \underset{k_2}{\overset{k_1}{\rightleftarrows}} P_N$$

$$\emptyset \xrightarrow{P_N \perp} M$$

$$M \longrightarrow \emptyset$$

Note that, in the scheme above, we used the horizontal arrow symbols $\longrightarrow$ and $\rightleftarrows$ to denote non-reversible and reversible processes, respectively, while we used the symbol Y$\perp$, over a horizontal arrow, to denote the repression effect on the process exerted by a species Y.

Before building the stochastic model, we write the deterministic full system by exploiting the mass conservation law to reduce the system complexity.

Indeed, though the total concentration of the protein PER $P_0 + P_1 + P_2 + P_N + C_1 + C_2 + C_3 + C_4 = P_t$ and of the mRNA are not conserved, due to the degradation, the conservation laws involve the enzymes E_i, F_i:

$$E_1 + C_1 = E_{10}\ , \quad F_1 + C_2 = F_{10}\ , \quad E_2 + C_3 = E_{20}\ , \quad F_2 + C_4 = F_{20} \tag{10}$$

so that it is possible to reduce the number of deterministic variables (species concentrations) to 9 independent ones. By taking, for example, $\{M, P_0, P_1, P_2, P_N, C_1, C_2, C_3, C_4\}$ as the set of the independent variables, using the law of mass action, the full system of equations governing the dynamics of the system is therefore:

$$\begin{cases} \dfrac{dP_0}{dt} = k_s M - a_1 P_0(E_{10} - C_1) + d_1 C_1 + b_2 C_2 \\ \dfrac{dP_1}{dt} = b_1 C_1 - a_2 P_1(F_{10} - C_2) + d_2 C_2 - a_3 P_1(E_{20} - C_3) + d_3 C_3 + b_4 C_4 \\ \dfrac{dP_2}{dt} = b_3 C_3 - a_4 P_2(F_{20} - C_4) + d_4 C_4 - \dfrac{v_d}{P_2 + K_d} P_2 - k_1 P_2 + k_2 P_N \\ \dfrac{dP_N}{dt} = k_1 P_2 - k_2 P_N \\ \dfrac{dC_1}{dt} = a_1[P_0(E_{10} - C_1) - K_{1M} C_1] \\ \dfrac{dC_2}{dt} = a_2[P_1(F_{10} - C_2) - K_{2M} C_2] \\ \dfrac{dC_3}{dt} = a_3[P_1(E_{20} - C_3) - K_{3M} C_3] \\ \dfrac{dC_4}{dt} = a_4[P_2(F_{20} - C_4)) - K_{4M} C_4] \\ \dfrac{dM}{dt} = v_s \dfrac{K_I^n}{K_I^n + P_N^n} - v_m \dfrac{M}{K_m + M} \end{cases}$$

where $K_{iM} = \frac{d_i + b_i}{a_i}$, with initial conditions

$$P_j(0) = P_{j0}, \; j = 0, 1, 2, \quad P_N(0) = P_{N0}, \quad M(0) = M_0,$$
$$C_i(0) = 0 \;, \; i = 1, \ldots, 4 \,. \tag{11}$$

The value of n should be determined experimentally. In absence of experimental data, the choice of n is somehow arbitrary. Following [25], we set in this case $n = 4$.

According to a large variety of literature that manages to reformulate the dynamics of a system from an ODE into a CME framework (see e.g. [61] or [63]), we treat the state variables as discrete copy numbers. In this context, we reinterpret the deterministic reaction rates as probabilities-per-unit-time (or propensities) of a properly defined Continuous-time Markov Chain (CTMC), i.e., a stochastic process whose trajectories evolve on a n-dimensional lattice, and whose dynamics (in terms of probability of being in a specific state of the CTMC) is described by the CME.

The CME dynamics is $\dot{\mathcal{P}} = G\mathcal{P}$, where G is the probability transition matrix of the CTMC, which is built according to the propensities in Table 1 (see [60] for further details), and $\mathcal{P}$ is the vector collecting the time-varying probabilities of all the states of the process. When the dimension of G is large enough to make computationally too demanding the exact solution of the equilibrium equation, $G\mathcal{P} = 0$, one can still employ the Gillespie Stochastic Simulation Algorithm (SSA) [49–53], returning one or more statistically correct trajectories of the process, which can be used in a Monte Carlo or in an ergodic setting to obtain a sampled (approximate) equilibrium distribution of the process.

4 QSSA of the PER Circadian Clock Stochastic Model

Similarly to the deterministic approach, where the complexes behave as fast variables, while the substrates are the slow variables, the double time scale affects as well the CMEs associated to the reset map detailed in Table 1, slowing down the computational efficiency. Therefore a need exists to introduce a QSSA also for the stochastic case. In [61, 64] it was shown a way to obtain both standard and total QSSA for the double phosphorylation/dephosphorylation (PDP) cycle. In both cases, we start from the ODE version of the QSSA and *stochastify* it by introducing suitably defined one-step processes. Note that, as already described at the end of Sect. 3, the stochastic approach considered in [61, 64] and in the present work does not consist in perturbing the ODE setting by means of noise terms, which is typical of the Stochastic Differential Equation (SDE)/Langevin approach, but in reinterpreting the deterministic reaction rates as probabilities-per-unit-time (or propensities) of a properly defined CTMC.

Table 1 Chemical processes, full system

Event no.	Description	Reset	Propensity
1	P_0 translation by M	$P_0 \to P_0 + 1$	$k_s \cdot M$
2	C_1 formation	$P_0 \to P_0 - 1$ $C_1 \to C_1 + 1$	$a_1 \cdot P_0 \cdot (E_{10} - C_1)$
3	C_1 dissociation	$P_0 \to P_0 + 1$ $C_1 \to C_1 - 1$	$d_1 \cdot C_1$
4	P_1 activation	$P_1 \to P_1 + 1$ $C_1 \to C_1 - 1$	$b_1 \cdot C_1$
5	C_2 formation	$P_1 \to P_1 - 1$ $C_2 \to C_2 + 1$	$a_2 \cdot P_1 \cdot (F_{10} - C_2)$
6	C_2 dissociation	$P_1 \to P_1 + 1$ $C_2 \to C_2 - 1$	$d_2 \cdot C_2$
7	P_1 deactivation	$P_0 \to P_0 + 1$ $C_2 \to C_2 - 1$	$b_2 \cdot C_2$
8	C_3 formation	$P_1 \to P_1 - 1$ $C_3 \to C_3 + 1$	$a_3 \cdot P_1 \cdot (E_{20} - C_3)$
9	C_3 dissociation	$P_1 \to P_1 + 1$ $C_3 \to C_3 - 1$	$d_3 \cdot C_3$
10	P_2 activation	$P_2 \to P_2 + 1$ $C_3 \to C_3 - 1$	$b_3 \cdot C_3$
11	C_4 formation	$P_2 \to P_2 - 1$ $C_4 \to C_4 + 1$	$a_4 \cdot P_2 \cdot (F_{20} - C_4)$
12	C_4 dissociation	$P_2 \to P_2 + 1$ $C_4 \to C_4 - 1$	$d_4 \cdot C_4$
13	P_2 deactivation	$P_1 \to P_1 + 1$ $C_4 \to C_4 - 1$	$b_4 \cdot C_4$
14	P_2 degradation	$P_2 \to P_2 - 1$	$v_d \cdot P_2/(K_d + P_2)$
15	P_2 transformation into P_N	$P_2 \to P_2 - 1$ $P_N \to P_N + 1$	$k_1 \cdot P_2$
16	P_N transformation into P_2	$P_2 \to P_2 + 1$ $P_N \to P_N - 1$	$k_2 \cdot P_N$
17	Nuclear M transcription (Hill)	$M \to M + 1$	$v_s \cdot {K_I}^n/({K_I}^n + {P_N}^n)$
18	M degradation (Michaelis-Menten)	$M \to M - 1$	$v_m \cdot M/(K_m + M)$

With respect to the ODE systems below reported, the continuous state-variables actually represent a first-order approximation of the expected value of the copy numbers, which are indeed stochastic variables [37]. Section 4.1 treats the QSSAs of the ODE system, whereas Sect. 4.2 concerns the CME-based-stochastic version of the QSSAs.

4.1 QSSA of the Deterministic Model of the PER Circadian Clock

With regard to the sQSSA, its ODE version is written by setting the complex dynamics at steady state:

$$\frac{dC_i}{dt} \approx 0$$

In this way complexes are related to substrate and enzyme concentrations by means of algebraic constraints and, after computations, the dynamics become

$$C_1 = \frac{P_0 E_{10}}{P_0 + K_{1M}} \quad , \quad C_2 = \frac{P_1 F_{10}}{P_1 + K_{2M}} \quad , \quad C_3 = \frac{P_1 E_{20}}{P_1 + K_{3M}}$$
$$C_4 = \frac{P_2 F_{20}}{P_2 + K_{4M}}$$

together with the system of 5 differential equations

$$\begin{cases} \dfrac{dP_0}{dt} = k_s M - \dfrac{V_1 P_0}{P_0 + K_{1M}} + \dfrac{V_2 P_1}{P_1 + K_{2M}} \\ \dfrac{dP_1}{dt} = \dfrac{V_1 P_0}{P_0 + K_{1M}} - \dfrac{V_2 P_1}{P_1 + K_{2M}} - \dfrac{V_3 P_1}{P_1 + K_{3M}} + \dfrac{V_4 P_2}{P_2 + K_{4M}} \\ \dfrac{dP_2}{dt} = \dfrac{V_3 P_1}{P_1 + K_{3M}} - \dfrac{V_4 P_2}{P_2 + K_{4M}} - \dfrac{v_d}{P_2 + K_d} P_2 - k_1 P_2 + k_2 P_N \\ \dfrac{dP_N}{dt} = k_1 P_2 - k_2 P_N \\ \dfrac{dM}{dt} = v_s \dfrac{K_I^n}{K_I^n + P_N^n} - v_m \dfrac{M}{K_m + M}, \end{cases} \tag{12}$$

(where $V_1 = b_1 E_{10}$, $V_2 = b_2 F_{10}$, $V_3 = b_3 E_{20}$, $V_4 = b_4 F_{20}$), which correspond to the equations reported in [25] and in [26] (formula 11.1, p. 476).

As already observed, the total concentration of the protein PER and of the *per* mRNA is not constant, due to the presence of feedback mechanisms and degradation phenomena.

For what concerns the tQSSA, we set the total substrates at a generic time instant t:

$$\overline{P_0} = P_0 + C_1 \quad ; \quad \overline{P_1} = P_1 + C_2 + C_3 \quad ; \quad \overline{P_2} = P_2 + C_4$$

In terms of these new variables, the dynamics of the total substrates are given by

$$\begin{cases} \dfrac{d\overline{P_0}}{dt} = k_s M - b_1 C_1 + b_2 C_2, \\ \dfrac{d\overline{P_1}}{dt} = b_1 C_1 - b_2 C_2 - b_3 C_3 + b_4 C_4, \\ \dfrac{d\overline{P_2}}{dt} = b_3 C_3 - b_4 C_4 - \left(\dfrac{v_d}{\overline{P_2} - C_4 + K_d} + k_1 \right) (\overline{P_2} - C_4) + k_2 P_N, \\ \dfrac{dP_N}{dt} = k_1 (\overline{P_2} - C_4) - k_2 P_N, \\ \dfrac{dC_1}{dt} = a_1 [C_1^2 - (\overline{P_0} + E_{10} + K_{1M}) C_1 + \overline{P_0} E_{10}], \\ \dfrac{dC_2}{dt} = a_2 [C_2^2 - (\overline{P_1} - C_3 + F_{10} + K_{2M}) C_2 + (\overline{P_1} - C_3) F_{10}], \\ \dfrac{dC_3}{dt} = a_3 [C_3^2 - (\overline{P_1} - C_2 + E_{20} + K_{3M}) C_3 + (\overline{P_1} - C_2) E_{20}], \\ \dfrac{dC_4}{dt} = a_4 [C_4^2 - (\overline{P_2} + F_{20} + K_{4M}) C_4 + \overline{P_2} F_{20}], \\ \dfrac{dM}{dt} = v_s \dfrac{K_I^n}{K_I^n + P_N^n} - v_m \dfrac{M}{K_m + M}, \end{cases} \tag{13}$$

Applying also in this case the QSSA $\left(\dfrac{dC_i}{dt} \approx 0 \right)$, we obtain a system of 4 coupled algebraic equations in C_i

$$\begin{cases} C_1^2 - (\overline{P_0} + E_{10} + K_{1M}) C_1 + \overline{P_0} E_{10} = 0, \\ C_2^2 - (\overline{P_1} - C_3 + F_{10} + K_{2M}) C_2 + (\overline{P_1} - C_3) F_{10} = 0, \\ C_3^2 - (\overline{P_1} - C_2 + E_{20} + K_{3M}) C_3 + (\overline{P_1} - C_2) E_{20} = 0, \\ C_4^2 - (\overline{P_2} + F_{20} + K_{4M}) C_4 + \overline{P_2} F_{20} = 0, \end{cases} \tag{14}$$

which bring to the unique biologically consistent solutions

$$\begin{cases} C_1^- = \dfrac{(\overline{P_0} + E_{10} + K_{1M}) - \sqrt{(\overline{P_0} + E_{10} + K_{1M})^2 - 4\overline{P_0} E_{10}}}{2}, \\ C_2^- = \dfrac{(\overline{P_1} - C_3^- + F_{10} + K_{2M}) - \sqrt{(\overline{P_1} - C_3^- + F_{10} + K_{2M})^2 - 4(\overline{P_1} - C_3^-) F_{10}}}{2}, \\ C_3^- = \dfrac{(\overline{P_1} - C_2^- + E_{20} + K_{3M}) - \sqrt{(\overline{P_1} - C_2^- + E_{20} + K_{3M})^2 - 4(\overline{P_1} - C_2^-) E_{20}}}{2}, \\ C_4^- = \dfrac{(\overline{P_2} + F_{20} + K_{4M}) - \sqrt{(\overline{P_2} + F_{20} + K_{4M})^2 - 4\overline{P_2} F_{20}}}{2}, \end{cases} \tag{15}$$

and to the system of 5 differential equations

$$\begin{cases} \dfrac{d\overline{P_0}}{dt} = k_s M - b_1 C_1^- + b_2 C_2^-, \\ \dfrac{d\overline{P_1}}{dt} = b_1 C_1^- - b_2 C_2^- - b_3 C_3^- + b_4 C_4^-, \\ \dfrac{d\overline{P_2}}{dt} = b_3 C_3^- - b_4 C_4^- - \left(\dfrac{v_d}{\overline{P_2} - C_4^- + K_d} + k_1 \right) (\overline{P_2} - C_4^-) + k_2 P_N, \\ \dfrac{dP_N}{dt} = k_1 (\overline{P_2} - C_4^-) - k_2 P_N, \\ \dfrac{dM}{dt} = v_s \dfrac{K_I^n}{K_I^n + P_N^n} - v_m \dfrac{M}{K_m + M}, \end{cases} \tag{16}$$

with initial conditions

$$\overline{P_i}(0) = P_{i0} \, , i = 0, 1, 2 \quad , \quad P_N(0) = P_{N0} \quad , \quad M(0) = M_0 \, .$$

Remark 2 As already observed, the value of P_t is determined by the sum of all the reactants containing the protein PER. Thus, in the full system we have $P_t = P_0 + P_1 + P_2 + P_N + C_1 + C_2 + C_3 + C_4$. In the tQSSA this law is conserved, because we use the concentrations of the total substrates, thus the equation for P_t becomes $P_t = \overline{P_0} + \overline{P_1} + \overline{P_2} + P_N$. In [25], as in several papers applying the sQSSA, the complexes are supposed not only to be constant, but also to be almost zero, in the conservation laws. Thus, in this case, $P_t = P_0 + P_1 + P_2 + P_N$. This is why the complex depletion paradox emerging for the sQSSA is not present for the total, since no violation of the mass conservation law occurs. Thus the tQSSA yields the same asymptotic values for all the reactants (complexes included) as in the full system.

4.2 QSSA of the CME-Based Stochastic Model of the PER Circadian Clock

In agreement with [61] or [63], we treat the state variables in (12) and (16) as discrete copy numbers that increase by one or decrease by one (according to a one-step process approach [37]), with propensity provided by the sum of the production (resp. clearance) rates of the ODE dynamics for the one-step increase (resp. decrease) reaction. In this way, the reset maps associated to the standard and total QSSA are reported in Tables 2 and 3, respectively.

Table 2 Chemical processes, sQSSA

Event no.	Description	Reset	Propensity
1	P_0 translation by M	$P_0 \to P_0 + 1$	$k_s \cdot M$
2	P_1 activation	$P_0 \to P_0 - 1$ $P_1 \to P_1 + 1$	$b_1 \cdot E_{10} \cdot P_0/(P_0 + K_{1M})$
3	P_1 deactivation	$P_0 \to P_0 + 1$ $P_1 \to P_1 - 1$	$b_2 \cdot F_{10} \cdot P_1/(P_1 + K_{2M})$
4	P_2 activation	$P_1 \to P_1 - 1$ $P_2 \to P_2 + 1$	$b_3 \cdot E_{20} \cdot P_1/(P_1 + K_{3M})$
5	P_2 deactivation	$P_1 \to P_1 + 1$ $P_2 \to P_2 - 1$	$b_4 \cdot F_{20} \cdot P_2/(P_2 + K_{4M})$
6	P_2 degradation	$P_2 \to P_2 - 1$	$v_d \cdot P_2/(K_d + P_2)$
7	P_2 transformation into P_N	$P_2 \to P_2 - 1$ $P_N \to P_N + 1$	$k_1 \cdot P_2$
8	P_N transformation into P_2	$P_2 \to P_2 + 1$ $P_N \to P_N - 1$	$k_2 \cdot P_N$
9	Nuclear M transcription (Hill)	$M \to M + 1$	$v_s \cdot {K_I}^n/({K_I}^n + {P_N}^n)$
10	M degradation (Michaelis-Menten)	$M \to M - 1$	$v_m \cdot M/(K_m + M)$

Table 3 Chemical processes, tQSSA

Event no.	Description	Reset	Propensity
1	$\overline{P_0}$ translation by M	$\overline{P_0} \to \overline{P_0} + 1$	$k_s \cdot M$
2	$\overline{P_1}$ activation	$\overline{P_0} \to \overline{P_0} - 1$ $\overline{P_1} \to \overline{P_1} + 1$	$b_1 \cdot C_1$
3	$\overline{P_1}$ deactivation	$\overline{P_0} \to \overline{P_0} + 1$ $\overline{P_1} \to \overline{P_1} - 1$	$b_2 \cdot C_2$
4	$\overline{P_2}$ activation	$\overline{P_1} \to \overline{P_1} - 1$ $\overline{P_2} \to \overline{P_2} + 1$	$b_3 \cdot C_3$
5	$\overline{P_2}$ deactivation	$\overline{P_1} \to \overline{P_1} + 1$ $\overline{P_2} \to \overline{P_2} - 1$	$b_4 \cdot C_4$
6	$\overline{P_2}$ degradation	$\overline{P_2} \to \overline{P_2} - 1$	$v_d \frac{(\bar{P_2} - C_4)}{K_d + (\bar{P_2} - C_4)}$
7	$\overline{P_2}$ transformation into P_N	$\overline{P_2} \to \overline{P_2} - 1$ $P_N \to P_N + 1$	$k_1 \cdot (\overline{P_2} - C_4)$
8	P_N transformation into $\overline{P_2}$	$\overline{P_2} \to \overline{P_2} + 1$ $P_N \to P_N - 1$	$k_2 \cdot P_N$
9	Nuclear M transcription (Hill)	$M \to M + 1$	$v_s \cdot {K_I}^n/({K_I}^n + {P_N}^n)$
10	M degradation (Michaelis-Menten)	$M \to M - 1$	$v_m \cdot M/(K_m + M)$

5 Numerical Simulations

We here provide simulation results for the PER circadian clock model, following both the deterministic and the stochastic approach and the full/sQSSA/tQSSA formulations presented in the previous sections.

In the deterministic approach, the PER circadian clock confirms the superior reliability of the total QSSA with respect to the standard QSSA. Indeed, although both approximations provide stable limit cycles, according to meaningful parameter settings, only the total is able to catch with very good precision the limit cycle emerging from the full system, see Fig. 1, where the phase-plane of M versus P_t is reported.

Following the stochastic approach, simulations are obtained by means of an ergodic implementation of the Gillespie algorithm [49]. Although strict periodicity is prevented in the stochastic framework, random paths approximately resemble the corresponding deterministic limit cycles, see e.g. Fig. 2 referred to the full system.

In spite of the noisy oscillations exhibited by the random paths, stochastic convergence is expected with respect to the probability distributions. For simplicity, we restrict our attention to the stationay behaviour of species M and P_t.

Figures 3 and 4 provide two different (3D and planar) views of the stationary joint probability distribution M vs. P_t, confirming the better matching of tQSSA to the full system, compared to the sQSSA (in spite of the same computational complexity of the underlying CTMC model), also in the stochastic framework.

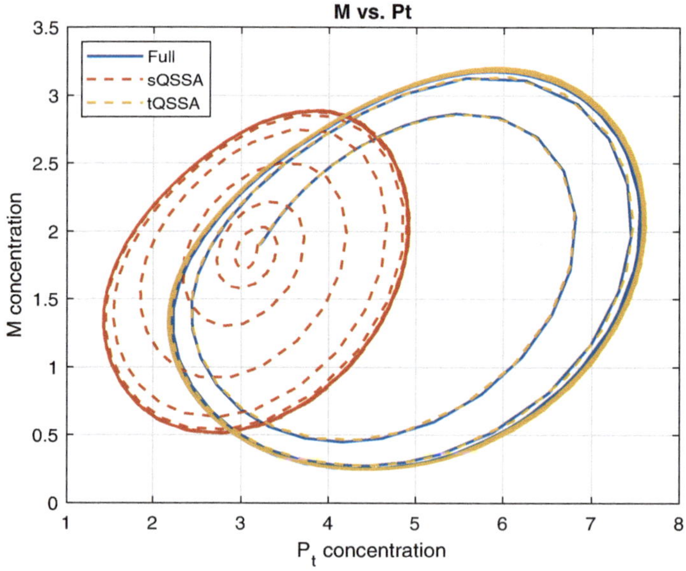

Fig. 1 Trajectories in the phase-plane M-versus-P_t for the full, sQSSA and tQSSA deterministic systems

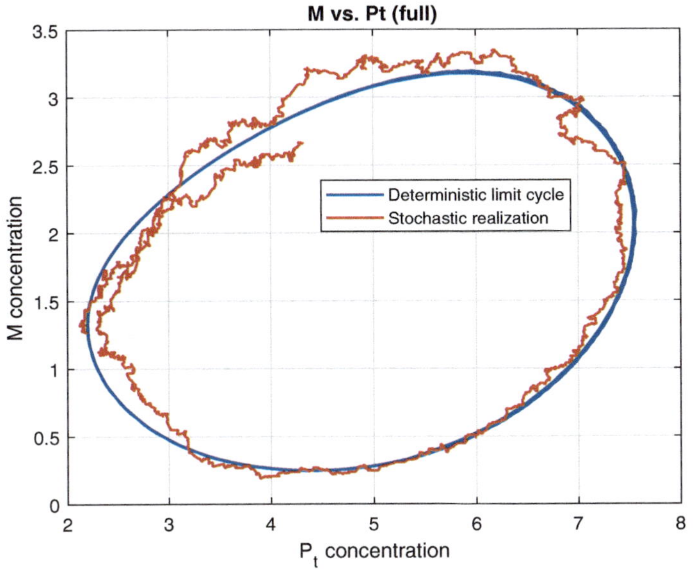

Fig. 2 Random path compared to the deterministic limit cycle (full system)

Note that the discrete states of the Markov chains are not visited with the same frequency/probability (which would result in a uniform distribution), meaning that along each noisy (approximately periodic) random trajectory, transitions occur with different propensities (or average rates).

The 3D plot in the top panel of Fig. 3 reveals the presence of two modes in the joint equilibrium distribution of the full system. Multimodality is an important feature of the stochastic models which was deeply investigated, among others, by Gandolfi and collaborators (see, for example [65, 66]), in a different biological context. We would like to stress that in the sQSSA there is a substantial loss of bimodality (see middle panel of Fig. 3), whereas this is preserved by the tQSSA, as shown in the bottom panel of the same figure. The fact that the sQSSA shows substantially unimodality was already observed in the stochastic treatment of the double phosphorylation-dephosphorylation cycle, in [64]. Here we show in particular that in some cases the sQSSA produced monostability in the deterministic framework and unimodality in the stochastic one, while both the full system and the tQSSA yielded bistability and bimodality in the deterministic and stochastic setting respectively. The results published in [64] show moreover that, in some cases, deterministic bistability does not provide a stationary bimodal distribution. It is well known that deterministic models could produce misleading results if not properly accounted for in the wider setting of the stochastic approach, according to which ODE models can be thought of as a first-order approximation of the average dynamics. The disappearance of one mode in the stationary bimodal distribution could occur when the probability of leaving one basin of attraction is so low that

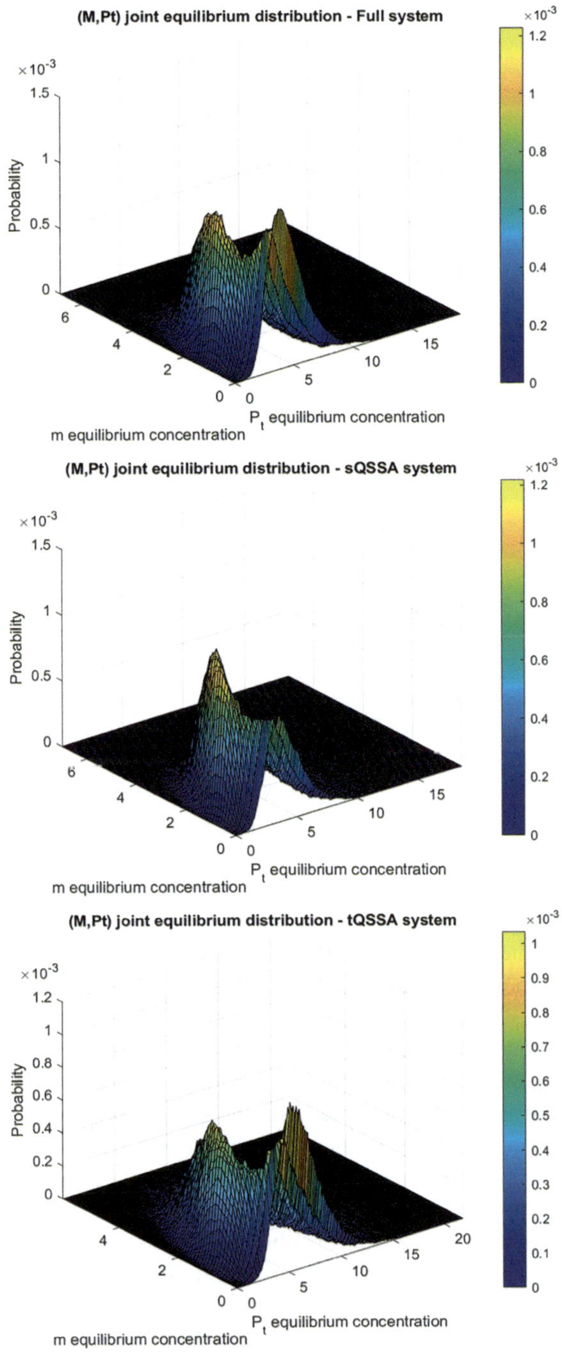

Fig. 3 Joint distribution M vs. P_t for full system (top panel), sQSSA system (middle panel), tQSSA system (bottom panel)

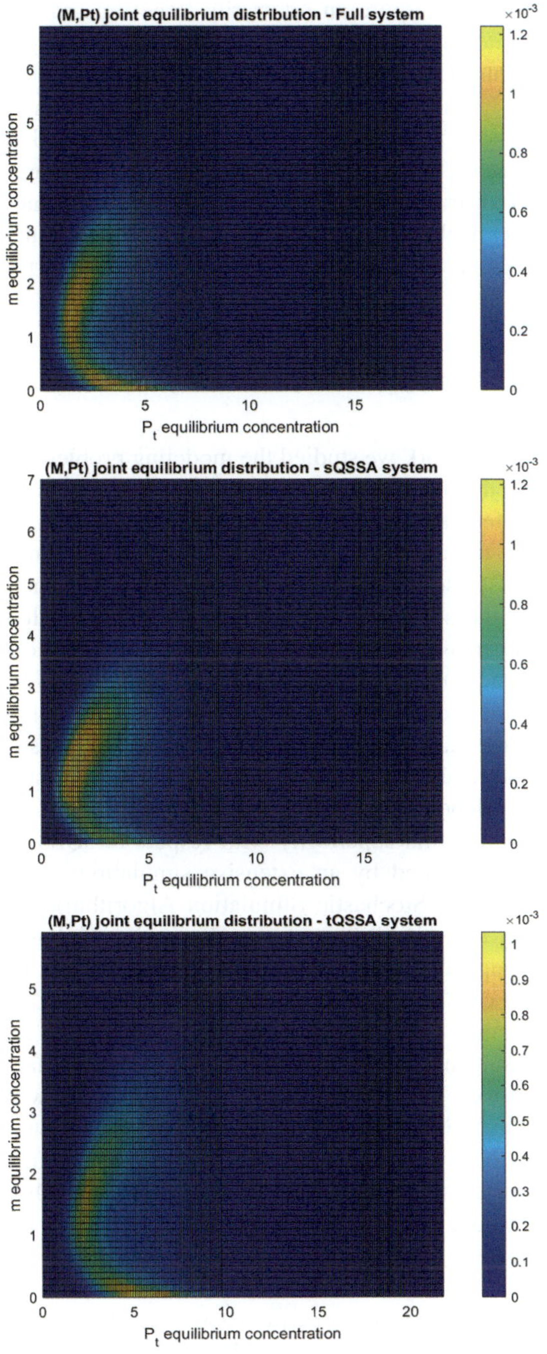

Fig. 4 Joint distribution (rotated view) M vs. P_t for full system (top panel), sQSSA system (middle panel), tQSSA system (bottom panel)

it might require an arbitrarily large time to occur. In this case, we can expect a switching phenomenon from one state to the other one after a very long time, in contrast with what occurs in the deterministic case.

Further qualitative explorations can be inferred by the banana-shaped distributions in Fig. 4, in particular the non-trivial correlations among the considered species, with a qualitative transition from unimodal to bimodal behavior in the M distribution conditioned to increasing values of P_t, with the two modes progressively drifting apart for larger P_t concentrations, up to the far ends of the (banana-shaped) non-trivial portions of the distributions.

6 Conclusions

In this work we studied the modeling problem of the biochemical reaction network of the PER circadian clock, following a discrete stochastic approach, which, differently from a purely deterministic framework, is able to jointly account for the discrete nature of the underlying biochemical species and to characterize the noisy oscillations of the considered reactions in addition to the mean-field behavior. In the stochastic formulation, deterministic long-term oscillations (limit cycles) are translated into noisy random paths, but convergence is preserved in terms of (unique) stationary probability distributions. In order to overcome the curse of dimensionality of the discrete stochastic model, formulated in terms of Continuous-Time Markov Chains (CTMCs), stochastic versions of the well-established standard and total quasi-steady-state approximations (sQSSA and tQSSA, respectively) are developed. In spite of the equal complexity in terms of species and reactions, tQSSA confirms its superiority with respect to sQSSA also in the stochastic framework, as confirmed by an extensive simulation campaign (performed by applying the Gillespie Stochastic Simulation Algorithm), with a very accurate matching with the original (full) system. Qualitative biological insights are inferred from the sampled joint stationary probability distributions regarding the correlations among the considered species. In particular, it can be observed that the discrete states of the Markov chains are not visited with the same frequency/probability (which would result in a uniform distribution), meaning that along each noisy (approximately periodic) random trajectory, transitions occur with different propensities (or average rates). Further studies will be devoted to the study of the full PER-TIM model and to the extension of the stochastic approach by using stochastic differential equations (for example, the Langevin equation, which however must be used with caution, because this approximation could fail, due the unbounded nature of Gaussian white noise).

Acknowledgments Alessandro Borri is member of the Gruppo Nazionale per l'Analisi Matematica e le sue Applicazioni (GNAMPA) of the Istituto Nazionale di Alta Matematica (INdAM, Italy). Alberto Maria Bersani is member of the Gruppo Nazionale per la Fisica Matematica (GNFM) of the Istituto Nazionale di Alta Matematica (INdAM, Italy). Gabriella Mavelli is member of SYSBIO - Centre of Systems Biology, since 2012.

References

1. Henri, V.: Recherches sur la loi de l'action de la sucrase. C. R. Hebd. Acad. Sci. **133**, 891–899 (1901)
2. Henri, V.: Über das gesetz der wirkung des invertins. Z. Phys. Chem. **39**, 194–196 (1901)
3. Henri, V.: Théorie générale de l'action de quelques diastases. C. R. Hebd. Acad. Sci. **135**, 916–919 (1902)
4. Michaelis, L., Menten, M.: Kinetics of invertase action. Biochem. Z **49**, 333–369 (1913) . Translation by Roger S. Goody, and Kenneth A. Johnson
5. Briggs, G.E., Haldane, J.B.S.: A note on the kinetics of enzyme action. Biochem. J. **19**, 338–339 (1925)
6. Lin, C.C., Segel, L.A.: Mathematics applied to deterministic problems in the natural sciences. Society for Industrial and Applied Mathematics SIAM, Philadelphia (1988)
7. Segel, L.: On the validity of the steady state assumption of enzyme kinetics. Bull. Math. Biol. **50**, 579–593 (1988)
8. Segel, L.A., Slemrod, M.: The quasi-steady state assumption: a case study in perturbation. SIAM Rev. **31**, 446–477 (1989)
9. Bersani, A.M., Bersani, E., Dell'Acqua, G., Pedersen, M.G.: New trends and perspectives in nonlinear intracellular dynamics: one century from Michaelis-Menten paper. Continuum Mech. Therm. **22**, 659–684 (2015)
10. Laidler, K.J.: Theory of the transient phase in kinetics, with special reference to enzyme systems. Can. J. Chem. **33**, 1614–1624 (1955)
11. Borghans, J., de Boer, R., Segel, L.: Extending the quasi-steady state approximation by changing variables. Bull. Math. Biol. **22**, 43–63 (1996)
12. Tzafriri, R.A.: Michaelis-Menten kinetics at high enzyme concentrations. Bull. Math. Biol. **65**, 1111–1129 (2003)
13. Dell'Acqua, G., Bersani, A.M.: A perturbation solution of Michaelis-Menten kinetics in a total framework. J. Math. Chem. **50**, 1136–1148 (2012)
14. Bersani, A.M., Dell'Acqua, G.: Is there anything left to say on enzyme kinetic constants and quasi-steady state approximation? J. Math. Chem. **50**, 335–344 (2012)
15. Pedersen, M.G., Bersani,A.M., Bersani, E.: Quasi-steady state approximations in intracellular signal transduction – a word of caution. J. Math. Chem. **43**, 1318–1344 (2008)
16. Flach, E. H., Schnell, S.: Use and abuse of the quasi-steady-state approximation. IEE Proceed. Syst. Biol. **153**, 187–191 (2006)
17. Ciliberto, A., Capuani, F., Tyson, J.J.: Modeling networks of coupled enzymatic reactions using the total quasi-steady state approximation. PLoS Comput. Biol. **22**, 463–472 (2007)
18. Sabouri-Ghomi, M., Ciliberto, A., Kar, S., Novak, B., Tyson, J.J.: Antagonism and bistability in protein interaction networks. J. Theor. Biol. **250**, 209–218 (2008)
19. Erneux, T., Goldbeter, A.: Rescue of the quasi-steady-state approximation in a model for oscillations in an enzymatic cascade. SIAM J. Appl. Math. **67**(2), 305–320 (2006)
20. Kim, J.K., Tyson, J.J.: Misuse of the Michaelis–Menten rate law for protein interaction networks and its remedy. PLOS Comput. Biol. **16**(10), e1008258 (2020)
21. Dunlap, J.C.: Molecular bases for circadian clocks. Cell **96**, 271–290 (1999)
22. Young, M.W., Kay, S.A.: Time zones: a comparative genetics of circadian clocks. Nat. Rev. Genet. **2**, 702–715 (2001)
23. Williams, J.A., Sehgal, A.: Molecular components of the circadian system in drosophila. Ann. Rev. Physiol. **63**, 729–755 (2001)
24. Reppert, S.M., Weaver, D.R.: Molecular analysis of mammalian circadian rhythms. Ann. Rev. Physiol. **63**, 647–676 (2001)
25. Goldbeter, A.: A model for circadian oscillations in the Drosophila period protein (PER). Proc. R. Soc. Lond. Ser. B **261**, 319–324 (1995)
26. Goldbeter, A.: Biochemical Oscillations and Cellular Rhythms: The Molecular Bases of Periodic and Chaotic Behavior. Cambridge University Press, Cambridge (1996)

27. Leloup, J.-C., Goldbeter, A.: A model for circadian rhythms in Drosophila incorporating the formation of a complex between the PER and TIM proteins. J. Biol. Rhythms **13**, 70–87 (1998)
28. Leloup, J.-C., Gonze, D., Goldbeter, A.: Limit cycle models for circadian rhythms based on transcriptional regulation in Drosophila and Neurospora. J. Biol. Rhythms **14**, 433–448 (1999)
29. Leloup, J.-C., Goldbeter, A.: Modeling the molecular regulatory mechanism of circadian rhythms in Drosophila. BioEssays **22**, 84–93 (2000)
30. Ueda, H.R., Hagiwara, M., Kitano, H.: Robust oscillations within the interlocked feedback model of Drosophila circadian rhythm. J. Theor. Biol. **210**, 401–406 (2001)
31. Smolen, P., Baxter, D.A., Byrne, J.H.: Modeling circadian oscillations with interlocking positive and negative feedback loops. J. Neurosci. **21**, 6644–6656 (2001)
32. Forger, D.B., Peskin, C.S.: A detailed predictive model of the mammalian circadian clock. Proc. Natl. Acad. Sci. USA **100**(25), 14806–14811 (2003)
33. Leloup, J.-C., Goldbeter, A.: Toward a detailed computational model for the mammalian circadian clock. Proc. Natl. Acad. Sci. USA **100**(12), 7051–7056 (2003)
34. Leloup, J.-C., Goldbeter, A.: Modeling the mammalian circadian clock: sensitivity analysis and multiplicity of oscillatory mechanisms. J. Theor. Biol. **230**(4), 541–62 (2004)
35. Becker-Weimann, S., Wolf, J., Herzel, H., Kramer, A.: Modeling feedback loops of the Mammalian circadian oscillator. Biophys. J. **87**(5), 3023–3034 (2004)
36. Barkai, N., Leibler, S.: Circadian clocks limited by noise. Nature (London) **403**, 267–268 (2000)
37. Van Kampen, N.G.: Stochastic Processes in Physics and Chemistry, 3rd edn. North Holland, Amsterdam (2007)
38. Gonze, D., Halloy, J., Goldbeter, A.: Robustness of circadian rhythms with respect to molecular noise. Proc. Natl. Acad. Sci. **99**, 673–678 (2002)
39. Gonze, D., Halloy, J., Goldbeter, A.: Deterministic versus stochastic models for circadian rhythms. J. Biol. Phys. **28**, 637–653 (2002)
40. Gonze, D., Halloy, J., Goldbeter, A.: Emergence of coherent oscillations in stochastic models for circadian rhythms. Physica A **342**, 221–233 (2004)
41. Gonze, D., Goldbeter, A.: Circadian rhythms and molecular noise. Chaos **16**, 026110 (2006)
42. Gonze, D., Halloy, J., Leloup, J.-C., Goldbeter, A.: Stochastic models for circadian rhythms: effect of molecular noise on periodic and chaotic behaviour. C. R. Biol. **326**, 189–203 (2003)
43. Denault, D.L., Loros, J.J., Dunlap, J.C.: WC-2 mediates WC-1-FRQ interaction within the PAS protein-linked circadian feedback loop of Neurospora. EMBO J. **20**, 109–117 (2001)
44. Cheng, P., Yang, Y., Heintzen, C., Liu, Y.: Coiled-coil domain-mediated FRQ-FRQ interaction is essential for its circadian clock function in Neurospora. EMBO J. **20**, 101–108 (2001)
45. Zhang, R., Gonze D.: Stochastic simulation of a model for circadian rhythms in plants. J. Theor. Biol. **527**, 110790 (2021)
46. Reppert, S.M., Weaver, D.R.: Coordination of circadian timing in mammals. Nature (London) **418**, 935–941 (2002)
47. Kunz, H., Achermann, P.: Simulation of circadian rhythm generation in the suprachiasmatic nucleus with locally coupled self-sustained oscillators. J. Theor. Biol. **224**, 63–78 (2003)
48. Cao, Y., Gillespie, D.T., Petzold, L.R.: The slow-scale stochastic simulation algorithm. J. Chem. Phys. **122**(1), 14116 (2005)
49. Gillespie, D.T.: Exact stochastic simulation of coupled chemical reactions. J. Phys. Chem. **81**(25), 2340–2361 (1977)
50. Gillespie, D.T.: Approximate accelerated stochastic simulation of chemically reacting systems. J. Phys. Chem. **115**, 1716–1733 (2001)
51. Gillespie, D.T.: Deterministic limit of stochastic chemical kinetics. J. Phys. Chem. B **113**(6), 1640–1644 (2009)
52. Gillespie, D.T.: A diffusional bimolecular propensity function. J. Chem. Phys. **131**(16), 164109 (2009)
53. Gillespie, D.T., Cao, Y., Petzold, L.: Multicale stochastic simulation algorithm with stochastic partial equilibrium assumption for chemically reacting systems. J. Comput. Phys. **206**(2), 395–411 (2005)

54. Mastny, E.A., Haseltine, E.L., Rawlings, J.B.: Two classes of quasi-steady state model reductions for stochastic kinetics. J. Chem. Phys. **127**(9), 094106 (2007)
55. Rao, C.V., Arkin, A.P.: Stochastic chemical kinetics and the quasi-steady state assumption: application to the Gillespie algorithm. J. Chem. Phys. **118**(11), 4999–5010 (2003)
56. Khoo, C., Hegland, M.: The total quasi-steady state assumption: its justification by singular perturbation theory and its application to the chemical master equation. ANZIAM J. **50**, C429–C443 (2008)
57. Szekely, T., Burrage, K.: Stochastic simulation in systems biology. Comput. Struct. Biotechnol. J. **12**(20), 14–25 (2014)
58. Thomas, P., Straube, A.V., Grima, R.: Communication: limitations of the stochastic quasi-steady state approximation in open biochemical reaction networks. J. Chem. Phys. **135**(18), 181103 (2011)
59. Borri, A., Carravetta, F., Mavelli, G., Palumbo, P.: Some results on the structural properties and the solution of the chemical master equation. Proceedings of the 2013 American Control Conference (ACC 2013) Washington, DC, USA, pp. 3777–3782 (2013)
60. Borri, A., Carravetta, F., Mavelli, G., Palumbo, P.: Block-tridiagonal state-space realization of chemical master equations: a tool to compute explicit solutions. J. Comput. Appl. Math. **296**, 410–426 (2016)
61. Bersani, A.M., Borri, A., Carravetta, F., Mavelli, G., Palumbo, P.: Quasi-steady state approximations of the chemical master equation in enzyme kinetics - application to the double phosphorylation/ dephosphorylation cycle. IEEE Conference on Decision and Control (CDC 2014), Los Angeles, CA, USA, pp. 3053–3058 (2014)
62. Dell'Acqua, G., Bersani, A.M.: Bistability and the complex depletion paradox in the double phosphorylation-dephosphorylation cycle. BIOINFORMATICS 2011 - International Conference on Bioinformatics Models, Methods and Algorithms, Rome, pp. 55–65 (2011)
63. Bazzani, A., Castellani, G., Giampieri, E., Remondini, D., Cooper, L.N.: Bistability in the chemical master equation for dual phosphorylation cycles. J. Chem. Phys. **136**(23), 235102 (2012)
64. Bersani, A.M., Borri, A., Carravetta, F., Mavelli, G., Palumbo, P.: On a stochastic approach to model the double phosphorylation/dephosphorylation cycle. M&MoCS J. **8**(4), 261–285 (2020)
65. d'Onofrio, A., Gandolfi, A.: Resistance to antitumor chemotherapy due to bounded-noise-induced transitions. Phys. Rev. E **82**(6), 061901 (2010)
66. d'Onofrio, A., Gandolfi, A., Gattoni, S.: The Norton–Simon hypothesis and the onset of non-genetic resistance to chemotherapy induced by stochastic fluctuations. Physica A **391**(24), 6484–6496 (2012)

What Is a Good Model?

With Examples from Endocrinology

Simona Panunzi, Marcello Pompa, Alessandro Borri, Laura D'Orsi, and Andrea De Gaetano

Abstract Mathematical models are simplified representations of reality, including only a subset of features or behaviors that are of interest. This paper reviews the most distinguishing properties of good models and of correct system design, with a focus on models describing the functioning of a system in terms of subcomponents and relationships (mechanistic models). Ingredients of good models include usefulness, mathematical coherence, statistical robustness, meaningfulness and interpretability, generalization capabilility, parsimony, tractability, falsifiability. Deterministic and stochastic mathematical models are now of widespread use for different goals: analysis or interpretation of the past, prediction of future behavior, numerical simulation, diagnosis of dangerous/critical events, control and decision making. Within the many modern applications of mathematical models, we highlight their increasing role in biology and medicine, and we support our discussion with examples from endocrinology.

1 Introduction

What is a mathematical model? Mathematical models can be defined as the representations of essential aspects of existing systems, systems which are so complicated that a first level of compromise is that of identifying only their most relevant parts. When approaching reality with mathematics we therefore already know that we are only approximating our understanding of the real world. Starting

S. Panunzi · M. Pompa · A. Borri (✉) · L. D'Orsi
CNR-IASI Biomathematics Laboratory, National Research Council of Italy, Rome, Italy
e-mail: simona.panunzi@biomatematica.it; marcello.pompa@biomatematica.it; alessandro.borri@biomatematica.it; laura.dorsi@biomatematica.it

A. De Gaetano
CNR-IASI Biomathematics Laboratory, Institute for Biomedical Research and Innovation (CNR-IRIB), National Research Council of Italy, Palermo, Italy

Department of Biomatics, Óbuda University, Budapest, Hungary
e-mail: andrea.degaetano@biomatematica.it

A. d'Onofrio et al. (eds.), *Problems in Mathematical Biophysics*, SEMA SIMAI Springer Series 38, https://doi.org/10.1007/978-3-031-60773-8_10

from a "conceptual" model, where the knowledge of the system is organized and integrated in a qualitative form on the basis of observations and some hypotheses [15], the next important aspect thus lies in deciding which elements of the system to retain and which to discard in the mathematical description. To cite Einstein: "As far as the laws of mathematics refer to reality, they are not certain; and as far as they are certain, they do not refer to reality". The second question to be answered is: what are mathematical models for? Mathematical models are not only useful to describe what we can perceive with our senses, but also to infer what we cannot perceive or observe, that is for phenomena we can only hypothesize to exist. The objectives mathematical models can achieve can vary. Models can be useful for developing new knowledge: by the quantification of what we know it is possible to point out precisely what we do not know; to test the effects of particular changes in a system: changing some model parameter values, when parameters have explicit physical meaning, could lead to understanding the causes of observed behavior [2]; to make decisions on the basis of the model's predictive ability: in biomedicine, for example, by testing the effect of new drugs or calibrating the right drug dosage to administer on the basis of the predicted organism response. An example is given by the amount of insulin to be delivered by an artificial pancreas on the basis of future predicted glycaemias. Another important use of a mathematical model lies in its interpretative power: by replicating quantitatively the (main features of the) hypothesized structure of a system, if the model predictions disagree with experimental results we know that the hypothesized structure is incorrect. This is in fact the Galileian method, as recently formalized by K. Popper's falsificationism [12].

Mathematical models can take different forms but a first important classification distinguishes mechanistic models from empirical models. A mechanistic model attempts a description of the inner functioning of a system by identifying its (main) sub-components and their relationships. In an empirical model, the mechanisms responsible for the observed behavior are not represented: the model attempts to capture quantitatively only the experimental observations, possibly with some input-output numerical relationship. While empirical models may be useful for interpolation, their use for extrapolation, i.e. for making predictions outside of the range of the available observations, is extremely risky. While, as mentioned above, there may be different purposes for the use of mathematical models, we believe that their usefulness for the investigating scientist resides mainly in their ability to force the investigator to make assumptions explicit and hence to formalize quantitatively the consequences of such assumptions, thereby revealing the possible incompatibility of the assumptions with observed reality. In this respect, models are central to the advancement of science by progressive elimination of untrue hypotheses. Conversely, capitalizing on robustly verified models may make experimental observations unnecessary. Indeed, following the evidence-based medicine approach according to which, for example, two separate prospective randomized clinical trials are required by the FDA to accept that a medical device is effective, Yeh et al. [18] lament the dearth of studies comparing randomized volunteers jumping from aircrafts with either a parachute or an empty backpack. According to standard statistical requirements, therefore, we ought not to be able to assert that a parachute is a life-saving device under the circumstances. The widespread use of parachutes, in fact, has to do with understanding the consequences from a purely physical model of the experience. In this case, at least, a good model makes painful experimental testing unnecessary.

This enjoyable example however focuses our attention on the main point of the present article: what is a good model? If we are to use a model to understand reality, and possibly to help us decide whether or not to invest money, take medications or use a parachute, what criteria should we use to decide that the model we are about to employ exhibits a reasonable degree of reliability? One often quoted aphorism from the statistician George E. P. Box is: "all models are wrong, but some are useful.". While we do agree with this aphorism to the letter, it appears to us that quoting Box has often the effect of justifying the use of **any** model or to make all models equivalent. This is so much more pernicious, because the judgment on the reliability of the model is typically left to the (hopefully technically prepared) modeler, many model users not being able to make rational sense of the squabble of different Authors defending their respective formulations.

Although the goal of a mathematical model is not to be a complete representation of the system under consideration, nevertheless it must be "useful" to help our understanding of the system itself. To be "useful" a model should be "good" and to be "good" it must have some essential characteristics. We list here a few features, which a potentially good model should exhibit.

A model should be *reasonable*, its behavior should not conflict with established knowledge. For example, a model which would predict concentrations to be negative would not be reasonable. The degree to which a model formulation conforms to established knowledge can be investigated by studying its qualitative behavior: existence and unicity of the solutions; positivity and boundedness (where appropriate); boundedness and possibly continuity of the derivatives; continuous dependence of the model output with respect to parameters and initial conditions; periodicity where expected; equilibria, with local and global stability; these are all features that may be deduced mathematically from the structure and properties of the model before considering any experimental observations.

A model should be easily *interpretable*, its functional form should mirror the structure of the system being modeled and its parameters should be clearly meaningful to the domain expert. For example, lack of interpretability of Artificial Intelligence (AI) methods, however successful in classification or recognition, is determining the current trend towards Explainable AI.

A model must be *robust*, that is the identified values of its parameters should not vary substantially for small sampling variations of the observed data from which identification is carried out, nor should they be unduly affected by a small number of missing data.

A model however should be sufficiently *sensitive* to allow detection of systematic differences in behavior (e.g. as determined by the presence of disease or by the effect of treatment), thus being useful in the discrimination between different conditions.

It is clear that robustness and sensitivity are somewhat conflicting requirements: the parameters of a maximally robust model (e.g. an arbitrary constant) will not change irrespective of changes in the dataset, making the identification of the model, if any, of very questionable usefulness. Similarly, the parameters of a maximally sensitive model would change drastically with every small change in the data, also making the model useless. Notice however that some model pathologies can be assessed considering (the lack of) robustness: an extreme example being that of a model where two parameters are linearly dependent, in which case an infinite number of combinations of values of these parameters would produce identical model results and allow the same fit to the available data.

The model parameters ought to be precisely *identifiable* from standard or appropriate experimental data sets affected by the usual levels of noise: the variance of the parameter estimates should be sufficiently small to afford a reasonable level of confidence that the estimated values are approximately correct.

A model should be *parsimonious*: it should not be more complicated than appropriate, which when expressed in quantitative terms suggests that a model should not include too many parameters or predictors. The introduction of a large number of parameters and variables necessarily requires the analysis of all their interactions, as well as the ability to interpret each individual relationship. While it is generally true that adding terms and parameters to a model makes the corresponding goodness of fit better, a more complicated model may lose predictive capacity on data sets different from the one on which fitting has been carried out. In this respect a parsimonious model is typically robust and its parameters estimated precisely.

However, Occam's razor [4, 5] is often blithely invoked to the detriment of structural insight. Suppose we are able to *a-posteriori* identify the parameters of both a simpler and of a more complex model. Stalwart supporters of Occam's razor would say that the simpler model ought to be retained because of simplicity. In fact, the identifiable more complex model reliably offers more information about the inner structure of the modeled phenomenon than the simpler one, and should in general be preferred.

A model should be *tractable*, that is, it should be computable (either analytically or numerically) at reasonable cost and it should be feasible to estimate its parameters and their variability from routine experimental results.

Finally, in accord with Popper's theory, a model should be *falsifiable*, that is it should be possible to design executable experiments capable of proving that the model is wrong (falsification principle). According to our current understanding, this is the very basis of the scientific method.

Mathematical models (deterministic or stochastic) are used to an ever larger extend in many fields, not only in the "hard" sciences but also in the social sciences. In particular, mathematical models are seeing an ever increasing application in biology and medicine. We focus our discussion on dynamical mathematical models in physiology, which integrate experimental observations and theoretical knowledge in a mechanistic framework, with the explicit purpose of helping to understand the structure of the phenomena under study and not just making forecasts. Such models can be used to make diagnoses, determine treatment regimens, assess personalized treatment options or safely test experimental procedures *in-silico*. The considerations that follow refer therefore to this limited interpretation of mathematical modeling in biomedicine.

2 Empirical (Data-Driven) Versus Mechanistic Models

This might be the right place to expound our very personal views on what intelligence is. Children of a mechanistic, scientific culture, we do believe that "thinking" is the result of the physical interconnections among a very large number of signal processing units. In this respect, there would be no *a-priori* impediment to the construction of machines developing an intelligence similar to or greater than human beings. However, technology, while wondrous, is still rather insufficient to provide the necessary numbers, dimensions and level of complexity. If we then add education (the transmission of previous experiences from generation to generation), and the fact that thinking individuals number in the billions on this planet, it appears rather evident that human scientists carry with them a distillate of many observations and associations. While the very concept of causality could be debated in a more philosophical arena than this, for the purposes of the practical modeler we may distinguish between constructions benefiting from this socially inherited very large set of observations, and models attempting to derive connections between, say, predictors and criteria, only from some limited experimental dataset at hand. We call the first models "mechanistic" or "structural" and the second ones "empirical" or "statistical" or sometimes "data-driven". It might be objected that if the 'limited' experimental dataset at hand is composed of millions or billions of automatically obtained observations ('Big Data') then the divide between the two categories ought to blur. In particular, we might surmise that a machine-learning algorithm trained on a specific problem for which millions of training data exist would outperform tradition-based but computationally inefficient human brains, and this is in fact the case. The problem of empirical models, however, even in this extreme case, is extrapolation. An Artificial Intelligence classifier or regressor, same as an interpolating polynomial or spline, may be optimal *within* the training domain, but may (and it usually does) provide nonsensical predictions if we stray from such training domain. In this respect mechanistic models, relying on possibly simplistic structural hypotheses drawn from traditional observation of many diverse but functionally homogeneous systems (think, for example, compartmental systems) have a much better chance to make meaningful extrapolations. Further, insofar as structural models mimic what we believe to be the internal organization of the system under investigation, they provide a piece-by-piece analogic representation of the system's components and allow us to make direct inferences from the abstract model into the workings of such components.

Just to make a simple example of the risks of using empirical models for extrapolation, let suppose to have observed the data of a metabolite that, starting from some steady-state initial value decreases over time after a perturbation, until it reaches a lower plateau value. Possible observations are reported in Fig. 1 as red points from time −40 min (forty minutes before the perturbation) to time 360 min. Using the mathematical solution derived from assuming equilibrium between constant production and linear elimination of the metabolite (elimination reputed to be faster after perturbation), the phenomenon could be interpreted using

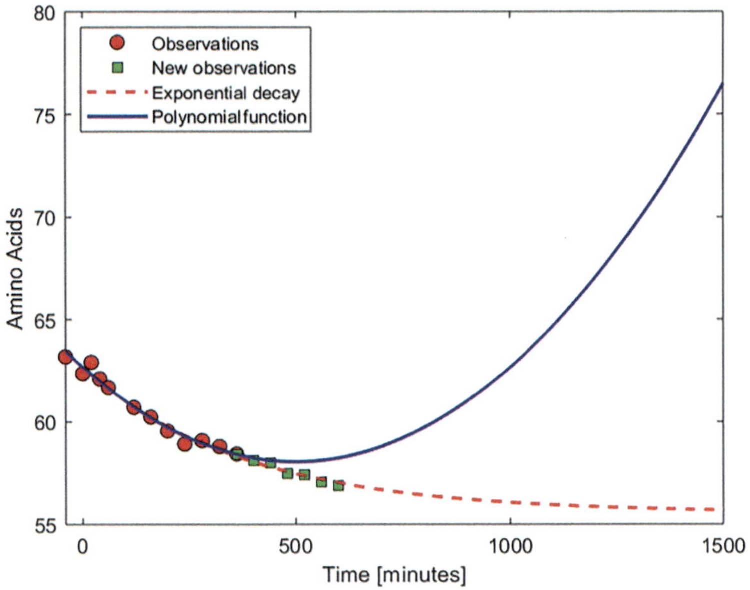

Fig. 1 Empirical versus mechanistic predictions

a decreasing exponential function with three parameters: λ, θ_1, θ_2:

$$y(t) = \theta_2 + (\theta_1 - \theta_2)e^{-\lambda t}. \quad (1)$$

The parameter λ represents the decay rate of the metabolite from its starting value θ_1 towards its lower level θ_2, decreasing its concentration by a delta equal to $(\theta_1 - \theta_2)$. The dashed red line in the figure represents the fitting obtained with the model of Eq. (1) and its predictions at future times. In the same figure, the blue line represents instead the fitting obtained with a second order polynomial function, a common interpolator which makes no assumptions on the mechanics giving rise to the observations:

$$p(t) = \alpha_1 + \alpha_2 t + \alpha_3 t^2. \quad (2)$$

As we can see from the graph, for late times the empirical model of Eq. (2) and the presumed mechanistic model predict very different concentrations. In particular the empirical model would predict a steep rise in concentrations, apparently unjustified by any structural consideration. Should we continue to observe the system and should our structural, mechanistic model be correct, then we would perform more observations which are in agreement with it, like the green squares in the figure. It may however happen that newer observations would not agree with the extrapolated time-course of the structural model: this would be undesirable from the point of view of accurate predictions, but it would be very positive from the point of view of our understanding of how the real underlying system works: in this case new experimental observations would have *falsified* the model, the model would have been proven wrong and we would have learned something about the real world. We could at this point devise another model, which is in accord with all hitherto available observations, and proceed to make new observations with the goal of possibly falsifying it.

In conclusion, empirical models may be of value when attempting to parsimoniously and effectively predict system behavior within a tested range of variable and parameter values. Structural models are of value when trying to understand the inner workings of the physical system and when extrapolation beyond the tested range is desirable, based upon assumed knowledge of such inner workings.

3 Studying a Model

In the present section we will make use of the physiological model developed by Bolie [1] for the representation of the glucose/insulin metabolism. This was the first, or one of the first mathematical models developed with the aim of representing glucose homeostasis. It was based on a system of simple ordinary differential equations, reducing to a minimum the complexity of the representation and limiting the number of functions representing the action of the liver, pancreas and peripheral tissues.

The model equations are reported below:

$$\frac{d\bar{G}(t)}{dt} = -k_1\bar{G}(t) - k_2\bar{I}(t) + \frac{k_3(t)}{V_G}, \qquad \bar{G}(0) = G_0, \tag{3}$$

$$\frac{d\bar{I}(t)}{dt} = -k_4\bar{I}(t) + k_5\bar{G}(t) + \frac{k_6(t)}{V_I}, \qquad \bar{I}(0) = I_0, \tag{4}$$

where:

- the variables $\bar{G}(t)$ and $\bar{I}(t)$ represent glucose and insulin supra-basal plasma concentrations respectively
- the parameter k_1 is the insulin-independent glucose elimination rate; k_2 is the rate of insulin-dependent glucose elimination (combining liver glycogen storage and muscle glucose utilization responding to elevated insulin concentrations); k_4 is the apparent first order insulin disappearance rate (insulin, a peptide, is eliminated from the bloodstream in the same fashion as all other peptides and proteins); k_5 is the sensitivity of pancreatic β-cells to glucose concentrations, whereby insulin secretion is driven by increasing glycemia; V_G and V_I are the apparent volumes of distribution for glucose and insulin respectively;

- the functions $k_3(t)$ and $k_6(t)$ represent the increments of glucose and insulin respectively, due to liver secretion of glucose and to external (enteral or parenteral) glucose intake and parenteral (injective) insulin administration.

When contemplating this model, we may say at the outset that it is both *parsimonious* and *tractable*. The system appears to be reduced at a minimum, with few parameters to be estimated.

We may however wonder whether such model does capture the relevant underlying physiology.

The model, in its original formulation, was designed to represent deviations in glucose and insulin concentrations from their average physiological values, as they occur in intravenous tolerance experiments. Other interpretations of the Bolie model [8] make use of the above equations to represent the change of glucose and insulin concentrations in plasma. Regardless of its interpretation, which determines the range over which the state variables are defined (raw concentrations or concentrations above some physiological value), the focus of our examination here is on the characteristic of being *reasonable*: does the model mirror the fundamental behavior of the system under study?

The major shortcoming of the system reported in the Eqs. (3) and (4) lies in the formulation of the insulin-dependent glucose elimination term $k_2\bar{I}(t)$, which describes the glucose elimination mechanism as dependent only on insulin levels: regardless of the amount of glucose present in the blood stream, as much glucose can be extracted from blood as desired if insulin concentration is sufficiently high. Elevated levels of insulin thus determine high rates of glucose uptake, even in the extreme case where glucose concentrations are close to zero. This mathematical formulation clearly does not conform with what we know of the implicated physiological mechanisms.

Figure 2a and b show the time course of glucose and insulin concentrations following an instantaneous infusion of 128.33 mmol of glucose and 4000 pmol of insulin, with parameters set to the values reported in Table 1, column "Parameter Set 1". The chosen values represent plausible values set on the basis of physiological considerations: k_1 has been set to 0.001 (in the physiological range as reported in [3, 9]); k_2 is approximated by the product of an average glucose concentration compatible with an IVGTT and an insulin sensitivity index of 1×10^{-4}, characteristic of subjects with normal glucose tolerance; k_4, the first-order insulin elimination rate, was set to a value compatible with what has been found in the literature; k_5 was determined by approximating the mean trend of the observations of glycemia and insulinemia resulting from a standard IVGTT on normal subjects (see [13]). Values for V_G and V_I were taken from [10], while G_0 and I_0 are mean blood glucose and insulin values observed during the fasting state in individuals with normal glucose tolerance.

Figure 2c and d are obtained by halving the values of parameters k_4 and k_5 (last column of Table 1). The first parameter produces a slower insulin elimination with a value that still lies in the acceptability range, considering that Bolie set the elimination of insulin at 0.92 h^{-1} which corresponds to a value of 0.015 min^{-1}. A

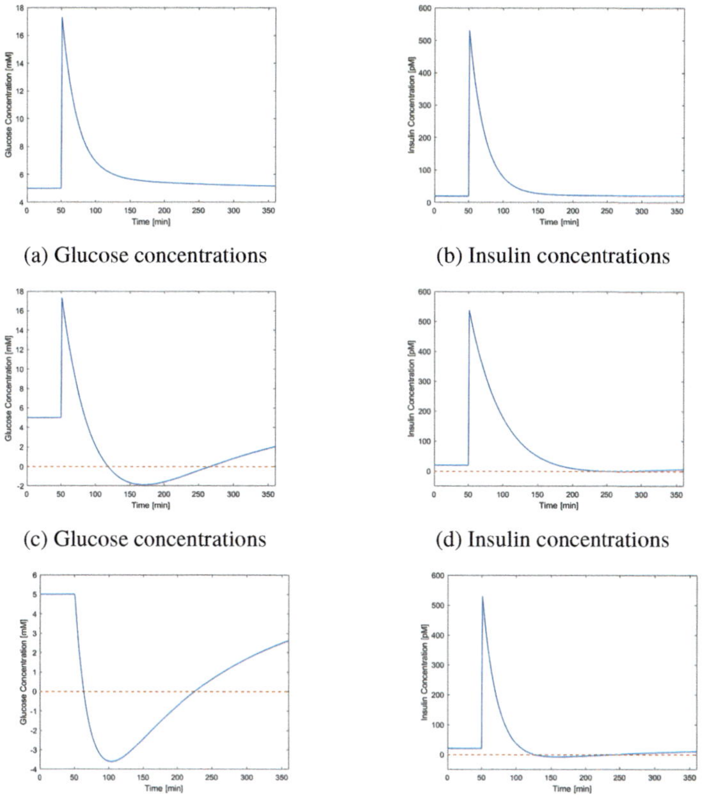

(a) Glucose concentrations (b) Insulin concentrations

(c) Glucose concentrations (d) Insulin concentrations

(e) Glucose concentrations (f) Insulin concentrations

Fig. 2 (a)–(b) Glucose and insulin concentrations over time after IVGTT (128.33 mmol) and IVITT (4000 pmol) with parameters in Table 1, column "Parameter Set 1"; (c)–(d) glucose and insulin concentrations over time after IVGTT (128.33 mmol) and IVITT (4000 pmol) with parameters in Table 1, column "Parameter Set 2"; (e)–(f) Glucose and insulin concentrations over time after IVITT (4000 pmol) with parameters in Table 1, column "Parameter Set 1"

Table 1 Bolie model parameters

Parameters	Parameter set 1	Parameter set 2
k_1 [1/min]	0.001	0.001
k_2 [1/min/pM]	0.001	0.001
k_4 [1/min]	0.05	0.025
k_5 [pM/min/mM]	0.2	0.1
V_G [L]	11.2	11.2
V_I [L]	8.4	8.4
G_0 [mM]	5	5
I_0 [pM]	20	20

lower value of the parameter k_5 produces a lower release of insulin in response to the same circulating glucose concentrations, in part contrasting smaller elimination values. Under these circumstances, we get negative glucose concentrations. Glucose values below zero are also observed when the same insulin infusion used in the previous simulation is administered with the first set of parameter values but in the absence of a glucose infusion (Fig. 2e–f). It is clear that glucose concentrations below zero are inconsistent with reality: finding such negative solutions for glycemia points to the fact that the model does not adequately represent the underlying functioning of the real system.

Some obstinate defenders of the use of insufficiently studied models might object at this point that the model as formulated produces plausible predictions if we restrict it to sufficiently circumscribed state variable and parameter regions. In our opinion such a viewpoint fails to recognize that the qualitatively aberrant behavior of the solutions indicates a fundamental flaw in the comprehension of the system's dynamics. The physiological assumptions (that one can extract as much glucose as desired from plasma without glucose in it) are wrong. Parameter estimates, even if obtained from observations within the circumscribed "model validity" range are going to be misleading because the estimation is carried out on a model with a recognizable structural inconsistency: in the specific example, any estimate of the k_2 parameter would quantify the insulin-dependent extraction rate of glucose *irrespective* of glucose concentrations, a clearly delusional idea. The incorrectness of a model is reflected in the estimate of its parameters of interest. In mechanistic models each parameter represents a particular aspect of the phenomenon, as described in the previous example, but if the model is wrong the resulting physiological reasoning is wrong.

This problem is far from being of mere academic interest: some models in widespread use suffer from such structural weaknesses. This is the case, for example, of the quantification of the insulin sensitivity index derived from historically established and insufficiently challenged models. The structural problems of the model formulation translate in practice in the limited ability to estimate the parameter of interest (insulin sensitivity in this case): the parameter estimates obtained by adapting the model to experimental observations exhibit extreme oscillations, with either such large values as to have no physiological sense or values near zero (the "zero-value" problem being actually discussed as such in the literature), which fail to quantify the reduced, but measurable parameter in the class of subjects (pre-diabetic) where this measurement would be of greatest use [11, 13].

We believe that such blunders (at least the most obvious ones) may be avoided by the careful collaboration between the mathematician deducing the qualitative properties of the model and the physiologist checking such qualitative properties against the accumulated knowledge about the underlying system.

Ideally therefore a model should be mathematically coherent, statistically robust and physiologically meaningful. In other words, the model should exhibit satisfactory properties of its solutions, its parameters should be statistically estimable with sufficient accuracy from data sets obtained from standard experimental procedures, and it should conform to well-established physiological concepts.

4 A-Posteriori Identifiability

In this section we move from rather philosophical (if important!) considerations to a more pragmatic suggestion about how to operate in practice.

Before delving into the details of a possible identification procedure, we should mention that correct statistical reasoning is necessary in order not to fall into parameter estimation pitfalls. One such, which has been thoroughly discussed elsewhere [13], is the mistake of using a "decoupling" procedure, fitting separately the arms of a feedback or control loop. This approach leads to completely erroneous conclusions regarding not only the values of the parameters but also regarding the qualitative nature of the process. Similarly, a bayesian approach which makes use of highly informative *a-priori* distributions on the model parameters, constraining parameters to assume values in a very restricted and predefined domain, often hides structural shortcomings in the model formulation. Oftentimes in these cases, if the parameters were left free to vary in a sufficiently wide admissible domain, values far too high or far too small, inconsistent with their physiological meaning, would be obtained. In the absence of coercing priors, extreme values of the model parameters obtained from fitting the available data would indicate that the model formulation fails to represent intrinsic and important features of the system.

Suppose now we have finally fitted our model to available data and we have found an optimal model parameter (vector). If the sample size is large we might make recourse to asymptotic normality of the parameter estimates, compute the inverse Hessian of the likelihood at the optimum and invoke Cràmer-Rao theorem to deliver a lower bound of the parameter dispersion matrix (which is typically used in the literature *as if* it were the actual dispersion!). Alternatively, we might choose to be more robust and use a Monte Carlo procedure to evaluate the a-posteriori identifiability of the model parameters. This approach however hinges on a few assumptions:

1. that the correct functional form of the model under analysis is known and that errors are normally distributed with zero mean around the model predictions;

$$y_j = f(t_j, \theta) + \varepsilon_j, \ \ \varepsilon_j \sim N(0, \Sigma(j, j)) \tag{5}$$

   where y_j is the observation at time t_j, the errors ε_j, $j = 1, \dots, n$ are assumed to be uncorrelated and where for example $\Sigma(j, j) = \sigma^2[f(t_j, \theta)]^2$.
2. that the point parameter estimate is close to the true parameter value;
3. that the observed variance of the residuals is close to the error variance.

Under these assumptions, the empirical distribution of a large number of parameter estimates obtained from samples randomly generated from the model (with additive error of zero mean and the above variance) will converge to their true distribution as the sample size grows.

In case of high coefficients of variation of the estimates of the model parameters, the use of Bayesian methods offers another way to evaluate parameter a-posteriori identifiability. A possible approach consists in the use of Monte Carlo Markov Chains (MCMC) with a Metropolis-Hastings [14] and Gibbs sampling algorithms [6, 7, 16, 17], with samples from full conditional distributions. Such an approach still requires the specification of a distribution of the observation error with the specification of its variance-covariance structure. The approach also requires the specification of suitable a-priori distributions for the meta-parameters (σ in Eq. (5)) and for the model structural parameters. As an example of the application of a MCMC approach, we can retain the assumption of normally distributed errors, assuming *a-priori* an inverse gamma distribution for the parameter σ^2 ($\sigma^2 \sim \mathrm{GI}(\frac{\nu}{2}, \frac{\tau}{2})$), while the parameter θ may be log-transformed and assumed to be uniformly distributed within reasonable regions. Under the above assumptions, the full conditional distribution of the scale parameter σ^2 is still an inverse gamma while the full conditional distributions of the log-transformed model parameters can be computed up to a proportionality constant. The Metropolis-Hastings algorithm provides samples from the full conditional distribution of the log-transformed parameter vector. At each iteration of the MCMC, a multinormal distribution could be used as proposal distribution with appropriate mean and variance-covariance matrix. At each iteration of the MCMC algorithm the proposal distribution will be centered on the values obtained from the preceding iteration. The MCMC algorithm is implemented by building independent chains of many (say 10,000) samples starting from different parameter values. The first (1000 or 2000) realizations are in general disregarded for the final computations to allow the Markov chains to reach equilibrium and get sufficiently close to the stationary distribution. The obtained empirical distributions of the model parameters will approximate the relative a-posteriori distributions and will be used to obtain credibility intervals or Credibility Regions (CR) from bivariate empirical distributions.

5 Conclusion

The use of mathematical models as a tool for quantitatively describing a complex system, for simplifying its representation and for drawing quantitative inferences on its component mechanisms, is of great and now generally recognized importance for the advancement of knowledge in many fields of research. Even in biomedicine this formal and quantitative approach is becoming indispensable.

We have examined the fundamental distinction between empirical and mechanistic models, and the use to which the latter can be put in the search for an in-depth understanding of real systems.

While it is true that “all models are wrong", in order to be useful a mathematical model must show some properties that cannot be overlooked. Otherwise there is a concrete risk of incurring into misleading and potentially dangerous conclusions when the model is used to make predictions or perform controls. A series of such properties can be proactively established on the model mathematical formulation even before data are analyzed.

The quality and the ultimate validity of a model depend primarily on the correctness of the assumptions on which it is based. The close collaboration and the continuous confrontation and dialogue between mathematicians and field experts is of paramount importance: it helps at the outset to select the appropriate level of simplification; it is then crucial in order to assure conformity of the model behavior to established knowledge.

Rigorous statistical procedures must be followed in the process of parameter estimation: failure to do so exposes the investigator to the risk of attributing values or behaviors to the modeled phenomenon that are very far from reality or that simply do not exist.

Good model construction is clearly a team effort: it benefits from a collaborative, interdisciplinary approach where biologists, mathematicians, statisticians, even engineers contribute their diverse expertise.

References

1. Bolie, V.W.: Coefficients of normal blood glucose regulation. Am. Physiol. Soc. 783–788 (1961)
2. Cobelli, C., Carson, E.: Introduction to Modeling in Physiology and Medicine. Academic Press, Cambridge (2019)
3. Dalla Man, C., Rizza, R.A., Cobelli, C., IEEE Fellow: Meal simulation model of the glucose-insulin system. IEEE Trans. Biomed. Eng. **54**(10), 1740–1749 (2007)
4. de Ockham, G.: Quaestiones et decisiones in quattuor libros Sententiarum Petri Lombardi, i, dist. 27, qu. 2, (K), (1495)
5. Froidmont, L.: Philosophia Christiana de Anima, p. 110 (1649)
6. Gelfand, A.E., Smith, A.F.M.: Sampling-based approaches to calculating marginal densities. J. Am. Stat. Assoc. **85**(410), 398–409 (1990)
7. Gelfand, A.E., Hills, S.E., Racine-Poon, A., Smith, A.F.M.: Illustration of bayesian inference in normal data models using gibbs sampling. J. Am. Stat. Assoc. **85**(412), 972–985 (1990)
8. Hirayama, H., Kasuya, Y., Yagishita, A., Katakura, T.: Evaluation of the bolie model describing blood glucose response for the application to continuous subcutaneous insulin infusion. J. Pharmacobio-Dyn. **8**, 808–818 (1985)
9. Hovorka, R.: Closed-loop insulin delivery from bench to clinical practice. Nat. Rev. Endocrinol. **7**, 385–395 (2011)
10. Hovorka, R., Shojaee-Moradie, F., Caroll, P., Chassin, L., Gowrie, I., Jackson, N., Tudor, R., Umpleby, A., Jones, R.: Partitioning glucose dristribution/transport, disposal, and endogenous production during ivgtt. Am. J. Physiol. **282**, 992–1007 (2002)
11. Panunzi, S., De Gaetano, A., Mingrone, G.: Advantages of the single delay model for the assessment of insulin sensitivity from the intravenous glucose tolerance test. Theor. Biol. Med. Model. **7**, 9 (2013)
12. Popper, K.: Conjectures and Refutations. Routledge and Kegan Paul, London (1963)
13. Panunzi, S., Palumbo, P., De Gaetano, A.: A discrete single delay model for the intra-venous glucose tolerance test. Theor. Biol. Med. Model. **4**, 1–16 (2007)
14. Smith, A.F.M., Roberts, G.O.: Bayesian computation via the gibbs sampler and related markov chain monte carlo methods. J. R. Stat. Soc. Series B (Methodological) **55**(1), 3–23 (1993)
15. Torres, N.V., Santos, G.: The (mathematical) modeling process in biosciences. Front. Genet. **6**(354) (2015)

16. Wakefield, J.: The bayesian analysis of population pharmacokinetic models. J. Am. Stat. Assoc. **91**(433), 62–75 (1996)
17. Wakefield, J.C., Smith, A.F.M., Racine-Poon, A., Gelfand, A.E.: Bayesian analysis of linear and non-linear population models by using the gibbs sampler. J. R. Stat. Soc. Series C (Appl. Stat.) **43**(1), 201–221 (1994)
18. Yeh, R.W., et.al.: Parachute use to prevent death and major trauma when jumping from aircraft: randomized controlled trial. Br. Med. J. (363) (2018)

Within-Host and Between-Hosts Epidemic Dynamics: A Journey with Alberto

Andrea Pugliese

Abstract In the past 20 years several authors have analysed models in which within-host infection dynamics is coupled to between-host transmission. Frequently in such models within-host dynamics was used to build appropriate functions appearing in epidemic models in order to study evolutionary problems (Gilchrist and Sasaki, J. Theor. Biol. 218:289–308, 2002), or to better describe the dynamics of immune boosting and waning (Barbarossa and Röst, J. Math. Biol. 71:1737–1770, 2015). Alberto Gandolfi, Carmela Sinisgalli and I worked at a model (Gandolfi et al., J. Math. Biol. 70:399–435, 2015) in which epidemic dynamics could affect within-host dynamics because the initial viral inoculum depended on the state of the infector, and thus on the overall state of the epidemics; this two-way interaction yields new dynamical patterns, with subsequent epidemic waves in absence of external perturbations.

A shortcoming of that model is the fact that once a host is infected, s/he cannot be reinfected at any stage; Alberto was aware of the problem, and had prepared a manuscript draft in which reinfections were allowed at the beginning of an infection, where they can be relevant for the dynamics. Unfortunately, we were not able to complete the work before Alberto's premature death. Here I show a simple model that takes account the effect of multiple infectious contacts through a system of ODEs with unexpected behaviour, thus showing the potential relevance of the theoretical models analysed with Alberto.

A. Pugliese (✉)
Department of Mathematics, University of Trento, Trento, Italy
e-mail: andrea.pugliese@unitn.it

A. d'Onofrio et al. (eds.), *Problems in Mathematical Biophysics*, SEMA SIMAI Springer Series 38, https://doi.org/10.1007/978-3-031-60773-8_11

1 Introduction

In the past 20+ years several authors have attempted to connect epidemic dynamics with within-host processes of pathogen growth and immune response; in fact, infectiousness will depend on pathogen load, while the length of the infectious and recovery periods will depend on the dynamics of immune response.

Clearly the immunological system is very complex, and even the simplified models (see, e.g. [3, 14]) that have been proposed to outline the infection dynamics, involve several specific aspects of humoral and/or cellular arm of the immune system. However, if a within-host model is only an ingredient of a multi-scale epidemic model, it is natural to look for a simple within-host model that can still represent some relevant features of infection dynamics. The simplest possible within-host model will include two variables: x, level of pathogens, y, level of immunity; and the infected class will be structured by these two variables through a density function $i(t, x, y)$.

A typical assumption is that, within each infected host, x and y are solutions of a simple system of ordinary differential equation $x' = f(x, y)$, $y' = g(x, y)$ with initial values x_0 and y_0 representing the pathogen load and immune level when an infection occurs. If $i(t, x, y)$ represents the density of infected individuals, one obtains the transport equation with loss terms (see, e.g., Chapter 6 in [20])

$$\frac{\partial i}{\partial t} + \frac{\partial}{\partial x}(f(x, y)i(t, x, y)) + \frac{\partial}{\partial y}(g(x, y)i(t, x, y)) = -(\mu + \alpha(x, y) + \gamma(x, y))i \tag{1}$$

where μ represents the natural death rate, α the disease-induced death rate, and γ the recovery rate.

The equation must be complemented by boundary conditions (discussed in the following Sections) that represent the rate of new infections, and will be non-local, i.e. will depend on the overall densities of infected individuals. Models of this type have been named *nested* models (see, for instance, the relative chapter in the book by Martcheva [17]), in the sense that the immunological model describing within-host dynamis yields appropriate values of parameters for the epidemic model, but the epidemic situation does not affect the within-host dynamics.

Some authors [9] asked whether the prevalence of a disease in the population should play a role in influencing the pathogen behavior within individual hosts. A mechanism causing such a feedback was conceived by Dushoff [8], though described only verbally and translated into a model at the between-host scale without a specific within-host model. Feng et al. [9] wrote instead a model having as variable the mean viral load in the population, thus neglecting the differences within infected individuals, and between infected and not.

In the work with Alberto Gandolfi we considered two mechanisms through which a feedback loop (from the scale of between-host transmission to the within-host scale) could occur. In [11] it was assumed that the initial pathogen load (the *inoculum*) depended on the pathogen load of the infectant (and thus on the overall infected density at the moment). In the following years, Alberto was actively working into elaborating the ideas by Dushoff [8] into a consistent and tractable system of PDEs; this work was truncated by the untimely death of Alberto. In the following Sections, I summarise the journey with Alberto and the results reached, and present a simple ODE model inspired by the system at which we were working at the time.

2 Within-Host Models

As mentioned above, an essential ingredient of this approach is a simple within-host model, often based on two variables only, x, level of pathogens, y, level of immunity.

Three very simple within/host models had been proposed by Gilchrist and Sasaki [13], André and Gandon [1], de Graaf et al. [4]. As for pathogen density x, [13] and [1] share the same equation (based on the prey equation in a Lotka-Volterra predator-prey model) $x' = rx - cxy$.

As for the equation for y', in [13] it is assumed that the replication rate of the specific immune cells is proportional to the pathogen load, yielding $y' = axy$, while in [1] replication rate is constant, once the immune system has been stimulated by the pathogen presence, hence obtaining $y' = ay$; this second assumption is perhaps closer to reality and lets the system be explicitly solved.

Summarizing, the equations are (2) for [13] and (3) for [1]:

$$\begin{cases} x' = rx - cxy \\ y' = axy \end{cases} \qquad (2) \qquad\qquad \begin{cases} x' = rx - cxy \\ y' = ay \end{cases} \qquad (3)$$

In Fig. 1 we show the solutions of the two systems for parameter values that yield similar peak infection values. It can in fact be seen (and easily proved) that, in both models, pathogen load $x(t)$ reaches a maximum and then decays towards 0. In system (2), the immune level $y(t)$ increases up to a final value y_∞. On the other hand, in (3) $y(t)$ grows to infinity, as can be seen by the solution $y(t) = y_0 e^{at}$. To avoid this unrealistic feature, one has to truncate the system at some time T when $x(T)$ gets below some threshold value, although this makes the final immune level $y(T)$ somewhat arbitrary.

The model by de Graaf et al. [4] avoids the arbitrariness by choosing an even simpler (and perhaps far from realistic) equation for x: $x' = rx - cy$. It is easy to see that there exists a value T such that $x(T) = 0$, while $x(t)$ becomes negative for $t > T$; thus the system is truncated at time T, yielding a final immune level $y_\infty = y_0 e^{aT}$, a quantity exploited in the analysis in [7].

All three models share two features: the immune level can only increase; the pathogen, after reaching a peak density, is eventually completely cleared, whichever is the initial value of x_0.

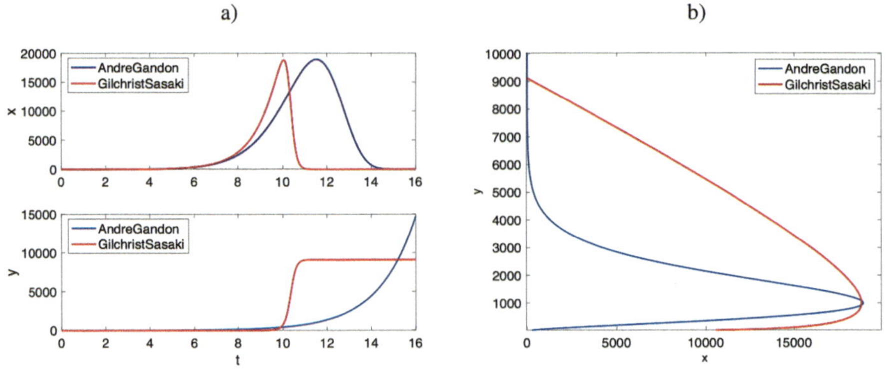

Fig. 1 (**a**) Solutions of system (2) (in red) and (3) (in blue) with x (or y) vs. t. (**b**) Same in the plane (x, y). Parameter values for both panels are $r = 1$, $c = 10^{-3}$, $x(0) = y(0) = 1$: $a = 3.14 \cdot 10^{-4}$ for (2); $a = 0.6$ for (3)

These models also neglect (since they are tailored to acute infections) the decay of specific immune cells, as well as their spontaneous production (at a very low rate). Such features have been condidered by Mohtashemi and Levins [18] and in several models presented by Nowak and May [19]. They all share one feature: either a pathogen is immediately cleared after the infection, or the system will settle to a positive equilibrium, where the pathogen is maintained at low density. It must be, however, remarked that, if the deterministic model is viewed as an approximation of a more realistic stochastic model, it is possible (or perhaps likely) that stochastic extinction of pathogen agents occurs when the pathogen load, as predicted by the deterministic model, is low.

In my first collaboration with Alberto [22], our aim was to slightly modify these models, so as to build a deterministic model with a less clear-cut behaviour: depending on initial values, a pathogen inoculum could give rise to immediate pathogen clerance, or to an acute infection followed by pathogen clearance, or to long-term persistence of the pathogen in presence of immune response.

We modified the equation for pathogens by assuming that, beyond the specific immune response triggered by a pathogen recognition, there exists also an innate immune response, modelled as a constant M. Moreover, we assumed that the pathogen is attacked according to a Holling-type 2 function (and not a bilinear law) by both specific and innate immune.

As for the equation for y' (change in immune response), we keep the spontaneous production and the decay of specific immune cells as in [18], but assume that replication of immune level follows a Michealis-Menten law, so that, at low pathogen density, is almost linear as in (2), but at high pathogen load, is approximately constant, as in (3).

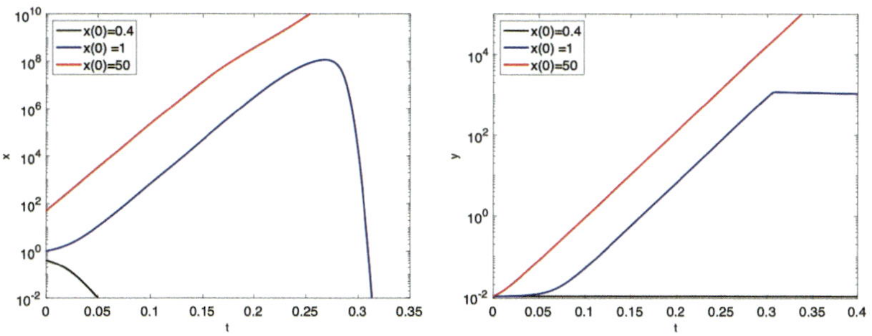

Fig. 2 Solutions of system (4)–(5) with parameter values $c_s = a = \delta = 1$, $r = 85$, $k_s = 10^{-8}$, $k_m = 2$, $c_m M = 200$, $k_a = 0.02$, $h = 0.05$. Left panel: pathogen load x (in log-scale) vs. time t; right panel: immune response y (in log-scale) vs. time t The lines correspond to $x(0) = 0.4$ (black), $x(0) = 1$ (blue), $x(0) = 100$ (red), while $y(0) = 0.05$. Note that the values $c_s = a = \delta = 1$ can always be obtained through appropriate rescalings of the variables

In summary, the equations that we used are

$$x' = rx - \frac{c_s x}{1 + k_s x} y - \frac{c_m x}{1 + k_m x} M \tag{4}$$

$$y' = \frac{ax}{1 + k_a x} y - \delta y + h. \tag{5}$$

From the quantitative point of view, we always assumed $k_m \gg k_s$ so that the innate response saturates at much lower pathogen densities than the specific response.

The model exhibits, according to parameter values, different dynamical behaviours, presented in [22]. One feature particularly interesting is the dependence of solutions of (4)–(5) on the initial value $x(0)$; as shown in Fig. 2, for $x(0)$ below a first threshold value, $x(t)$ decreases monotonically to 0, while $y(t)$ barely changes; for $x(0)$ above this threshold, $x(t)$ initially increases exponentially, with $y(t)$ increasing in response to that; if $x(0)$ is below a second threshold, the immune response controls pathogen growth, so that $x(t)$ reaches a peak and then decreases to 0, leading to infection clearance; when $x(0)$ is too large, both $x(t)$ and $y(t)$ continue growing exponentially, possibly simulating a critical infection.

This feature is particularly important for the immuno-epidemiological model developed in [11], as it fulfills one condition asked by [8] for coupled immuno-epidemiological models, namely that 'the intensity of initial exposure has an effect on immunological outcome'. This may be clearer by looking at the solutions in the phase-plane as in Fig. 3 where the two unstable equilibria of (4)–(5) are also shown.

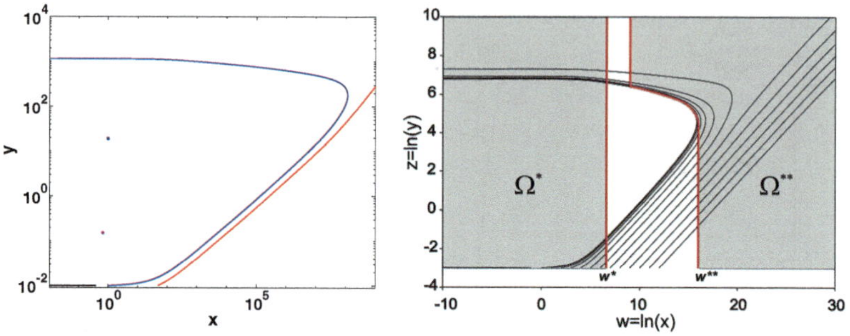

Fig. 3 Left panel: same solutions of (4)–(5) as in Fig. 2, here shown in the phase plane (x, y). The two dots correspond to unstable equilibria of the system, the magenta being a saddle point, while the blue one is completely unstable. Right panel: the regions where $\beta(x, y) \equiv 0$ (Ω^* and Ω^{**}) or equal to $\bar{\beta}$ (the remainder). Adapted from Figure 3 of [11]

3 An Immuno-Epidemiological Model with Dose-Dependent Outcome

In the previous Section, I presented a within-host model in which the outcome depends on the intensity of initial exposure. The other ingredient of the immuno-epidemiological model developed with Alberto and Carmela Sinisgalli [11] is the assumption that the initial pathogen load (the 'inoculum') from which an infection develops depends on the pathogen load of the infector; although we did not know of empirical studies in this sense, it appears natural assuming that, on average, an infected individual with higher pathogen load (and thus a higher concentration in the mucosae) will transfer a higher quantity of pathogens to an individual he/she infects.

We thus assumed that, when an infected individual with pathogen load ξ infects somebody, the 'inoculum' passed to the new infected will have density $p(\cdot, \xi)$; in other words, the probability that the new infected will have pathogen load between x_0 and x_1 is equal to $\int_{x_0}^{x_1} p(x, \xi)\, dx$. Necessarily

$$\int_0^{\infty} p(x, \xi)\, dx = 1 \quad \text{and a natural assumption is } p(x, \xi) \equiv 0 \text{ for } x \in [\xi, +\infty).$$

The boundary condition for (1) is built from this assumption, together with simplifying assumption that all susceptible individuals have the same immune level $\bar{y}$. Note, indeed, that if we assume that system (4)–(5) holds also for susceptible individuals that harbour no pathogens ($x \equiv 0$), the value of y will converge to $\bar{y} = h/\delta$; this will be assumed to be the level of y in all susceptibles.

Equation (1) will hold on $x > 0,\ y > \bar{y}$ while the boundary condition at $y = \bar{y}$ is

$$g(x, \bar{y})i(t, x, \bar{y}) = S(t) \int_0^\infty \int_{\bar{y}}^\infty \beta(\xi, y)i(t, \xi, y)p(x, \xi)\, dy\, d\xi \qquad t,\ x > 0 \tag{6}$$

where $\beta(\xi, y)$ represents the transmission rate of an infected individual whose internal variables are (ξ, y) and $S(t)$ are the susceptibles.

Integrating the right hand side of (6) over x, we have a corresponding exit rate from the susceptible class. Adding a constant recruitment and a constant death rate of susceptibles (equal to the natural death rate used in (1)), we obtain the equation

$$S'(t) = \Lambda - \mu S(t) - \lambda(t)S(t) \tag{7}$$

where the force of infection $\lambda(t)$ is given by

$$\lambda(t) = \int_0^\infty \int_{\bar{y}}^\infty \beta(\xi, y)i(t, \xi, y)\, dy\, d\xi. \tag{8}$$

It could be argued that it would be more appropriate dividing $i(t, \xi, y)$ by total population size $N(t)$ both in (6) and in (8), according to the 'true' mass-action law, as defined in [5]. In case $N(t)$ were constant, the denominator can simply be included in the term $\beta(\xi, y)$; if $N(t)$ were variable (as it would necessarily happen if the disease-induced death rate $\alpha(x, y) \not\equiv 0$), the analysis of the system with the 'true' mass-action law would be more complex. For the sake of simplicity, the system studied will be (1)–(6)–(7) together with appropriate initial conditions.

Actually, we found more convenient changing the variables from x and y to θ (time since infection) and x_0 (value of x at the infection). Precisely, letting $\xi(\theta, x_0)$, $\eta(\theta, x_0)$ be the solutions of

$$\begin{cases} \dfrac{\partial \xi}{d\theta} & = f(\xi, \eta) \\ \dfrac{\partial \eta}{d\theta} & = g(\xi, \eta) \\ \xi(0, x_0) & = x_0 \\ \eta(0, x_0) & = \bar{y} \end{cases} \tag{9}$$

we use as main variable

$$u(t, \theta, x_0) = i(t, \xi(\theta, x_0), \eta(\theta, x_0)) \left|\det \begin{pmatrix} \xi_\theta & \xi_{x_0} \\ \eta_\theta & \eta_{x_0} \end{pmatrix}\right|.$$

A simple calculation shows that $u(t, \theta, x_0)$ and $S(t)$ satisfy the equations

$$\begin{aligned} \frac{\partial u(t,\theta,x_0)}{\partial t} + \frac{\partial u(t,\theta,x_0)}{\partial \theta} &= -\tilde{\alpha}(\theta, x_0) u(t,\theta,x_0) \\ u(t,0,x_0) &= S(t) \int_0^{\infty}\int_0^{\infty} \tilde{\beta}(\theta,x)\tilde{p}(\theta,x_0,x) u(t,\theta,x)\,dx\,d\theta \\ S'(t) &= \Lambda - S(t)\int_0^{\infty}\int_0^{\infty} \tilde{\beta}(\theta,x) u(t,\theta,x)\,dx\,d\theta - \mu S(t) \end{aligned} \tag{10}$$

having defined

$$\begin{aligned} \tilde{\alpha}(\theta, x_0) &= \mu + \alpha(\xi(\theta,x_0), \eta(\theta,x_0)) \\ \tilde{\beta}(\theta, x_0) &= \beta(\xi(\theta,x_0), \eta(\theta,x_0)) \\ \tilde{p}(\theta, x_0, x) &= p(x_0, \xi(\theta, x)). \end{aligned} \tag{11}$$

4 A Numerical Example

Equations (10) have been the basis of the numerical method presented in [11]; the method has been used to simulate a specific example in which it is assumed that a new infected starts with an inoculum $x(0)$ equal to a fixed fraction $\rho \in (0, 1)$ of the infectant pathogen load; namely, the function $p(x_0, \xi) = \delta_{\rho\xi}(x_0)$. Of course, the Dirac's delta is not a proper function, and system was rewritten for this case through an appropriate limiting procedure.

Without going into the details (that can be found in [11]), I simply show here the ingredients (for which Alberto's ideas were crucial) and the results of the simulations. To reduce the number of parameters, it was assumed that the transmission coefficient $\beta(x, y)$ is either 0 (when $x < x^*$) or a constant $\bar{\beta}$ above x^*; furthemore, we also assumed that, when pathogen load went beyond x^{**}, an infected would be isolated, reducing again to 0 the value of $\beta(x, y)$; the assumption is presented graphically in the right panel of Fig. 3, where Ω^* represents the area in which an infected does not transmit the infection, because of low pathogen level, while Ω^{**} represents the region in which infected are isolated.

In the simulations one can then keep track of several quantities, represented in the left panel of Fig. 4: I (the individuals that have ever been infected), R (the individuals that have been infected, but have re-entered the region Ω^* and thus cannot transmit the infection any longer), I_{12} (the infected whose pathogen load is above the threshold of e^{12}, and are thus taken as symptomatic), H (the infected who are in isolation). Indeed, the complexity of the model allows to define, within the same structure, many different classes of infected individuals, as it has often been done in several models of the COVID-19 epidemic, see for instance [12].

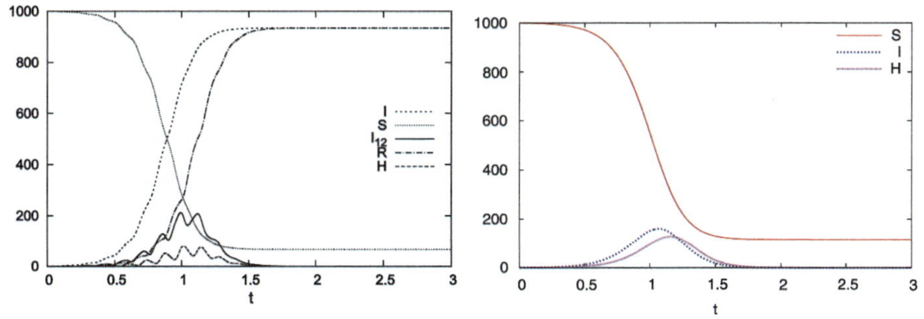

Fig. 4 Left panel: One solution of system (10) with f and g given by (4)–(5). Here $\Lambda = \mu = \alpha = 0$, β is shown in Fig. 4, right panel, $p(x_0, \xi) = \delta_{\rho\xi}(x_0)$; parameter values of (4)–(5) as in Fig. 3; other parameters are $\rho = 10^{-5}$, $\bar{\beta} = 0.035$, $x^* \approx 618$, $x^{**} = e^{16}$. Reprinted from Figure 4 of [11]. Right panel: simulations of an $SEI(H)R$ model where the distributions of the times spent in E, I and Q are as in Figure 8a) of [11], and the contact rate adjusted so as to reach approximately the same value of S_∞ as in the left panel. Reprinted from Figure 9 of [11]

The overall pattern is similar to what can be obtained with a classical SIR model: the susceptible fraction decreases from 1 to a value S_∞ (in the simulation around 5%), while the recovered fraction follows the complementary path; the acutely infected fractions reach a peak and then decline. However the details are rather different from a simple SIR model: in fact, one notices several minor oscillations in the acutely infected fractions.

The structure of the model implicitly subdivides the infection period in a latency period (before transmission starts), in an infectiousness period and, in some cases, in an isolation period, each of which has a non-exponential distribution. One may think that the features seen in the left panel of Fig. 4 depend on these distributions. However, a simulation of an SEI(H)R model, with the distribution of each stage according to the empirical distributions obtained in the simulations of (10), is much more similar to the standard SIR model: see the right panel in Fig. 4. It thus appears that the cause of the oscillations seen in the left panel of Fig. 4 is the feedback mechanism between the within-host and between-host scales

5 Analysis of the Immuno-Epidemiological Model

A definition of R_0 was given in [11] , but without an explicit connection with properties of the system (10). Furthermore, well-posedness of the system and existence of equilibria were not discussed. Here I sketch the main steps in the analysis, without striving for full mathematical rigour.

5.1 Reduction to an Integral Equation

First of all, one can integrate the first of (10) along the characteristics, as standard in equations for age-structured populations [15], obtaining

$$u(t,\theta,x_0) = \begin{cases} B(t-\theta,x_0)\tilde{\pi}(\theta,x_0) & t>\theta \\ \tilde{u}_0(\theta-t,x_0)\tilde{\pi}(\theta,x_0) & 0<t<\theta \end{cases} \tag{12}$$

with

$$B(t,x_0) = u(t,0,x_0) \quad \tilde{\pi}(\theta,x_0) = \exp\{-\int_0^\theta \tilde{\alpha}(s,x_0)\,ds\}, \tag{13}$$

while the second line in (12) depends on the initial condition

$$u(0,\theta,x_0) = u_0(\theta,x_0) \text{ having defined } \tilde{u}_0(s,x) = \frac{u_0(s,x)}{\tilde{\pi}(s,x)}.$$

Using the second of (10), one then obtains an integral equation for $B(t,x_0)$. It is convenient writing this as an abstract equation.

Thus we look for $B \in C([0,T];L^1(\mathbb{R}_+))$, defined as $B(t)(x_0)$ satisfying

$$B(t) = S(t)\left(\int_0^t A(\theta)B(t-\theta)\,d\theta + F(t)\right) \tag{14}$$

where $A(\theta) \in \mathcal{L}\left(L^1(\mathbb{R}_+)\right)$ and $F(t) \in L^1(\mathbb{R}_+)$ are defined by

$$(A(\theta)\varphi)(x_0) = \int_0^\infty \tilde{p}(\theta,x_0,x_0')\tilde{\beta}(\theta,x_0')\tilde{\pi}(\theta,x_0')\varphi(x_0')\,dx_0' \tag{15}$$

and

$$F(t)(x_0) = \int_t^\infty\int_0^\infty \tilde{p}(\theta,x_0,x_0')\tilde{\beta}(\theta,x_0')\tilde{\pi}(\theta,x_0')\tilde{u}_0(\theta-t,x_0')\,dx_0'\,d\theta. \tag{16}$$

Clearly, these definitions require some assumptions.

A simple minimal assumption is

(H1) $\beta(\cdot,\cdot) \in L^\infty(\mathbb{R}_+,\mathbb{R}_+)$ with $0 \le \beta(x,y) \le \beta_M$, $\alpha(x,y) \ge 0$, $\mu > 0$, $p(\cdot,\xi) \in L^1(0,\infty)$ with $p(\cdot,\cdot) \ge 0$ and $\int_0^\infty p(x,\xi)\,dx = 1$ a.e. in $\xi \in (0,+\infty)$, $u_0 \in L^1(\mathbb{R}_+ \times \mathbb{R}_+)$ with $u_0 \ge 0$.

Under (H1), it is easy to see that $0 \le \tilde{\beta}(\theta,x_0') \le \beta_M$, $0 \le \tilde{\pi}(\theta,x_0') \le e^{-\mu\theta}$, so that

$$\|A(\theta)\| \le \beta_M e^{-\mu\theta}. \tag{17}$$

Furthermore

$$\tilde{\pi}(\theta, x_0')\tilde{u}_0(\theta - t, x_0') \leq e^{-\mu t} u_0(\theta - t, x_0)$$

so that $\|F(t)\|_1 \leq \beta_M e^{-\mu t} \|u_0\|_1$.

The renewal Eq. (14) must be coupled with an equation for the scalar $S(t)$, namely the third of (10) that can be rewritten as

$$S'(t) = \Lambda - \mu S(t) - S(t) \left(\int_0^t Q(\theta) B(t - \theta)\, d\theta + G(t) \right) \tag{18}$$

with initial condition $S(0) = S_0 > 0$, where

$$Q(\theta) : L^1(\mathbb{R}_+) \to \mathbb{R} \quad \text{with } Q(\theta)\varphi = \|A(\theta)\varphi\|$$

and

$$G(t) = \int_t^\infty \int_0^\infty \tilde{\beta}(\theta, x_0')\tilde{\pi}(\theta, x_0')\tilde{u}_0(\theta - t, x_0')\, dx_0'\, d\theta.$$

Note that if we let $p(\cdot, \xi)$ to be a measure-valued function (as in the example where, formally, $p(\cdot, \xi) = \delta_{\rho\xi}(\cdot)$), one could look at (14) as an equation for $B(t)$ in the space of Borel measures on $\mathbb{R}_+$ (see, for instance, [10]). This generalization is left for future work.

5.2 *Well Posedness of the System*

Existence and uniqueness of solutions of (14)–(18) can be proved via a fixed point theorem. Let $\bar{S} = \max\{\Lambda/\mu, S_0\}$. Assigned a function $S \in C([0, T]; [0, \bar{S}])$, (14) is a linear problem for $B(\cdot)$, and, since $A(\vartheta)$ are bounded operators, a unique solution $B(S)(\cdot) \geq 0$ exists. Furthermore, one can prove that for any pair $S_1, \ S_2 \in C([0, T]; [0, \bar{S}])$, there exist C and a such that

$$\|B(S_2)(t) - B(S_1)(t)\| \leq Ce^{at} \max_{s \in [0,t]} |S_2(s) - S_1(s)|$$

Then, one defines a map $\tau : C([0, T]; [0, \bar{S}]) \to C([0, T]; [0, \bar{S}])$ as the solution of

$$(\tau(S))'(t) = \Lambda - (\tau(S))(t) \left(\mu + \int_0^t Q(t - \theta) B(S)(\theta)\, d\theta + G(t) \right)$$

with $(\tau(S))(0) = S_0$.

By standard methods, one shows that τ^n is a contraction if n is large enough.

5.3 Existence of Equilibria

In order to look for equilibria of (14)–(18), one needs first to assume that the first of (12) holds for all $\theta > 0$, or in other words $\tilde{u}_0(s, \xi) = B(-s)(\xi)$ for $s, \xi > 0$. Setting then $B(t) \equiv B^*$ and $S(t) \equiv S^*$ in (14)–(18), one obtains

$$B^* = S^* K B^* \qquad \Lambda = S^*(\mu + L B^*) \tag{19}$$

where

$$K = \int_0^\infty A(\vartheta)\, d\vartheta \qquad L = \int_0^\infty Q(\vartheta)\, d\vartheta \tag{20}$$

with the convergence ensured by (17).

One solution of (19) consists in the Disease-Free-Equilibrium (DFE) $E_0 = (\Lambda/\mu, 0)$.

If $B^* \neq 0$, it can be seen from the first of (19) that B^* has to be an eigenvector of K with eigenvalue $1/S^*$. Notice that K is a positive operator and is also compact under weak assumptions on the kernel. Certainly K is compact if (H2) holds.

(H2) there exists $x_M > 0$ such that $p(x, \xi) \equiv 0$ for $x > x_M$.

Since K is positive and compact, Krein-Rutman theorem states that $\rho(K)$ is an eigenvalue, corresponding to a positive eigenvector.

Clearly the same property holds for $\frac{\Lambda}{\mu}K$. Then we can define

Definition 1 The basic reproduction ratio $R_0 = \rho(\frac{\Lambda}{\mu}K)$; let v_0 be the corresponding positive eigenvector with $\|v_0\| = 1$.

Notice that $\dfrac{\Lambda}{\mu}K$ can be seen as the next-generation operator, as defined in [6]. Indeed, a newly infected with 'inoculum' x_0' will produce on average

$$\frac{\Lambda}{\mu}\tilde{p}(\theta, x_0, x_0')\tilde{\beta}(\theta, x_0')\tilde{\pi}(\theta, x_0')$$

new infected individuals of 'inoculum' x_0 at time of infection θ. If φ is the distribution of newly infected, multiplying the previous quantity by $\varphi(x_0')$ and integrating over all x_0' and $\theta \in (0, +\infty)$, we obtain the average density of new infected with 'inoculum' x_0 produced over all the period of infection by the initial distribution. The computation here described produces $(K\varphi)(x_0)$. Therefore, the definition of R_0 as the spectral radius of K is consistent with the definition in [6].

Using this definition in the first of (19), we see that $B^* = c v_0$, while $S^* = \dfrac{\Lambda}{\mu R_0}$.

From the second of (19), using the fact that for $v \geq 0$, $Lv = \|Kv\|$, we obtain $c = \mu\left(1 - \dfrac{1}{R_0}\right)$, which is positive if and only if $R_0 > 1$.

We have thus established

Proposition 1 *If* $R_0 > 1$, *system* (14)–(18) *has a positive equilibrium*

$$E^* = \left(\frac{\Lambda}{\mu R_0}, \mu\left(1 - \frac{1}{R_0}\right) v_0\right).$$

In order to exclude the existence of other equiilibria, it is convenient adding assumptions on the continuity and positivity of the functions in the kernel.

(H3) $p(\cdot, \xi)$ is continuous, uniformly in $\xi \in (0, +\infty)$; β is continuous in $\mathbb{R}_+ \times [\bar{y}, +\infty)$.
(H4) Letting

$$x_m = \inf\{\xi : \exists \tau > 0 \text{ s.t. } \tilde{\beta}(\tau, \xi)\tilde{\pi}(\tau, \xi) > 0\}$$

$x_m < x_M$ and for each pair x and ξ in (x_m, x_M) there exists τ such that $\tilde{p}(\tau, x, \xi)\tilde{\beta}(\tau, \xi)\tilde{\pi}(\tau, \xi) > 0$.

Under (H2) and (H3), it is easy to see that the image of $u \in L^1$ is a continuous function with support in $[0, x_M]$. Necessarily, any eigenvector of K must then be a continuous function with support in $[0, x_M]$.

Consider now K_m the restriction of K to the functions in $C([x_m, x_M])$, i.e.

$$(K_m f)(x) = \int_{x_m}^{x_M} \int_0^{\infty} \tilde{p}(\theta, x, x_0')\tilde{\beta}(\theta, x_0')\tilde{\pi}(\theta, x_0') f(x_0')\, d\theta\, dx_0' \quad x \in ([x_m, x_M]).$$

It follows from (H4) that for any continuous function f with support in $[0, x_M]$, $(K_m f)(x) = (K f)(x) \; \forall\, x \in [x_m, x_M]$.

Because of (H4), K_m is strongly positive; hence the conclusions of Krein-Rutman theorem can be strengthened to stating that K_m has a unique (up to multiples) nonnegative eigenvector, i.e. the positive eigenvector corresponding to $\rho(K_m)$.

As stated above, assumptions (H3)–(H4) imply that (the restriction to $[x_m, x_M]$ of) any eigenvector of K is also an eigenvector of K_m. We can then conclude

Proposition 2 *Assume (H1)–(H4). If* $R_0 > 1$, E^* *is the unique positive equilibrium of system* (14)–(18). *If* $R_0 \le 1$, *there are no positive equilibria.*

5.4 *Stability of the DFE*

Linearizing (14)–(18) around the DFE $E_0 = (\Lambda/\mu, 0)$, one obtain the linear renewal equation

$$B(t) = \frac{\Lambda}{\mu}\left(\int_0^t A(\theta)B(t-\theta)d\theta + F(t)\right). \tag{21}$$

It is well-known (see, for instance, Theorem 0.7 in [21]) that (21) admits a resolvent $R(\cdot) \in L^1(0, +\infty)$ if and only if

$$I - \hat{A}(\lambda) \text{ is invertible for each } \lambda \text{ s.t. } \mathrm{Re}\,(\lambda) \geq 0. \tag{22}$$

When $R(\cdot) \in L^1(0, +\infty)$, a simple computation using the variation of parameters formula [21] shows that, if $\lim_{t\to\infty} F(t) = 0$ (as in our case), then $\lim_{t\to+\infty} B(t) = 0$. Vice versa, if $I - \hat{A}(\lambda)$ is not invertible for some λ with $\mathrm{Re}\,(\lambda) > 0$ [$\mathrm{Re}\,(\lambda) = 0$], it is easy to find functions $F(t)$ such that $|B(t)|$ grows exponentially [is bounded away from 0].

Thus condition (22) is necessary and sufficient for 0 to be an asymptotically stable equilibrium of (21), in the sense that for each function F such that $\lim_{t\to\infty} F(t) = 0$, then $\lim_{t\to+\infty} B(t) = 0$.

The final step is to state (22) in terms of R_0. In order to do that, the following result is needed.

Lemma 1 *Let $A(\theta)$ be an integral operator in $L^1(\mu)$ with nonnegative kernel. Then* (22) *is equivalent to* $\rho\left(\int_0^\infty A(\theta)\,d\theta\right) < 1$.

This is presumably a well-known fact, but I could not find it in the literature; thus, I present a proof in the Appendix.

With the help of this result, one can then state

Proposition 3 0 *is an asymptotically stable equilibrium of* (21) *if and only if* $R_0 < 1$.

In order to obtain the asymptotic stability of E_0 for (14)–(18), one needs a linearization theorem. To this aim, it is probably better seeing (10) as a differential equation in $L^1(\mathbb{R}_+ \times \mathbb{R}_+)$ and use the theory of semigroups with non-densely defined generators [16], as the literature in that area is more complete. It seems likely that, using that theory, it is possible to prove rigorously that E_0 is asymptotically stable [unstable] if $R_0 < 1$ [$R_0 > 1$].

6 A Model with Reinfection of Infected Individuals

In the model presented in the previous Section it is assumed that infected individuals cannot be reinfected.

Several authors [2, 4] present models where reinfections may occur because of immunity waning.

On the other hand, with Alberto we were considering a model in which reinfections of actively infected individuals could affect the dynamics of within-host pathogen-immune interactions. The idea was that, when an infection is in the initial stages, receiving an extra pathogen dose through contact with another infected

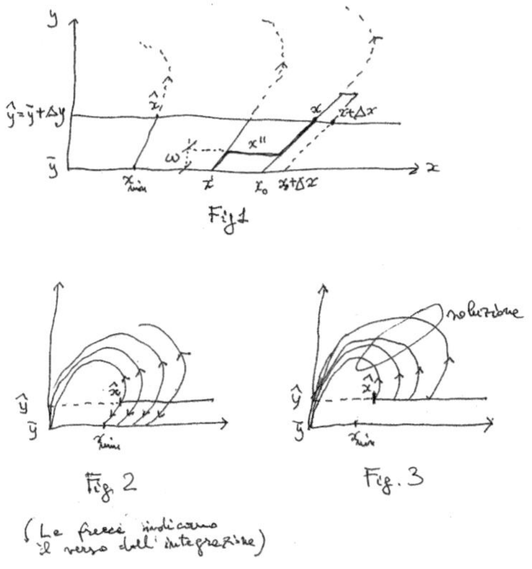

Fig. 5 A hand-drawn figure sketched by Alberto on the effect of a reinfection of an infected individual: an infected with pathogen load x' would reach load x in a small time interval, upon receiving an extra dose x'' in another infectious contact

could change the course of the infection, possibly switching between an infection controlled by the innate immune system to an infection becoming acute.

This would be another pathway through which the epidemic situation in the population could affect the within-host dynamics.

Alberto had proposed a possible model for this, outlined in the hand-drawn sketch shown in Fig. 5, also as a personal memory of him. An assumption was that reinfections are to be considered only when $y < \bar{y} + \Delta y$, where $\Delta y > 0$ has to be calibrated; the rationale is that when the specific immune response has already been primed, an extra inoculum is immediately blocked. He also translated the idea into a PDE system, that was rather complex; unfortunately, my slow answers to his suggestions and his untimely death have stopped this development.

In order to build on his unfinished work, I present here a much simpler model, translating the idea in terms of transitions between compartments. Thus I assume that, upon infection, a Susceptible enters the Exposed compartment; when an individual is in the Exposed compartment, the infection may be controlled by the innate immune response, so that the individual returns Susceptible, or the pathogen may succeed to grow, so that the individual becomes actively Infectious. If a further infectious contact occurs when an individual is in the Exposed compartment, the balance may turn in favour of the pathogen, so that the individual becomes Infectious. Finally, an Infectious individual recovers, and enters the Removed compartment. The scheme is presented in Fig. 6, where the dashed lines represent necessary factors for the transitions.

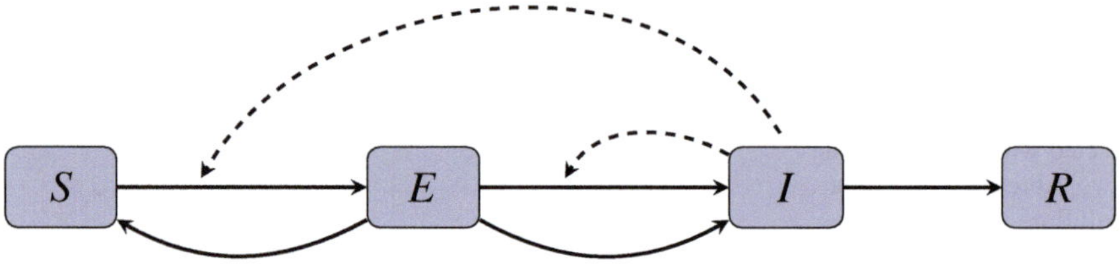

Fig. 6 The transitions between compartments modelled through system (23)

The scheme translates into the following system of ordinary differential equations

$$\begin{cases} S' = \mu - \beta SI + \delta(1-p)E - \mu S \\ E' = \beta SI - \delta E - q\beta EI - \mu E \\ I' = \delta pE + q\beta EI - \gamma I - \mu I \end{cases} . \tag{23}$$

Here S, E, I represent the fraction of the population in the respective class, and it is assumed that the population is in demographic equilibrium, so that the birth rate μ is equal to the mortality rate. The equation for R' is not written, as $R = 1 - S - E - I$.

It is simple to observe that the stability of the DFE $(1, 0, 0)$ depends on the value of

$$R_0 = \frac{p\delta}{\delta + \mu} \frac{\beta}{\gamma + \mu}$$

and that if $R_0 > 1$, when the initial conditions are close to the DFE, $E(t)$ and $I(t)$ initially grow exponentially at rate

$$r = -\frac{\delta + \gamma + 2\mu}{2} + \sqrt{\frac{(\gamma - \delta)^2}{4} + \beta p\delta}.$$

It may be noted that R_0 and r are the same as for an $SEIR$ model with contact rate $p\beta$. This is because the nonlinear term $q\beta EI$ is negligible relative to the linear terms, and does not enter the linearization. However, when the infectious and the exposed fractions grow, the nonlinear term may become dominant. A simulation of the system for a single epidemic ($\mu = 0$) showing this feature is shown in Fig. 7, where (when plotted in logarithmic scale) one can clearly see that the rate of exponential growth increases sharply before the epidemic peak.

Another feature that emerges from the analysis of the system with $\mu > 0$ is the presence of backward bifurcation, so that endemic equilibria may exist for $R_0 < 1$. A detailed analysis of the condition for backward bifurcation, as well as an exploration of the qualitative behaviour of system (23), is left for future work.

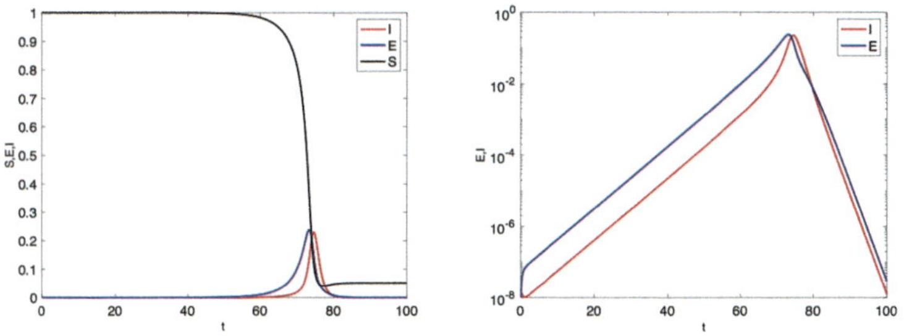

Fig. 7 A simulation of system (23). Parameter values are R0 = 1.5; $p = 0.2$, $q = 0.8$, $\gamma = 1$, $\delta = 0.8$, $\beta = 7.5$, $\mu = 0$, so that $R_0 = 1.5$. Initial values are $I(0) = 1.667 \cdot 10^{-8}$, $S(0) = 1 - I(0)$, $E(0) = 0$. Left panel: S, E and I in linear scale vs. t. Right panel: E and I in logarithmic scale vs. t

7 Concluding Remarks

In this chapter I have presented an overview of immuno-epidemiological models from the viewpoint of my collaboration with Alberto on the topic. He was particularly interested in suggesting and investigating mechanisms by which a feedback could arise from the epidemic to the immunological scale; in other words, whether and how within-host dynamics could depend on the epidemiological state of the population.

I hope I have been able to show how Alberto's ideas can give rise to new developments, even if, unfortunately, he could not complete the work.

Moving from the analysis of system (23), which is little more than a caricature, to properly set and analyse a multi-scale model with reinfection of infected individuals is definitely a big challenge. I hope this will be tackled in the future, possibly through different mathematical approaches.

Appendix

Proof of Lemma 1 I will use repeatedly the following property of compact positive operators.

Proposition ([23], Prop. 3) *If A is a positive compact operator on the ordered Banach space X, then*

$$\rho(A) = \sup\{r > 0 : \exists\, v \geq 0 \; s.t. \; Av \geq rv\}.$$

Now, let $\rho(K) \geq 1$; consider the operators $\hat{A}(\lambda) = \int_0^\infty A(\theta) e^{-\lambda\theta}\, d\theta$ for $\lambda \geq 0$. It is clear that they are all positive compact operators; moreover, if $0 \leq \lambda_1 < \lambda_2$, $0 \leq \hat{A}(\lambda_2)v \leq \hat{A}(\lambda_1)v$ for any $v \geq 0$. It follows that the function $h(\lambda) = \rho(\hat{A}(\lambda))$

is strictly decreasing on $[0, +\infty)$. It is also possible to prove that h is continuous; hence there exists $\bar{\lambda} \geq 0$ such that $h(\bar{\lambda}) = \rho(\hat{A}(\bar{\lambda})) = 1$, i.e. there exists $v > 0$ such that $\hat{A}(\bar{\lambda})v = v$; we thus obtain that (22) does not hold.

Vice versa, assume $\rho(K) < 1$ and assume (22) is violated. Since $\hat{A}(\bar{\lambda})$ is compact, there exists λ with $\mathrm{Re}\,(\lambda) \geq 0$ and $v \in L^1(\mathbb{R}_+)$, $v \neq 0$ such that

$$v = \int_0^\infty A(\theta)e^{-\lambda\theta} v \, d\theta, \text{ i.e. } v(x) = \int_0^\infty \int_{\mathbb{R}} a(\theta, x, y)e^{-\lambda\theta} v(y) \, \mu(dy) \, d\theta \; a.e. \tag{24}$$

Now, since the kernel of the integral operator $a \geq 0$, we have

$$|a(\theta, x, y)e^{-\lambda\theta} v(y)| = a(\theta, x, y)e^{-(\mathrm{Re}\,(\lambda))\theta} |v(y)| \leq a(\theta, x, y)|v(y)|.$$

Then, from (24), we obtain

$$\begin{aligned} |v(x)| &\leq \int_0^\infty \int_{\mathbb{R}} |a(\theta, x, y)e^{-\lambda\theta} v(y)| \, \mu(dy) \, d\theta \\ &\leq \int_0^\infty \int_{\mathbb{R}} a(\theta, x, y)|v(y)|\mu(dy) \, d\theta = (K|v|)(x). \end{aligned} \tag{25}$$

with $|v| \in L^1$ defined as $|v|(x) = |v(x)|$. We have thus obtained $K|v| \geq |v|$, which implies $\rho(K) \geq 1$, reaching a contradiction.

Acknowledgments This work was supported by the Italian Ministry of University and Research (MUR) through the PRIN 2020 project (No. 2020JLWP23) 'Integrated Mathematical Approaches to Socio-Epidemiological Dynamics' (CUP: E15F21005420006).

Andrea Pugliese is member of the "Gruppo Nazionale per l'Analisi Matematica e le sue Applicazioni" (GNAMPA) of the "Istituto Nazionale di Alta Matematica" (INdAM).

I especially thank Alberto D'Onofrio, Antonio Fasano, Federico Papa and Carmela Sinisgalli for having involved me in contributing to this book, that is a deserved homage to the scientific value and very nice and kind personality of Alberto. I also thank Carmela for help in retrieving and adjusting the original figures from [11].

Finally, I thank two anonymous reviewers of the manuscript for their careful reading and constructive comments.

References

1. André, J.-B., Gandon, S.: Vaccination, within-host dynamics, and virulence evolution. Evolution (NY) **60**, 13–23 (2006)
2. Barbarossa, M.V., Röst, G.: Immuno-epidemiology of a population structured by immune status: a mathematical study of waning immunity and immune system boosting. J. Math. Biol. **71**, 1737–1770 (2015)

3. Cao, P., Wang, Z., Yan, A.W.C., et al.: On the role of CD8+ T cells in determining recovery time from influenza virus infection. Front. Immunol. **7** (2016)
4. de Graaf, W.F., Kretzschmar, M.E.E., Teunis, P.F.M., Diekmann, O.: A two-phase within-host model for immune response and its application to serological profiles of pertussis. Epidemics **9**, 1–7 (2014)
5. de Jong, M., Diekmann, O., Heesterbeek, J.A.P.: How does transmission of infection depend on population size? In: Mollison, D. (ed.), Epidemic Models: Their Structure and Relation to Data, pp. 84–94. Cambridge University Press, Cambridge (1995)
6. Diekmann, O., Heesterbeek, J.A.P., Metz, J.A.J.: On the definition and the computation of the basic reproduction ratio R0 in models for infectious diseases in heterogeneous populations. J. Math. Biol. **28**, 365–382 (1990)
7. Diekmann, O., de Graaf, W.F., Kretzschmar, M.E.E., Teunis, P.F.M.: Waning and boosting: on the dynamics of immune status. J. Math. Biol. **77**, 2023–2048 (2018)
8. Dushoff, J.: Incorporating immunological ideas in epidemiological models. J. Theor. Biol. **180**, 181–187 (1996)
9. Feng, Z., Velasco-Hernandez, J., Tapia-Santos, B., Leite, M.C.A.: A model for coupling within-host and between-host dynamics in an infectious disease. Nonlinear Dyn. **68**, 401–411 (2012)
10. Franco, E., Gyllenberg, M., Diekmann, O.: One dimensional reduction of a renewal equation for a measure-valued function of time describing population dynamics. Acta Appl. Math. **175**, 12 (2021)
11. Gandolfi, A., Pugliese, A., Sinisgalli, C.: Epidemic dynamics and host immune response: a nested approach. J. Math. Biol. **70**, 399–435 (2015)
12. Gatto, M., Bertuzzo, E., Mari, L., et al.: Spread and dynamics of the COVID-19 epidemic in Italy: Effects of emergency containment measures. Proc. Natl. Acad. Sci. **117**, 10484–10491 (2020)
13. Gilchrist, M.A., Sasaki, A.: Modeling host-parasite coevolution. J. Theor. Biol. **218**, 289–308 (2002)
14. Handel, A., Longini, I.M., Antia, R.: Towards a quantitative understanding of the within-host dynamics of influenza a infections. J. R. Soc. Interface **7**, 35–47 (2010)
15. Iannelli, M., Milner, F.A.: The Basic Approach to Age-Structured Population Dynamics. Springer, Berlin (2017)
16. Magal, P., Ruan, S.: Theory and Applications of Abstract Semilinear Cauchy Problems. Springer, Berlin (2018)
17. Martcheva, M.: An Introduction to Mathematical Epidemiology. Springer, Berlin (2015)
18. Mohtashemi, M., Levins, R.: Transient dynamics and early diagnosis in infectious disease. J. Math. Biol. **43**, 446–470 (2001)
19. Nowak, M.A., May, R.M.: Virus dynamics: Mathematical principles of immunology and virology. Oxford University Press, Oxford (2000)
20. Perthame, B.: Transport Equations in Biology. Birkhäuser, Basel (2007)
21. Prüss, J.: Evolutionary Integral Equations and Applications. Birkhäuser, Basel (1993)
22. Pugliese, A., Gandolfi, A.: A simple model of pathogen-immune dynamics including specific and non-specific immunity. Math. Biosci. **214**, 73–80 (2008)
23. Schaefer, H.H.: Some spectral properties of positive linear operators. Pac. J. Math. **10**, 1009–1019 (1960)

Diffusion Phenomena on Metric Graphs

Alberto Tesei

Abstract We outline some recent results concerning blow-up of solutions to a model parabolic equation on metric graphs, pointing out the occurrence of the Fujita phenomenon.

1 Introduction

Many natural and social phenomena are studied using *discrete graphs* (e.g., see [23]). Discrete graphs consist of collections of *vertices* and *edges*, an edge being just an ordered couple of vertices (see Definition 5). On the other hand, edges of *metric graphs* (also called *quantum graphs* [5]) are thought of as line segments glued together at the vertices, thus giving rise to spatially continuous networks where dynamics occurs.

Therefore, metric graphs provide the proper framework to model spatial structures of interest in physics and biology, like river basins, cave systems and vascular networks [38–40]. More generally, metric graphs appear with different motivations in several areas of science, including quantum physics, chemistry, engineering, mathematical biology, and domains of mathematics like algebra and number theory (see [1, 5, 10, 30, 34, 38–41] and references therein).

In this general framework, dynamics is typically ruled by partial differential equations stated on the edges, and satisfying boundary conditions at the vertices of the graph. Obviously, the choice of the equations is dictated by the modelling of the phenomenon under consideration. For instance, conservation laws or wave equations on networks have been addressed for traffic flows in [16], or in [12] for flexible strings distributed along a planar graph. Semilinear hyperbolic chemotaxis models, motivated by tissue-engineering scaffolds used for improving wound

---

A. Tesei (✉)
Dipartimento di Matematica "G. Castelnuovo", Università Sapienza di Roma, Roma, Italy

Istituto per le Applicazioni del Calcolo "M. Picone", CNR, Roma, Italy
e-mail: alberto.tesei@lincei.it

A. d'Onofrio et al. (eds.), *Problems in Mathematical Biophysics*, SEMA SIMAI Springer Series 38, https://doi.org/10.1007/978-3-031-60773-8_12

healing, have been considered in [8, 18, 19] (see also [7, 9] and references therein). A discretized version of diffusion equations on weighted graphs has been recently proposed in the modelling of Alzheimer's disease (see [6] and references therein).

On the strength of the above motivations, many challenging mathematical problems arise. In particular, a major task is to understand how the geometry of the graph (i.e., features like branching degree of vertices, length of edges, or symmetry properties) influence the dynamics on it, compared to when the problem is studied in Euclidean space—an issue already addressed for Riemannian manifolds. In this connection, observe that metric graphs occupy an intermediate position between Riemannian manifolds and discrete graphs, since they can be viewed as one-dimensional manifolds with singularities. Further mathematical interest lies in both metric graphs and Riemannian manifolds being examples of *metric measure spaces* (see the Appendix).

Our interest is in diffusion phenomena on metric graphs. In the case of manifolds a good deal of information has been gathered for typical phenomena like blow-up, wave propagation and pattern formation, pointing out their dependence on manifold features like *curvature* or *volume doubling* properties (in particular, see [2, 3, 29] and references therein).

In this note we outline some recent results in the same spirit concerning blow-up on metric graphs [35]. Compared to the case of manifolds, the overall picture is less clear and the results more sparse. It is arguable whether an exhaustive theory can only be obtained for specific classes of graphs, depending on their structure; an indication of this comes from studying wave propagation on trees ([36]; see Definition 7).

2 Statement of the Problem and Outline of Results

Both for metric graphs and for Riemannian manifolds, addressing blow-up phenomena suggests to consider the Cauchy problem

$$\begin{cases} u_t = \Delta u + h(t)|u|^{p-1}u & \text{in } X \times (0,T) \\ u = u_0 & \text{in } X \times \{0\}, \end{cases} \tag{1}$$

where X is a metric measure space (see the Appendix). The following will be assumed:

$$\begin{cases} (i) \;\; h : (0,\infty) \mapsto (0,\infty), \; h \in C(0,\infty) \cap L^1_{loc}([0,\infty)), \\ \quad\;\; \text{for some} z \in (0,1) \text{ there exists } \lim_{t\to 0^+} t^z h(t) \in (0,\infty); \\ (ii) \;\; p > 1\,; \\ (iii) \; u_0 \in C(X), \; u_0 \geq 0 \;\; u_0 \not\equiv 0\,. \end{cases} \tag{H}$$

2.1 Outline of Results: Euclidean Space

Consider the Cauchy problem

$$\begin{cases} u_t = \Delta u + h(t)|u|^{p-1}u & \text{in } \mathbb{R}^N \times (0,T) \\ u = u_0 & \text{in } \mathbb{R}^N \times \{0\} \end{cases} \tag{2}$$

with h, p and $u_0 \in L^\infty(\mathbb{R}^N)$ satisfying assumption (H). Fujita proved (see [15]) that when $h \equiv 1$ there exists a critical value $p^* := 1+\frac{2}{N}$, called the *Fujita exponent*, with the following properties:

(i) if $1 < p < p^*$ nontrivial global solutions of (2) do not exist;
(ii) if $p > p^*$ global solutions exist for u_0 sufficiently small.

Nonexistence of nontrivial global solutions for $p = p^*$ was proven later [21, 25]. The same *Fujita phenomenon* has been observed in other cases (see [26] and references therein, [37, 43]). If $h(t) = t^q$ $(q > -1)$, the Fujita exponent becomes $p^* := 1 + \frac{2(q+1)}{N}$ (see [31]).

The situation is different when studying the same problem in a bounded domain $\Omega \subseteq \mathbb{R}^N$ with homogeneous Dirichlet boundary conditions. In this case for $h(t) = t^q$ $(q > -1)$ formally there holds $p^* = 1$—namely, global solutions exist for all $p > 1$ if u_0 is sufficiently small. The dichotomy observed for (2) only arises if $h(t) = e^{\beta t}$ $(\beta > 0)$, with critical exponent $p^* := 1 + \frac{\beta}{\lambda_0}$ (see [31]; here λ_0 denotes the first eigenvalue of $-\Delta$ with homogeneous Dirichlet boundary conditions). The heuristic explanation is that "the growth of h must compensate the strong diffusion effect due to homogeneous Dirichlet boundary conditions".

2.2 Outline of Results: Riemannian Manifolds

The situation depicted for the Euclidean case easily extends to complete Riemannian manifolds with nonnegative *Ricci curvature*, where the *heat kernel estimate* holds:

$$c_1 \frac{e^{-\frac{[c_2 d(x,y)]^2}{t}}}{V(x,\sqrt{t})} \le K(x,y,t) \le C_2 \frac{e^{-\frac{[C_2 d(x,y)]^2}{t}}}{V(x,\sqrt{t})}\,; \tag{HK}$$

here $d = d(x, y)$ is the distance on the manifold X, $(x, y, t) \in X \times X \times (0, \infty)$, $c_1, c_2, C_1, C_2 > 0$, and K is the heat kernel (see Sect. 4). Estimate (HK) follows from the *Li-Yau gradient estimate* [27] and implies *volume doubling*, in the sense of the following definition.

Definition 1 A metric measure space X has the *volume doubling property* if for some $C_D > 0$ there holds

$$V(x_0, 2r) \leq C_D V(x_0, r) \quad \text{for any } x_0 \in X, r > 0\,, \tag{VD}$$

$V(x_0, r)$ denoting the volume of the ball $B(x_0, r) := \{x \in X \mid d(x_0, x) < r\}$. The volume doubling property holds *locally* in X, if for any compact subset $K \subseteq X$ there exists $C_D = C_D(K)$ such that inequality (VD) holds for any $x_0 \in X$ and $r > 0$ with $B(x_0, 2r) \subseteq K$.

A related notion is as follows:

Definition 2 The volume growth in a metric measure space X is said to be:

(i) *(at most) polynomial*, if there exists $C \geq 1$ and $n \in \mathbb{N}$ such that

$$V(x_0, r) \leq C r^n V(x_0, 1) \quad \text{for any } x_0 \in X, r > 1\,;$$

(ii) *(at least) exponential*, if there exist $C, \epsilon > 0$ such that

$$V(x_0, r) \geq C e^{\epsilon r} \quad \text{for any } x_0 \in X, r > 0\,.$$

The link between the validity of (HK) and intrinsic features of the manifolds like volume doubling, Poincaré inequality and parabolic Harnack inequality is well understood. In particular, if X satisfies (HK), it also satisfies (VD) (see [17, Theorem 2.7]), whence plainly

$$V(x_0, r) \leq M r^{[s]+1} V(x_0, 1) \quad \text{for any } x_0 \in X, r > 1 \tag{3}$$

with a constant $M \geq 1$ only depending on C_D, and $s := \log_2 C_D = \frac{\log C_D}{\log 2}$. Hence X has polynomial volume growth. In this case a power $h(t) = t^q$ $(q > -1)$ is sufficient to produce the Fujita phenomenon, whereas the exponential $h(t) = e^{\alpha t}$ $(\alpha > 0)$ is "too strong", thus every nontrivial solution blows up in finite time (see [35, Proposition 8.1]).

Instead, for manifolds with constant negative curvature as the hyperbolic space $\mathbb{H}^N$ the volume growth is exponential (e.g., see [44, Theorem 0.2] and subsequent remarks), thus an exponential like $h(t) = e^{\alpha t}$ $(\alpha > 0)$ is needed to produce the Fujita phenomenon (see [35, Proposition 8.3]). It is worth pointing out the analogy between the situation in $\mathbb{H}^N$ $(N \geq 2)$ and that in bounded domains of $\mathbb{R}^N$ (in fact, the supremum of the L^2-spectrum of the Laplacian is in both cases strictly negative), or in compact Riemannian manifolds with homogeneous Dirichlet boundary conditions (see [35, Proposition 8.2]).

2.3 Outline of Results: Metric Graphs

As outlined above, blow-up on Riemannian manifolds are related to the characteristics of diffusion, which depends on properties like curvature and volume growth through a well-established spectral theory.

Unfortunately, for metric graphs a proper notion of curvature is cumbersome, whereas spectral theory is exhaustive only in special cases, like *finite graphs* or *regular trees* (see [5, 33, 42]). The situation is more settled for discrete graphs [28, 44], where blow-up results rely on discrete heat kernel estimates proven in [4, 22]. However, such results do not imply similar results for metric graphs, since in general the discrete Laplacian and the corresponding metric Laplacian have different spectral properties (see [5, Theorem 3.6.1], [42, Example 6.3]).

An additional difficulty is that for a general metric graph volume doubling and heat kernel estimates only have a local character (see [20, Corollaries 3.4-3.5, Theorem 4.2 and Proposition 4.13]). Yet in specific cases, such as the following three, the graph geometry allows the volume doubling property to hold globally.

Example 1 Let Γ be a generalized star graph with finite sets Γ_f, Γ_i (see Definition 6). It is easily seen that the graph is of linear volume growth. Furthermore,

$$V(x_0, r) \leq \sum_{e \in \Gamma_f} l(e) + 2D_i n_i r ,$$

where $D_i := \sup_{v \in \Gamma_i} d_v$ and $n_i \in \mathbb{N}$ is the cardinality of Γ_i . Hence there exist $\bar{r} > 0$ and $\bar{M} > 2D_i n_i$ such that for any $x_0 \in \Gamma$ and $r > \bar{r}$ there holds $V(x_0, r) \leq \bar{M} r$. Therefore, the volume doubling holds with $s = 1$ and $C_D = 2\bar{M}$.

Example 2 Let Γ be the metric graph associated with the Euclidean lattice $\mathbb{Z}^N$. Volume doubling holds globally for $\mathbb{Z}^N$ with $C_D = 2^N$, hence the volume growth is of order N. The same holds for Γ, since for any $x_0 \in \Gamma$ there exists $q_0 \in \mathbb{Z}^N$ such that $|q_0 - x_0| < 1$ and $V(q_0, [r] - 1) \leq V(x_0, r) < V(q_0, [r] + 2)$ $(R > 1)$.

Example 3 Let all edges of a regular tree (see Definition 7) be of the same length, say $l_e = 1$. Then the volume of the ball $B(O, r)$ (O denoting the root of the tree) satisfies the inequalities

$$\sum_{k=0}^{[r]} \beta(k) = V(O, [r]) \leq V(O, r) \leq V(O, [r] + 1) = \sum_{k=0}^{[r]+1} \beta(k) . \tag{4}$$

For a homogeneous tree (see Definition 8) with branching number b, by (4) there holds $V(O, r) \geq \sum_{k=0}^{[r]} b^k \geq \frac{[r]}{2} b^{\frac{[r]}{2}} + 1$. Hence the volume growth is (at least) exponential.

Expectedly, for metric graphs with polynomial volume growth (like those in Examples 1–2) the situation should be similar to that in $\mathbb{R}^N$, or in Riemannian manifolds with nonnegative Ricci curvature. This is indeed the case, as shown by the following two propositions, whose proof relies on Theorem 3.

Proposition 1 *Let Γ be a generalized star graph with finite sets Γ_f, Γ_i, and let (H) hold.*

(i) Let $h(t) = t^q$ $(q > -1)$ and set $p^*_\Gamma := 1 + 2(q+1)$. Then:

(a) if $1 < p < p^*_\Gamma$, every nontrivial solution of problem (1) *(with $X = \Gamma$) blows up in finite time;*
(b) if $p > p^*_\Gamma$, problem (1) *possesses global solutions for small initial data.*

(ii) Let $h(t) = e^{\alpha t}$ $(\alpha > 0)$. Then for all $p > 1$ every nontrivial solution of problem (1) *(with $X = \Gamma$) blows up in finite time.*

Observe that the exponent p^*_Γ in Proposition 1 is the same as the the Fujita exponent for $\mathbb{R}^N$ with $N = 1$, in agreement with being the growth linear in both cases.

Arguing as in the proof of Proposition 1, similar results are proven for the Euclidean lattice (see Example 2):

Proposition 2 *Let Γ be the metric graph associated with $\mathbb{Z}^N$, and let (H) hold. Then the same conclusions of Proposition 1 hold with $p^*_\Gamma := 1 + \frac{2(q+1)}{N}$.*

For regular metric trees satisfying suitable branching conditions the spectrum of the Laplacian is strictly negative (see Theorem 5) and the volume growth is exponential (see Example 3). Then, as in the case of the hyperbolic space $\mathbb{H}^N$, an exponential h is needed to produce the Fujita phenomenon (whereas for $h(t) = t^q$ global solutions with small initial data always exist). This is the content of the following proposition, which is a consequence of Theorem 4 and Proposition 9.

Proposition 3 *Let $X = \mathcal{T}$ be a regular metric tree of infinite height satisfying condition* (18). *Let $\Delta = \Delta_D$ be the Dirichlet Laplacian, and let (H) hold.*

(i) Let $h(t) = t^q$ $(q > -1)$. Then for small initial data u_0 there exist global solutions of problem (1).
(ii) Let $h(t) = e^{\alpha t}$ and $p^*_{\mathcal{T}} := 1 + \frac{\alpha}{E_0}$ $(\alpha > 0)$, where $E_0 := \min \sigma(-\Delta_D) > 0$. Then:

(a) if $1 < p < p^*_{\mathcal{T}}$, every nontrivial solution of problem (1) *blows up in finite time;*
(b) if $p > p^*_{\mathcal{T}}$, problem (1) *possesses global solutions for small initial data.*

If $\mathcal{T}$ is a homogeneous tree of infinite height, condition (18) is satisfied (see Remark 1), thus the results of Proposition 3 apply to this case. From (19)–(20) we get the inequality

$$\frac{(\log b)^2}{4b} \leq E_0 \leq b(\log b)^2 ,$$

which points out the dependence of the spectrum on the branching number b. Accordingly, there holds

$$1 + \frac{\alpha}{b(\log b)^2} \leq p^*_{\mathcal{T}} \leq 1 + \frac{4\alpha b}{(\log b)^2} .$$

Due to the strict negativity of the spectrum of the Laplacian, the same conclusions of Proposition 3 hold for compact metric graphs with homogeneous Dirichlet boundary conditions:

Proposition 4 *Let $X = \Gamma$ be a compact metric graph, let $\Delta = \Delta_D$ be the Dirichlet Laplacian, and let (H) hold.*

(i) Let $h(t) = t^q$ $(q > -1)$. Then for small initial data u_0 there exist global solutions of problem (1).
(ii) Let $h(t) = e^{\alpha t}$ and $p^*_\Gamma := 1 + \frac{\alpha}{\lambda_0}$ $(\alpha > 0)$, $-\lambda_0 < 0$ being the first eigenvalue of Δ_D. Then:

  (a) if $1 < p < p^*_\Gamma$, every nontrivial solution of problem (1) *blows up in finite time;*
  (b) if $p > p^*_\Gamma$, problem (1) *possesses global solutions for small initial data.*

3 Blow-Up Criteria

3.1 Local and Global Existence

Definition 3 Let (H) hold and $u_0 \in L^\infty(X)$. A function $u \in C(\overline{Q}_\tau) \cap C([0,\tau]; L^\infty(X))$ is called a *mild solution* of (1) in $[0,\tau]$, if for any $t \in [0,\tau]$

$$u(\cdot,t) = e^{\Delta t}u_0 + \int_0^t e^{\Delta(t-s)}h(s)(|u|^{p-1}u)(\cdot,s)\,ds \quad \text{in } X . \tag{5}$$

A mild solution of (1) is *global* if (5) holds in $[0,\tau]$ for any $\tau \in (0,\infty)$.

By semigroup theory and a standard application of the contraction principle (e.g., see the proof of [45, Theorem 4.1]) the following well-posedness result can be proven.

Theorem 1 *Let (H) hold and $u_0 \in L^\infty(X)$. Then problem* (1) *has a unique mild solution $u \geq 0$ in a maximal existence interval* $[0,T)$ $(T \in (0,\infty])$. *Either $T = \infty$ and the solution is global, or $T < \infty$ and $\|u\|_{C([0,\tau];L^\infty(X))} \to \infty$ as $\tau \to T^-$.*

Sufficient conditions for global existence can be proven (see [35, Proposition 6.1]):

Proposition 5 *Let* (H) *hold with* h *nondecreasing, and let* $G \in C([0,\infty)) \cap C^1(0,\infty)$, $G > 0$ *in* $(0,\infty)$ *be nondecreasing. Let there exist* $y_0 \in X$ *such that for all* $x \in X$

$$\liminf_{t\to\infty} \left(\frac{G'(t)}{G^p(t)h(t)}\right)^{\frac{1}{p-1}} \frac{1}{K(x,y_0,t)} > 1\,. \tag{6}$$

Then problem (1) *possesses global solutions for small initial data.*

Set

$$H(t) := \int_0^t h(s)\,ds \quad (t \in [0,\infty))\,, \qquad H_\infty := \lim_{t\to\infty} H(t)\,. \tag{7}$$

Since $\{e^{\Delta t}\}_{t\geq 0}$ is contracting on $L^\infty(X)$ (see (23)), from (5) by Gronwall's inequality we obtain

$$\|u(\cdot,t)\|_\infty \leq \|u_0\|_\infty \left\{1-(p-1)\|u_0\|_\infty^{p-1} H(t)\right\}^{-\frac{1}{p-1}}, \qquad (t \in [0,\tau)) \tag{8}$$

whence the subsequent result immediately follows:

Theorem 2 *Let* (H) *hold and* $u_0 \in L^\infty(X)$.

(i) If $H_\infty \leq \frac{1}{(p-1)\|u_0\|_\infty^{p-1}}$, *the unique nonnegative mild solution of problem* (1) *is global.*
(ii) If $H_\infty > \frac{1}{(p-1)\|u_0\|_\infty^{p-1}}$, *the maximal time of existence of* u *is not less than* $H^{-1}\left(\frac{1}{(p-1)\|u_0\|_\infty^{p-1}}\right)$.

3.2 Finite Time Blow-Up

Theorem 1 suggests the following definition.

Definition 4 Let $T > 0$ be the maximal time of existence of a mild solution u of problem (1). If $T < \infty$, T is called the *blow-up time* of u, and u *blows up in finite time* as $t \to T^-$.

Let u be a mild solution of problem (1) in $[0,T]$. Set

$$U(x,t) := \int_X K(x,z,T-t)u(z,t)\,d\mu(z)\,, \tag{9}$$

thus in particular

$$U(x,0) = \int_X K(x,z,T)u_0(z)\,d\mu(z) = \left(e^{\Delta T}u_0\right)(x)\,. \tag{10}$$

The following lemma is well-known (see [31, 43]).

Lemma 1 *Let (H) hold. Let $u \geq 0$ be a mild solution of problem* (1) *in* $[0, T]$, *and let H, U be as in* (7) *and* (9). *Then for any $x \in X$ and $t \in [0, T]$*

$$(p-1)H(t) \leq \frac{1}{[U(x,0)]^{p-1}}\,. \tag{11}$$

From Lemma 1 we get the following general blow-up criterion.

Proposition 6 *Let (H) hold, and let $g \in C(X \times [0,\infty))$, $g > 0$. Let the following assumptions be satisfied:*

(i) for any $\tau > 0$ there exists $f \in C(X)$ (only depending on τ and u_0), $f(x) > 0$ for any $x \in X$, such that

$$\left(e^{\Delta t}u_0\right)(x) \geq \frac{f(x)}{g(x,t)} \quad \text{for any } t \in [\tau,\infty) \text{ and } x \in X; \tag{12}$$

(ii) there exists $x_0 \in X$ such that

$$\limsup_{t\to\infty} \frac{[H(t)]^{\frac{1}{p-1}}}{g(x_0,t)} = \infty\,. \tag{13}$$

Then the corresponding mild solution of problem (1) *blows up in finite time.*

Proof Let $u \geq 0$, $u \not\equiv 0$ be the unique mild solution of problem (1) in $[0, T]$. By (10)–(12) there holds

$$\frac{f(x)}{g(x,T)} \leq U(x,0) \leq \left(\frac{1}{p-1}\right)^{\frac{1}{p-1}} [H(T)]^{-\frac{1}{p-1}}\,,$$

whence

$$\frac{[H(T)]^{\frac{1}{p-1}}}{g(x,T)} \leq \left(\frac{1}{p-1}\right)^{\frac{1}{p-1}} \frac{1}{f(x)} \qquad (x \in X)\,.$$

If u is a global solution of (1), the above inequality holds for all $x \in X$ and $T > 0$ sufficiently large. However, for $x = x_0$ letting $T \to \infty$ gives a contradiction with assumption (13). Then the conclusion follows. □

The proof of the following variant of Proposition 6 is almost identical, thus is omitted.

Proposition 7 *Let (H) hold, and let $g \in C(X \times [0, \infty))$, $g > 0$. Let the following assumptions be satisfied:*

(i) there exist $x_0 \in X$, $\tau > 0$ and $C > 0$ such that

$$\left(e^{\Delta t} u_0\right)(x_0) \ \geq \ \frac{C}{g(x_0, t)} \quad \text{for any } t \in [\tau, \infty);$$

(ii) for any $x \in X$ there holds

$$\limsup_{t \to \infty} \frac{[H(t)]^{\frac{1}{p-1}}}{g(x, t)} = \infty .$$

Then the corresponding mild solution of problem (1) *blows up in finite time.*

Specific blow-up criteria follow from Propositions 6–7 by a proper choice of the function g. If the bilateral estimate (HK) holds, a natural candidate for the function g is $g(x, t) = V(x, t^{\frac{1}{2}})$. Then by Proposition 6 we get the following result:

Theorem 3 *Let (H) hold, and let the heat kernel estimate (HK) be satisfied. Let there exist $x_0 \in X$ such that*

$$\limsup_{t \to \infty} \frac{[H(t)]^{\frac{1}{p-1}}}{V(x_0, t^{\frac{1}{2}})} = \infty . \tag{14}$$

Then every nontrivial nonnegative mild solution of problem (1) *blows up in finite time.*

Proof The result will follow from Proposition 6 with $g(x, t) = V(x, t^{\frac{1}{2}})$, if we prove the following

Claim For any $\tau > 0$ and u_0 as in (H)–(iii) there exists $f \in C(X)$ (only depending on τ and u_0) such that $f(x) > 0$ for any $x \in X$, and

$$\left(e^{\Delta t} u_0\right)(x) \ \geq \ \frac{f(x)}{V(x, t^{\frac{1}{2}})} \quad \text{for any } t \in [\tau, \infty) \text{ and } x \in X. \tag{15}$$

By assumption (H)–(iii) there exist $x_0 \in X$, $r > 0$ and $\delta > 0$ such that $u_0 \geq \delta$ in the ball $B(x_0, r)$. Then by (22) and the estimate from below in (HK) we get

$$\left(e^{\Delta t} u_0\right)(x) \ \geq \ \delta c_1 \frac{1}{V(x, t^{\frac{1}{2}})} \int_{B(x_0, r)} g_\beta(c_2\, d(x, y), \tau)\, d\mu(y)$$

for any $t \in [\tau, \infty)$ and $x \in X$. Setting

$$f(x) := \delta c_1 \int_{B(x_0,r)} g_\beta(c_2\, d(x, y), \tau)\, d\mu(y)$$

the Claim follows.

□

A different choice of the function g makes use of information concerning the spectrum of the Laplacian. This gives the following result:

Theorem 4 *Let (H) hold, let the kernel K of $\{e^{\Delta t}\}_{t\geq 0}$ exist, and let $E_0 := \min \sigma(-\Delta) > 0$. Let the following assumptions be satisfied:*

(i) there holds

$$\lim_{t\to\infty} \frac{\log K(x, y, t)}{t} = -E_0$$

locally uniformly for $x, y \in X$;

(ii) for some $\bar{\epsilon} > 0$ there holds

$$\limsup_{t\to\infty} \frac{[H(t)]^{\frac{1}{p-1}}}{e^{(E_0+\bar{\epsilon})t}} = \infty\,.$$

Then every mild solution of problem (1) *blows up in finite time.*

Proof The conclusion follows from Proposition 7 with $g(x, t) \equiv g(t) = e^{(E_0+\bar{\epsilon})t}$, if we prove the following

Claim There exist $x_0 \in X$ and $C > 0$ (depending on u_0) with the following property: for any $\epsilon > 0$ there exists $\tau > 0$ such that

$$\left(e^{\Delta t} u_0\right)(x_0) \geq \frac{C}{e^{(E_0+\epsilon)t}} \quad \text{for any } t \in [\tau, \infty)\,.$$

By assumption (H)–(iii) there exist $x_0 \in X$ and $r > 0$ such that $C \equiv C(u_0) := \int_{B(x_0,r)} u_0\, d\mu > 0$. On the other hand, by (i) for every $\epsilon > 0$ there exists $\tau > 0$ such that for any $t \geq \tau$ and $x, y \in B(x_0, r)$ there holds $K(x, y, t) > e^{-(E_0+\epsilon)t}$. It follows that

$$\begin{aligned}\left(e^{\Delta t} u_0\right)(x_0) &= \int_X K(x_0, y, t) u_0(y)\, d\mu(y) \\ &\geq \int_{B(x_0,r)} K(x_0, y, t) u_0(y)\, d\mu(y) \geq \frac{C}{e^{(E_0+\epsilon)t}}\,.\end{aligned}$$

This proves the Claim.

□

4 Proof of Blow-Up Results

Proof of Proposition 1 As pointed out in Example 1, in this case there exists $\bar{M} > 0$ such that $V(x_0, r) \leq \bar{M} r$ for any $x_0 \in \Gamma$ and r sufficiently large; moreover, estimate (HK) holds globally. Hence part (a) of claim (i) and claim (ii) follow immediately from Theorem 3, since in both cases equality (14) holds.

To prove part (b) of claim (i), let $p > p_\Gamma^*$ be fixed. Let us apply Proposition 5 with $G(t) = t^r$ $(r > 0)$. Since $V(x_0, r)$ has linear growth and (HK) holds, for some constant $\tilde{M} > 0$ and all t sufficiently large there holds

$$\left(\frac{G'(t)}{G^p(t)h(t)}\right)^{\frac{1}{p-1}} \frac{1}{K(x, y_0, t)} \geq \tilde{M}\, t^{-r-\left(\frac{q+1}{p-1}-\frac{1}{2}\right)} = \tilde{M}\, t^{-r+\frac{p-p_\Gamma^*}{2(p-1)}}\,.$$

Choosing $r \in \left(0, \frac{p-p_\Gamma^*}{2(p-1)}\right)$ we have that $\lim_{t\to\infty} t^{-r+\frac{p-p_\Gamma^*}{2(p-1)}} = \infty$, thus inequality (6) is satisfied. Hence the result follows. □

Using Theorem 4 and Proposition 9 we can prove Proposition 4. We omit the proofs of Propositions 2–3 since they are analogous to those of Propositions 1 and 4, respectively.

Proof of Proposition 4 Claim (ii)–(a) follows from Theorem 4 and Proposition 9, since for any $\epsilon \in \left(0, -\lambda_0 + \frac{\alpha}{p-1}\right)$ there holds

$$\lim_{t\to\infty} \frac{[H(t)]^{\frac{1}{p-1}}}{e^{(\lambda_0+\epsilon)t}} = \lim_{t\to\infty} \frac{1}{e^{(\lambda_0+\epsilon)t}} \left(\frac{e^{\alpha t}-1}{\alpha}\right)^{\frac{1}{p-1}} = \infty\,.$$

To prove claims (i) and (ii)–(b) we make use of Proposition 5 with $G(t) = e^{\lambda t}$ $(\lambda > 0)$. By (25) for large values of t there holds $K_D(x, y, t) \sim e^{-\lambda_0 t}$ $(x, y \in \Gamma)$, thus inequality (6) is satisfied if

$$\lim_{t\to\infty} \frac{e^{(\lambda_0-\lambda)t}}{[h(t)]^{\frac{1}{p-1}}} = \infty\,. \tag{16}$$

If $h(t) = t^q$, equality (16) holds for any $\lambda \in (0, \lambda_0)$, thus claim (i) follows. If $h(t) = e^{\alpha t}$ and $p > p_\Gamma^*$, equality (16) holds if $\lambda_0 - \lambda - \frac{\alpha}{p-1} > 0$, namely if $\lambda \in \left(0, \frac{p-p_\Gamma^*}{p-1}\lambda_0\right)$. Hence part (b) of claim (ii) follows. □

Appendix

Let us recall some notions relative to metric graphs. As outlined below, metric graphs are *metric measure spaces*, since they are endowed with a geodesic distance d, a resulting Borel σ-algebra $\mathcal{B}$ and a Radon measure $\mu : \mathcal{B} \mapsto [0, \infty]$. Metric measure spaces provide a unified framework for both metric graphs and Riemannian manifolds.

Metric Graphs

The following definitions are well known (e.g., see [5, 32]).

Definition 5

(i) A *directed graph* is a couple (V, E), where V is a countable set and $E \subseteq V \times V$. The elements of V are called *vertices*, those of E *edges*. The edges are *directed*, since each $e \in E$ is an ordered couple (u, v) (written $u \in e$, $v \in e$). We say that u is the *initial endpoint*, v is the *terminal endpoint* of e, and that the edge $e = (u, v)$ *emanates* from u.
The *degree* $d_v \in \mathbb{N}$ of a vertex v is the number of edges for which v is an endpoint (either initial or terminal). The set $\partial V := \{v \in V \mid d_v = 1\}$ is called the *boundary* of the graph.

(ii) A directed graph is *simple* (or *oriented*) if for all $u, v \in V$ there holds $(v, v) \notin E$, and $(u, v) \in E \Rightarrow (v, u) \notin E$.

Definition 6

(i) An oriented graph (V, E) is called a *metric graph* (denoted by Γ) if:

  (a) each $e \in E$ is associated with an interval $I_e := [0, l(e)]$, $l(e) \in (0, \infty]$ being the *length* of e;

  (b) each $e \in E$ is assigned a coordinate $x_e \in I_e$ increasing in the direction of e.

  An edge $e \in E$ with $l(e) = \infty$ is called *ray*. A metric graph is called *of bounded geometry* if $D := \sup_{v \in V} d_v < \infty$ and $l := \inf_{e \in E} l(e) > 0$. A finite metric graph whose edges all have finite lengths is called *compact*.

(ii) A metric graph Γ is called a *generalized star graph* if $\Gamma \equiv \Gamma_f \cup \Gamma_i$, where $\Gamma_f, \Gamma_i \neq \emptyset$, the edges of Γ_f are of finite length and those of Γ_i are rays.

A connected metric graph Γ is a metric measure space. In fact, the distance function is first defined for couples of vertices by minimizing the length of the paths connecting them. Namely, for some $n \in \mathbb{N}$ there exist $\{e_1, \ldots, e_n\} \in E^n$, $\{x_1, \ldots, x_{n+1}\} \in V^{n+1}$ such that $x_1 = u$, $x_{n+1} = v$ and either $e_k = (x_k, x_{k+1})$ or $e_k = (x_{k+1}, x_k)$ ($k = 1, \ldots, n$); then the length of the path is $l(P) := \sum_{k=1}^{n} l(e_k)$, and the distance between u and v is $d(u, v) := \inf\{l(P) \mid P \text{ connects} u \text{and } v\}$. This notion is then extended to the intervals I_e in the natural way. This defines a distance on Γ (identified with the disjoint union of the intervals I_e), which generates a Borel σ-algebra by standard arguments.

Therefore Γ is a metric space, with Borel σ-algebra $\mathcal{B} = \mathcal{B}(\Gamma)$. A Radon measure $\mu : \mathcal{B} \mapsto [0, \infty]$ is induced on Γ by the Lebesgue measure λ on each interval I_e, namely

$$\mu(G) := \sum_{e \in E} \lambda(I_e \cap G) \quad \text{for any } G \in \mathcal{B}.$$

The way to define Lebesgue and Sobolev spaces on Γ is now paved. We refer the reader e.g. to [32] for details.

Metric Trees

An interesting class of metric graphs is pointed out by the following definition.

Definition 7

(i) A metric graph $\mathcal{T}$ associated with a tree, with a *root* vertex O singled out, is called a *metric tree*. The quantity $h(\mathcal{T}) := \sup_{x \in \mathcal{T}} \rho(x)$, where $\rho(x) := d(x, O)$ $(x \in \mathcal{T})$, is called the *height* of $\mathcal{T}$.

(ii) The *generation* $\text{gen}(v)$ of a vertex $v \in \mathcal{T}$ is the number of vertices which lie on the unique path connecting v with O (including the starting point but excluding the end point). The *branching number* $b(v)$ of $v \in \mathcal{T}$ is the number of edges emanating from v.

(iii) A tree $\mathcal{T}$ is *regular*, if all vertices of the same generation have equal branching numbers, and all edges emanating from them are of the same length.

Clearly,

(a) in a regular tree all vertices of the same generation have the same distance from the root;

(b) a regular tree is uniquely determined by two *generating sequences* $\{b_n\} \equiv \{b_n(\mathcal{T})\} \subseteq \{0\} \cup \mathbb{N}$, $\{\rho_n\} \equiv \{\rho_n(\mathcal{T})\} \subseteq \overline{\mathbb{R}}_+$, with $\{\rho_n\}$ increasing and $\rho_0 = 0$, such that

$$n = \text{gen}(v)\,, \quad b_n = b(v)\,, \quad \rho_n = \rho(v) \quad \text{for any vertex } v \text{ of } \mathcal{T}.$$

We always suppose that $b_0 = 1$, $b_n \geq 2$ for any $n \in \mathbb{N}$, $\text{gen}(v) < \infty$ for all $v \in \mathcal{T}$, and $h(\mathcal{T}) = \infty$. In particular, there holds $\partial\mathcal{T} = O$, i.e. the boundary of a metric tree consists of its root. Observe that $\bigcup_{n \in \mathbb{N}} I_n = (0, h(\mathcal{T})) = \mathbb{R}_+$.

Definition 8 A regular tree $\mathcal{T}$ is *homogeneous*, if the branching number $b(v)$ is the same for all $v \in \mathcal{T} \setminus O$, and all edges are of the same length.

If $\mathcal{T}$ is homogeneous, for any vertex v of $\mathcal{T}$ with $\mathrm{gen}(v) = n$ there holds $\rho_n = nr$, where $r > 0$ is the length of any edge ($n \in \mathbb{N}$). If $\mathcal{T}$ is regular, we denote by b_j the branching number of any vertex of generation j ($j \in \mathbb{N} \cup \{0\}$), and we introduce the *branching function* $\beta : [0, h(\mathcal{T})) \mapsto \mathbb{N}$,

$$\beta(\rho) := \mathrm{card}\{x \in \mathcal{T} \,|\, \rho(x) = \rho\} = \sum_{n=1}^{\infty} (b_0 b_1 \dots b_{n-1}) \chi_{(\rho_{n-1}, \rho_n]}(\rho) . \tag{17}$$

Clearly, β is an increasing step function continuous from the left, with jump $\beta(\rho_n^+) - \beta(\rho_n^-) = b_0 b_1 \dots b_{n-1}(b_n - 1)$ at any point $\rho = \rho_n$ ($n \in \mathbb{N}$).

The *reduced height* of $\mathcal{T}$ is

$$L(\mathcal{T}) := \int_0^\infty \frac{d\rho}{\beta(\rho)} .$$

For regular trees following holds (see [42, Theorem 5.2 and Subsection 7.1].

Theorem 5 *Let $\mathcal{T}$ be a regular metric tree of infinite height, $h(\mathcal{T}) = \infty$. Then the following statements are equivalent:*

(i) $E_0 := \min \sigma(-\Delta) > 0$;
(ii) there holds

$$L(\mathcal{T}) < \infty, \qquad B(\mathcal{T}) := \sup_{t>0} \left(\int_0^t \beta(\rho)\, d\rho \int_t^\infty \frac{d\rho}{\beta(\rho)} \right) < \infty . \tag{18}$$

Moreover,

$$[4B(\mathcal{T})]^{-1} \le E_0 \le B(\mathcal{T})^{-1}. \tag{19}$$

Remark 1 Let $\mathcal{T}$ be a homogeneous tree of infinite height, with branching number $b(v) \equiv b \ge 2$ for all vertices $v \neq O$, and all edges of the same length $l_e = 1$. From (17) we obtain $\beta(r) = b^{[r]}$, thus plainly

$$H(\mathcal{T}) \le \frac{b}{\log b}, \qquad \frac{1}{b(\log b)^2} \le B(\mathcal{T}) \le \frac{b}{(\log b)^2} . \tag{20}$$

Laplacian on Metric Graphs

A metric graph Γ can be endowed in a standard way with a *Dirichlet* (or *energy*) *form*. The most obvious choice is the energy form

$$\mathfrak{a}(f,g) := \int_\Gamma f'g'\,d\mu\,, \qquad f,g \in \mathfrak{D}(\mathfrak{a}) := H^1(\Gamma)\,.$$

The operator associated with $\mathfrak{a}$ is the *Laplacian* Δ, with domain

$$D(\Delta) := \Big\{ f \in H^1(\Gamma) \mid f_e \in H^2(0,l(e))\,\forall e \in E,$$

$$\times \sum_{e\in E}\int_0^{l(e)} |f_e''|^2\,d\lambda < \infty,\ \sum_{e\ni v}\frac{df_e}{dn}(v) = 0\,\forall\, v \in V \Big\} \tag{21a}$$

(here $\frac{df_e}{dn}(v)$ denotes the outer derivative of f_e at the vertex v). The Laplacian acts on $D(\Delta)$ in the natural way:

$$(\Delta f)_e := f_e'' \quad \text{for any } f \in D(\Delta) \qquad (e \in E)\,. \tag{21b}$$

Since $H^1(\Gamma) \subseteq C(\Gamma)$, every $f \in D(\Delta)$ is continuous at any vertex $v \in V \setminus \partial V$. The condition $\sum_{e\ni v}\frac{df_e}{dn}(v) = 0$ is the *Kirchhoff transmission condition* if $v \in V \setminus \partial V$, and the *Neumann homogeneous boundary condition* if $v \in \partial V$. Hence the Laplacian defined in (21) is called *Neumann Laplacian* and denoted by Δ_N, if clarity so requires.

If $V_D \subseteq V$, $V_D \neq \emptyset$, a different notion of Laplacian is obtained considering the Dirichlet form

$$\mathfrak{a}_{V_D}(f,g) := \int_\Gamma f'g'\,d\mu\,, \qquad f,g \in \mathfrak{D}(\mathfrak{a}_{V_D}) := H^1_1(\Gamma, V_D)$$

(see [14]). The Laplacian Δ_{V_D} associated with $\mathfrak{a}_{V_D}$ is the operator defined as follows:

$$(\Delta_{V_D} f)_e := f_e'' \quad \text{for any } f \in D(\Delta_{V_D}) \quad (e \in E)\,,$$

where

$$D(\Delta_{V_D}) := \Big\{ f \in H^1_1(\Gamma, V_D) \mid f_e \in H^2(0,l(e))\,\forall e \in E,$$

$$\times \sum_{e\in E}\int_0^{l(e)} |f_e''|^2\,d\lambda < \infty,\ \sum_{e\ni v}\frac{df_e}{dn}(v) = 0\,\forall\, v \in V \setminus V_D \Big\}.$$

More explicitly, the vertex conditions satisfied by all $f \in D(\Delta_{V_D}) \subseteq C(\Gamma)$ are:

$$\begin{cases} f(v) = 0 & \forall v \in V_D\,, \\ \sum_{e \ni v} \frac{df_e}{dn}(v) = 0 & \forall v \in V \setminus V_D\ . \end{cases}$$

Therefore, it is natural to call *Dirichlet Laplacian* the Laplacian $\Delta_D \equiv \Delta_{\partial V}$ (observe that the Neumann Laplacian satisfies $\Delta_N \equiv \Delta_\emptyset$). In particular, for a metric tree the vertex conditions satisfied by any $f \in D(\Delta_D) \subseteq C(\mathcal{T})$ explicitly read:

$$\begin{cases} f(O) = 0\,, \\ \sum_{e \ni v} \frac{df_e}{dn}(v) = 0 \quad \forall v \in \mathcal{T} \setminus O\,. \end{cases}$$

Any realization of the Laplacian on a metric graph Γ ia a self-adjoint nonpositive operator, thus the spectrum $\sigma(-\Delta)$ is contained in $\overline{\mathbb{R}}_+$. We denote by $\{e^{\Delta t}\}_{t \geq 0}$ the analytic semigroup generated by Δ in $L^2(\Gamma)$, which exists by classical results (e.g., see [45]). As in [13, Theorems 1.4.1–1.4.2], the following holds:

Proposition 8 *The semigroup* $\{e^{\Delta t}\}_{t \geq 0}$ *may be extended to a positivity preserving contraction semigroup* $\{T_p(t)\}_{t \geq 0}$ *on* $L^p(\Gamma) \equiv L^p(\Gamma, \mu)$ *for all* $p \in [1, \infty]$. *Moreover, there holds*

$$T_p(t)u = T_q(t)u \quad \text{for all } p, q \in [1, \infty] \text{ and } u \in L^p(\Gamma) \cap L^q(\Gamma)\,.$$

The semigroup $\{T_p(t)\}_{t \geq 0}$ *is strongly continuous if* $p \in [1, \infty)$ *and holomorphic if* $p \in (1, \infty)$.

The notation $e^{\Delta t} \equiv T_p(t)$ ($p \in [1, \infty]$) is currently used.

If for some smooth function $K : \Gamma \times \Gamma \times \overline{\mathbb{R}}_+ \mapsto \overline{\mathbb{R}}_+$ there holds

$$e^{\Delta t} u = \int_\Gamma K(\cdot, y, t)\, u(y)\, d\mu(y) \qquad (u \in \bigcup_{p=1}^{\infty} L^p(\Gamma)), \tag{22}$$

we say that the *heat kernel* K *exists on* Γ. Fundamental properties of the kernel are:

(a) K is symmetric, $K(x, y, t) = K(y, x, t)$, and smooth;
(b) for any fixed $y \in \Gamma$ $K(\cdot, y, \cdot)$ satisfies the heat equation in the classical sense,

$$K_t(x, y, t) = \Delta_x K(x, y, t)\,;$$

(c) for any $x, y, z \in \Gamma$ and $s, t \in \overline{\mathbb{R}}_+$

$$\int_\Gamma K(x, y, t)\, K(y, z, s)\, d\mu(y) = K(x, z, s + t)\,;$$

(d) for any $x \in \Gamma$ there holds

$$\lim_{t\to 0^+} \int_\Gamma K(x,y,t)\,u(y)\,d\mu(y) = u(x) \quad \text{for all } u \in C(\Gamma)\cap L^\infty(\Gamma).$$

The existence of the heat kernel on metric graphs is ensured by [24, Proposition 3.3]. Moreover, there holds

$$\int_\Gamma K(\cdot,y,\cdot)\,d\mu(y) = 1 \quad \text{in } \Gamma\times\overline{\mathbb{R}}_+ \tag{23}$$

—namely, for all $x \in \Gamma$ and $t \in \overline{\mathbb{R}}_+$ the measure $K(x,y,t)\,d\mu(y)$ is a probability measure.

If Γ is compact, due to the compact embedding of $H^1(\Gamma)$ in $L^2(\Gamma)$, $\sigma(-\Delta)$ consists of eigenvalues $\lambda_j \geq 0$ ($j \in \{0\}\cup\mathbb{N}$) of finite multiplicity. The first eigenvalue λ_0 is 0 for $-\Delta_N$ and strictly positive for $-\Delta_D$. In both cases there exists an orthonormal basis $\{\varphi_j\} \subseteq L^2(\Gamma)$ of eigenfunctions such that

$$D(-\Delta) = \left\{ f \in L^2(\Gamma) \;\middle|\; \sum_{j=0}^{\infty} \lambda_j^2 \left(\int_\Gamma f\varphi_j\,d\mu \right)^2 < \infty \right\}, \tag{24a}$$

$$-\Delta f = \sum_{j=0}^{\infty} \lambda_j \left(\int_\Gamma f\varphi_j\,d\mu \right) \varphi_j \qquad (f \in D(-\Delta)). \tag{24b}$$

By (24) and the spectral theorem there holds

$$K(x,y,t) = \sum_{j=0}^{\infty} e^{-\lambda_j t}\varphi_j(x)\varphi_j(y), \tag{25}$$

where K denotes the heat kernel either of Δ_N or of Δ_D, and the series is uniformly convergent on $\Gamma\times\Gamma\times[\tau,\infty)$ for any $\tau > 0$. Observe that for Δ_N equality (25) reads

$$K(x,y,t) = \frac{1}{\mu(\Gamma)} + \sum_{j=1}^{\infty} e^{-\lambda_j t}\varphi_j(x)\varphi_j(y).$$

Let us mention the following result, which is analogous to [11, Corollary 1]. We refer the reader to [35] for the proof.

Proposition 9 *Let Γ be a metric graph with Dirichlet Laplacian Δ_D and associated heat kernel K_D, and let $E_0 := \min \sigma(-\Delta_D)$. Then there holds*

$$\lim_{t\to\infty} \frac{\log K_D(x, y, t)}{t} = -E_0$$

locally uniformly for $x, y \in \Gamma$.

References

1. Baker, M., Faber, X.: Metric graphs, Laplacian operators, and electrical networks. In: Quantum Graphs and Their Applications. Contemporary Mathematics, vol. 415, pp. 15–33. American Mathematical Society, Providence (2006)
2. Bandle, C., Pozio, M.A., Tesei, A.: The Fujita exponent in the hyperbolic space. J. Differ. Equ. **251**, 2143–2163 (2011)
3. Bandle, C., Punzo, F., Tesei, A.: Existence and nonexistence of patterns on Riemannian manifolds. J. Math. Anal. Appl. **387**, 33–47 (2012)
4. Bauer, F., Horn, P., Lin, Y., Lippner, G., Mangoubi, D., Yau, S.-T.: Li-Yau inequality on graphs. J. Differ. Geom. **99**, 359–405 (2015)
5. Berkolaiko, G., Kuchment, P.: Introduction to Quantum Graphs. American Mathematical Society, Providence (2013)
6. Bertsch, M., Franchi, B., Raj, A., Tesi, M.C.: Macroscopic modelling of Alzheimer's disease: difficulties and challenges. Brain Multiphys. **2**, 100040 (2021)
7. Borsche, R., Klar, A., Ha Pham, T.N.: Nonlinear flux-limited models for chemotaxis on networks. Netw. Heterog. Media **12**, 381–401 (2017)
8. Bretti, G., Natalini, R., Ribot, M.: A hyperbolic model of chemotaxis on a network: a numerical study. ESAIM: Math. Mod. Numer. Anal. **48**, 231–258 (2014)
9. Camilli, F., Corrias, L.: Parabolic models for chemotaxis on weighted networks. J. Math. Pures Appl. **108**, 459–480 (2017)
10. Carlson, R.: Linear network models related to blood flow. In: Quantum Graphs and Their Applications. Contemporary Mathematics, vol. 415, pp. 65–80. American Mathematical Society, Providence (2006)
11. Chavel, I., Karp, L.: Large time behavior of the heat kernel: the parabolic λ-potential alternative. Comment. Math. Helvetici **66**, 541–556 (1991)
12. Dáger, R., Zuazua, E.: Wave propagation, observation and control in 1-d flexible multi-structures. Mathématiques et Applications, vol. 50. Springer, Berlin (2006)
13. Davies, E.B.: Heat Kernel and Spectral Theory. Cambridge University Press, Cambridge (1989)
14. Del Pezzo, L.M., Rossi, J.D.: The first eigenvalue of the p-laplacian on quantum graphs. Anal. Math. Phys. **6**, 365–391 (2016)
15. Fujita, H.: On the blowing up of solutions of the Cauchy problem for $u_t = \Delta u + f^{1+\alpha}$. J. Fac. Sci. Tokyo Sect. IA Math. **13**, 109–124 (1966)
16. Garavello, M., Piccoli, B.: Traffic flow on networks. Conservation Laws Models. AIMS Series on Applied Mathematics, vol. 1. AIMS Press, Springfield (2006)
17. Grigoryan, A., Saloff-Coste, L.: Stability results for Harnack inequalities. Ann. Inst. Fourier (Grenoble) **55**, 825–890 (2005)
18. Guarguaglini, F.R., Natalini, R.: Global smooth solutions for a hyperbolic chemotaxis model on a network. SIAM J. Math. Anal. **47**, 4652–4671 (2015)
19. Guarguaglini, F.R., Natalini, R.: Vanishing viscosity approximation for linear transport equations on finite star-shaped networks. J. Evol. Equ. **21**, 2413–2447 (2021)

20. Häseler, S.: Heat kernel estimates and related inequalities on metric graphs. arXiv:1101.3010v1
21. Hayakawa, K.: On nonexistence of global solutions of some semilinear parabolic differential equations. Proc. Jpn. Acad. Ser. A Math. Sci. **49**, 503–505 (1973)
22. Horn, P., Lin, Y., Liu, S., Yau, S.-T.: Volume doubling, Poincaré inequality and Gaussian heat kernel estimate for non-negatively curved graphs. J. Reine Angew. Math. **2019**, 89–30 (2021)
23. Jackson, M.O.: Social and Economic Networks. Princeton University Press, Princeton (2008)
24. Keller, M., Lenz, D., Vogt, H., Wojciechowski, R.: Note on basic features of large time behaviour of heat kernels. J. Reine Angew. Math. **708**, 73–95 (2015)
25. Kobayashi, K., Sirao, T., Tanaka, H.: On the growing up problem for semilinear heat equations. J. Math. Soc. Jpn. **29**, 407–424 (1977)
26. Levine, H.A.: The role of critical exponents in blowup theorems. SIAM Rev. **32**, 262–288 (1990)
27. Li, P., Yau, S.-T.: On the parabolic kernel of the Schrödinger operator. Acta Math. **156**, 153–201 (1986)
28. Lin, Y., Wu, Y.: The existence and nonexistence of global solutions for a semilinear heat equation on graphs. Calc. Var. Partial Differ. Equ. **56**, 101–122 (2017)
29. Matano, H., Punzo, F., Tesei, A.: Front propagation for nonlinear diffusion equations on the hyperbolic space. J. Eur. Math. Soc. **17**, 1199–1227 (2015)
30. Maury, B., Salort, D., Vannier, C.: Trace theorem for trees and application to the human lungs. Netw. Heterog. Media **4**, 469–500 (2009)
31. Meier, P.: On the critical exponent for reaction-diffusion equations. Arch. Rat. Mech. Anal. **109**, 63–71 (1990)
32. Mugnolo, D.: Semigroup Methods for Evolution Equations on Networks. Springer, Berlin (2014)
33. Naimark, K., Solomyak, M.: Geometry of Sobolev spaces on regular trees and the Hardy inequalities. Russ. J. Math. Phys. **8**, 322–335 (2001)
34. Nicaise, S.: Some results on spectral theory over networks applied to nerve impulse transmission. In: Brezinski, C., Draux, A., Magnus, A.P., Maroni, P., Ronveaux, A. (eds.) Polynômes Orthogonaux et Applications. Lecture Notes in Mathematics, vol. 1171, pp 532–541. Springer, Berlin (1985)
35. Punzo, F., Tesei, A.: Blow-up on metric graphs and Riemannian manifolds. Discrete Contin. Dyn. Syst. Ser. B 28, 6362–6392 (2023)
36. Punzo, F., Tesei, A.: Wave propagation on regular trees (forthcoming)
37. Qi, Y.-W.: The critical exponents of parabolic equations and blow-up in $\mathbb{R}^n$. Proc. Roy. Soc. Edinb. Sect. A **128**, 123–136 (1998)
38. Ramirez, J.M.: Population persistence under advection-diffusion in river networks. J. Math. Biol. **65**, 919–942 (2012)
39. Sarhad, J., Carlson, R., Anderson, K.E.: Population persistence in river networks. J. Math. Biol. **69**, 401–448 (2014)
40. Sarhad, J., Manifold, S., Anderson, K.E.: Geometric indicators of population persistence in branching continuous-space networks. J. Math. Biol. **74**, 981–1009 (2017)
41. Sherwin, S., Franke, V., Peiro, J., Parker, K.: One-dimensional modeling of a vascular network in space-time variables. J. Eng. Math. **47**, 217–250 (2003)
42. Solomyak, M.: On the spectrum of the Laplacian on regular metric trees. Waves Random Complex Media **14**, 155–171 (2004). Special section on quantum graphs
43. Weissler, F.B.: L^p energy and blow-up for a semilinear heat equation. Proc. Symp. Pure Math. **45**, 545–551 (1986)
44. Wu, Y.: On nonexistence of global solutions for a semilinear heat equation on graphs. Nonlinear Anal. **171**, 73–84 (2018)
45. Yagi, A.: Abstract Parabolic Equations and their Applications. Springer, Berlin (2010)

Challenging Age-Structured and First Order Transition Cell Cycle Models of Cell Proliferation

Paolo Ubezio

Abstract Uncontrolled cell proliferation is the key feature of tumours. Because experimental measures provide only a partial view to the underlying proliferative processes, such as cell cycling, cell quiescence and cell death, mathematical modelling aims to provide a unifying view of the data with a quantitative description of the contributing basic processes. Modelling approaches to proliferation of cell populations can be divided in two main categories: those based on first order transitions between successive compartments and those including a structure of the cells' life cycle. Here we challenge basic models belonging to the two categories to fit time course data sets, from our laboratory experience, obtained observing the proliferative phenomenon with different experimental techniques in a cancer cell line. We disclose the limitations of too simple models. At the minimal complexity level accounting for all available data the two approaches converge and suggest similar scenarios for the underlying proliferation process, in both untreated conditions and after treatment.

1 Introduction

Cell proliferation is driven by cell cycling processes through G_1, S, G_2 and M phases, by the quiescence-proliferation interplay and by cell death processes. The increase of the number of individuals of a cell population originates from the cell cycle, where each cell progressively increases its mass, doubles its DNA content in phase S and finally divides in mitosis (M), giving birth to two daughter cells. This expansive mechanism is counterbalanced by quiescence and death processes, arresting cell cycle progression in a G_0 phase, usually at the beginning of the cycle, where cells can stop for an indefinite time or be committed to die.

P. Ubezio (✉)
Unit of Biophysics, Laboratory of Cancer Pharmacology, Istituto di Ricerche Farmacologiche Mario Negri IRCCS, Milan, Italy
e-mail: paolo.ubezio@marionegri.it

A. d'Onofrio et al. (eds.), *Problems in Mathematical Biophysics*, SEMA SIMAI Springer Series 38, https://doi.org/10.1007/978-3-031-60773-8_13

Cell cycle progression, at the cell population level, has been modelled since the sixties of the last century with balance equations [1–3], either assuming that the progression occurs through subsequent compartments or phases and the transit from a compartment to the next is a Poisson process ("first order transition" models), or using structured models, with structures based on cell age [4, 5], "maturation" [6], size [7], DNA content [8, 9]. First order transition models are described by a set of ordinary differential equations or even a single equation as in the core of most PK/PD models, where a cell population is represented with a single compartment. Structured models are instead described by partial differential equations like the Von Foerster equation [4] for an age-structured population, taking advantage of the fact that age increases exactly with time. The corresponding equations with other structures have a limited use because they include the rate at which cells increase the structuring variable, which is generally unknown and is expected to be highly variable among cells. The models are usually implemented in their discrete-time form with difference equations, for computing convenience and because the meaningful variations of the state variables, such as numbers of cells, typically occur over hours. Thus, difference equations with a not too short time step go possibly closer to the real world than differential equations, which treat the cell number or cell "density" as continuous variables subject to infinitesimal variations.

Although the adoption of simple models may be justified by lack of experimental data available to the modeler, excessive simplification may lead to unrealistic or wrong deductions. Several experimental methods can be used to obtain information on the kinetics of progression within the cell cycle, and rich datasets exist, particularly for tumour cell proliferation and the effects of treatment, that can be used to challenge the mathematical models. A direct comparison of the different modelling approaches has not been made before, in our knowledge, and is pursued in the present work, exploiting one of the databases previously published by our laboratory, and exploring the predictions and limitations of models at increasing complexity levels.

2 The Database

The dataset used here comes from our studies in vitro on the effects of anticancer treatments on the ovarian cancer cell line IGROV-1 [10–14], mainly from the work on the effects of X-rays published in Plos Computational Biology, where the details of the experimental methods are reported [15]. We will consider first untreated cells, which were in exponential phase at the beginning of the experiment ($t = 0$), and then examples of treatment modelling. The IGROV-1 cell population was observed with independent experimental platforms, measuring: 1) cell cycle percentages by DNA flow cytometry (DNA_FC) (pooling G_2 and mitotic cells as G_2M), 2) the genealogical trees of individual cells by time lapse microscopy (TLM) and 3) absolute numbers of cells by coulter counter (ACC).

Main data are time courses up to t = 72 h or 96 h of:

- the overall cell number (N(t))
- the percent of cells in G_1, S and G_2M cell cycle phases (%G_1(t), %S(t), %G_2M(t)),
- the percent of "generation 0" (gen0) cells, i.e. cells observed at t = 0 and tracked until they divide or die (%Ngen0(t)), and of those in subsequent generations: %Ngen1(t), i.e. siblings of gen0 cells, %Ngen2(t), %Ngen3(t), etc.

Moreover, the intermitotic time of thousands of cells was individually directly recorded by TLM, giving the frequency distribution of the cell cycle time (F(T_c)).

In addition, we considered short-time bi-parametric flow cytometric measures of a pulse-chase experiment with bromodeoxyuridine (BrdU), where the cell cycle progression of cells that were in S-phase at t = 0 (keeping the BrdU label and thus named BrdU+ cells) was followed up to 12 h, giving the overall percentage of BrdU+ cells (%BrdU+(t)) and the percentage of residual undivided BrdU+ cells (%Res+(t)).

3 Results

Aim of the present study is a conceptual comparison of the capability of the first order transition and age-structured models to reproduce the observations. We will challenge the models with the data of IGROV-1 cells, to obtain for each model a set of best-fit parameter values for the unperturbed growth of this tumour cell line and then for X-rays treatments.

3.1 *The Growth Curve and the Steady State of Balanced Growth*

The experimental growth curve of an untreated cell population of IGROV-1 cells was exponential in the first 48 h with negligible cell death, and thus easily fitted also by a simple one-compartment first order model without consideration of cell cycle phases (Model A, Fig. 1). Subsequent decrease of growth rate due to reduction of nutrients/space availability could be considered introducing a time dependence of parameters, or using a traditional Gompertz equation, but we will suppose for the moment to limit our observation to a time period where the growth curve is exponential with good approximation.

Obviously Model A allows to deduce the "growth rate" ($\alpha = k\text{-}\mu$) from the growth curve alone, while would require an independent measure of the death rate μ in order to specify the proliferation rate *k*. Because there are several methods to evaluate cell death, one may expect that the death rates could be directly measured.

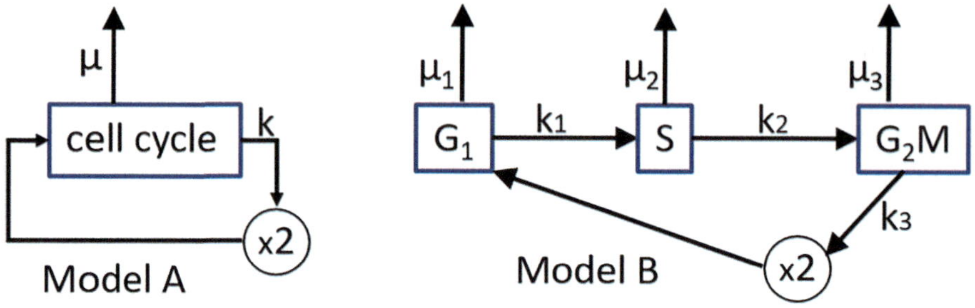

Fig. 1 Model A. One-compartment first order model: $dN(t)/dt = (k\text{-}\mu)\cdot N(t)$, where k is the proliferation rate and μ the death rate. Model B. Model with first order transitions between cell cycle phases: $dN_{G1}/dt = 2k_3\cdot N_{G2M}\text{-}(k_1 + \mu_1)\cdot N_{G1}$; $dN_S/dt = k_1\cdot N_{G1}\text{-}(k_2 + \mu_2)\cdot N_S$; $dN_{G2M}/dt = k_2\cdot N_S\text{-}(k_3 + \mu_3)\cdot N_{G2M}$ where k_1, k_2, k_3 are exit rates from G_1, S and G_2M respectively and μ_1, μ_2, μ_3 the corresponding death rates. The state variables $N_{G1}(t)$, $N_S(t)$ and $N_{G2M}(t)$ are the number of cells in cell cycle phases, directly related to the observable quantities: $N(t) = N_{G1}(t) + N_S(t) + N_{G2M}(t)$, $\%G_1(t) = 100\cdot N_{G1}(t)/N(t)$, $\%S(t) = 100\cdot N_S(t)/N(t)$ and $\%G_2M(t) = 100\cdot N_{G2M}(t)/N(t)$

Unfortunately, this is not the case. The popular methods are based on probes of cell death, detecting the cells which are in a window of the dying process, thus giving the percent of cells in that time-window (of unspecified width) at the particular time of the measure. The relationship of this percentage with the death rate or the overall number of dead cells in a time interval remains elusive. That's why a proper use of these markers is limited to a qualitative evaluation of the presence of death. In our experiments we considered cell death negligible when measuring <3% positive cells detected by one of these markers. Nevertheless, the information of presence/absence of cell death is important to steer the choice of considering or not the death parameter μ.

Although model A is often adopted in proliferation modelling, particularly when only the overall cell number or mass is available (e.g. in the PK/PD field), its description of the phenomenon is clearly too simplistic. It is enough to consider a simple measure of cell cycle percentages $\%G_1$, %S $\%G_2M$, to be forced to abandon model A for a more complex one, such as model B, including consideration of G_1, S and G_2M phases.

Coupling measures of the growth curve with cell cycle percentages imposes a strong constrain, which allows to deal several situations with model B avoiding over-parametrization, and in particular to make a deeper analysis of the exponential growth. In experimental research a balanced steady state of exponential growth is usually (often implicitly) assumed when measuring the characteristic properties of a cell line and it is the starting point in most experiments on effects of treatments.

Model B fits properly contemporaneously the number of cells and cell cycle percentages of the IGROV-1 cell line in the exponential growth (Fig. 2), as can be expected because, assuming the steady state and uniform death rate ($\mu = \mu_1 = \mu_2 = \mu_3$), the equations can be solved knowing the growth rate (directly calculated from the growth curve) and the constant values of $\%G_1$, %S and $\%G_2M$.

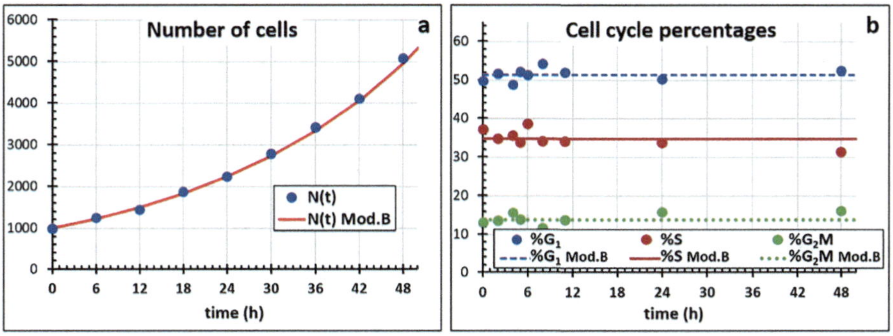

Fig. 2 Contemporary fit of the number of cells N(t) (panel **a**) and cell cycle percentages $\%G_1(t)$, S(t) and $G_2M(t)$ (panel **b**) in the exponential phase of growth with Model B with three parameters: k_1, k_2, k_3, (with fixed death rates: $\mu_1 = \mu_2 = \mu_3 = 0$), assuming the desynchronised starting distribution associated with parameters' values. (N(t) normalised to N(0) = 1000)

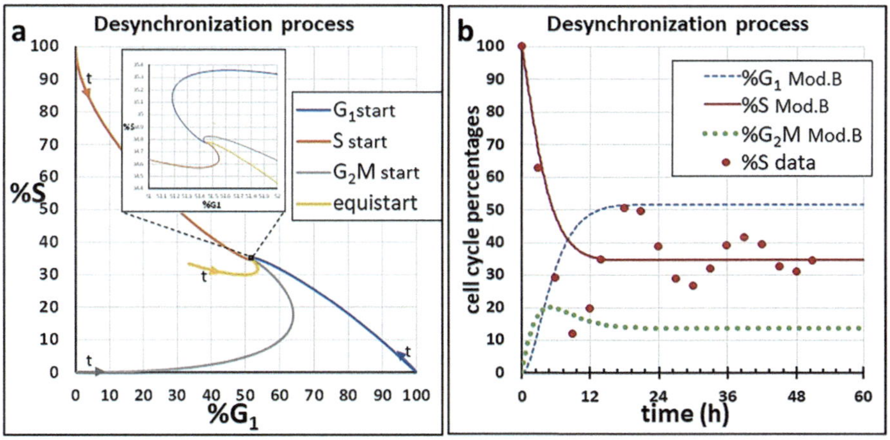

Fig. 3 The desynchronization process in IGROV-1 cells as simulated by the best fit model B with parameters k_1, k_2, k_3 as in Fig. 2. (**a**) Approach to the unique steady state starting with 100% cells in G_1 ("G1 start"), in S ("S start"), in G_2M ("G2M start") or 33.33% in each phase ("equistart"), in a %S vs %G1 plot with time as parameter of the curves. (**b**) Time course of cell cycle percentages starting with 100% S phase cells, as simulated by the model ("$\%G_1$ Mod.B", "%S Mod.B" and "$\%G_2M$ Mod.B") compared with experimental data ("%S") [16]

It is interesting to consider briefly the desynchronization process, which is also used in the fitting procedure at each variation of the parameters to achieve the starting cell cycle distribution. Figure 3a shows how model B realizes the desynchronization to the unique balanced steady state for the IGROV-1 cell line from different starting distributions. Even assuming that at a given time the cells are all in a single cell cycle phase, the model B desynchronizes the cell population in a very short time (variation of cell cycle percentages become less than 1% in 20 h, less than one doubling time). This can be compared with published data on

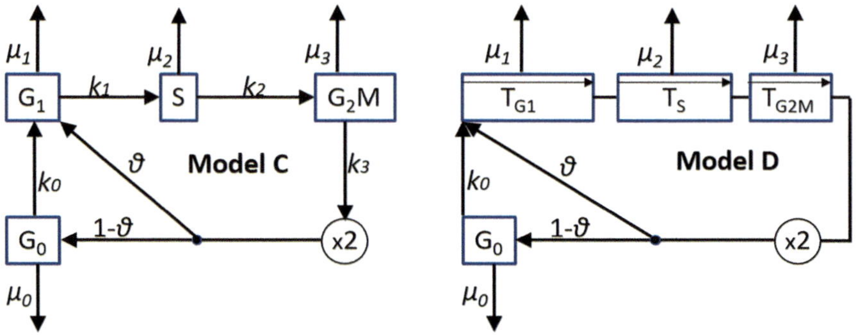

Fig. 4 Models including the quiescence compartment. Model C: first order transition model. Model D: Bertuzzi-Gandolfi age-structured model. In both models, quiescence is rendered with a similar G_0 compartment. The parameters of the G_0 compartment are: θ (fraction of cells bypassing G_0), k_O (rate of re-entering in cycle of quiescent cells), μ_O (rate of death of quiescent cells). G_1, S G_2M parameters are phase exit rates (k_1, k_2, k_3) in model C, phase durations (T_{G1}, T_S, T_{G2M}) in model D, phase death rates (μ_1, μ_2, μ_3). Simpler models without G_0, as model B, are obtained by setting $\theta = 1$

the same cell line [16] for cells initially in phase S (Fig. 3b). Data show a much slower desynchronization rate than predicted with model B, suggesting that the model does not catch properly the actual flow of the cells through cell cycle phases. In fact, according to models A and B, cells can complete the whole cell cycle in an infinitesimal time, while in reality it takes at least hours only for DNA replication and mitosis.

In the case of IGROV-1 cells in the exponential growth phase, independent experiments demonstrated that both quiescence and cell death were negligible, so it was acceptable to assume $\mu_1 = \mu_2 = \mu_3 = 0$ and not to consider quiescence, but in general this is not true, particularly *in vivo*.

Model C (Fig. 4) is a minimal model including all three pillars concurring to proliferation, adding a quiescence ("G_0") compartment [6] to the model B, which already considered cell cycle and cell death (or, more generally, "cell loss"). In this model, newborn cells may remain proliferating with probability θ or are committed to become quiescent (probability 1- θ); then quiescent cells die with rate μ_O or re-enter the cell cycle, with rate k_O.

Similarly, model D (Bertuzzi-Gandolfi model [17]) is a basic age-structured model with fixed phase durations T_{G1}, T_S, T_{G2M}. In model D all cells entering G_1 at t = t' will exit G_1 at t = t' + T_{G1}, if they do not die in the meantime, and similarly for S and G_2M phases. Within each phase, the ages a_{G1}, a_S, a_{G2M} are measured from cell entry in the corresponding phase. In the discrete-time formulation, each phase is divided in small age intervals Δa equal to the step time Δt. Considering that the cells of age between a_{G1} and $a_{G1}+\Delta a$ (with $\leq a_{G1} < T_{G1}$) at time t are those in the previous age interval in the previous step time which do not die in Δt, the simple balance equation holds: $N_{G1}(a_{G1}, t)=(1- \mu_{G1}\cdot\Delta t)\cdot N_{G1}(a_{G1}-\Delta a, t-\Delta t)$. Cells exiting

G_1 enter S at $a_S = 0$ while cells entering at $a_{G1} = 0$ come from G_2M exit and division. S and G_2M age structures act in the same way, while G_0 is modelled as a single compartment, with first-order exit rate (k_O) and without age structure, as in model C.

The introduction of G_0 makes cell cycle models flexible to account for different biological hypotheses. For instance, it is possible to consider the case in which all cells necessarily pass through G_0 by setting $\theta = 0$ and $k_O > 0$. Conversely, with $\theta > 0$ and $k_O = 0$ models C and D interpret G_0 as a compartment of definitively quiescent cells out of cycle. Moreover, G_0 compartment can be viewed as a compartment of quiescent stem cells, recruited to proliferation with rate k_O.

It should point out that a univocal balanced steady state of exponential growth exists also in these and in more complex models derived from them, as far as parameters are constant in time (see reviews [18, 19] of the mathematical literature on this topic). A variation of some of the parameters, possibly caused by variation of environmental factors, would lead the population towards a different steady state. As an example of this phenomenon we will consider in the next section the approach to confluence.

3.2 *The Approach to Confluence*

The steady state of exponential growth cannot be maintained indefinitely, both *in vitro* and *in vivo*, because nutrients are progressively consumed and/or the available space reduces. In this case we typically observe a slowdown of the proliferation with an increase of %G1, indicating that at least a fraction of the cells where delayed or arrested before the onset of DNA synthesis. This motivated the introduction in the models of a compartment (G_0) collecting cells committed to become quiescent before/within G_1. From the experimental point of view, if cell cycle percentages are calculated based on DNA cellular content only, proliferating (G_1) and quiescent (G_0) cells are indistinguishable and "%G_1" is renamed as "%G_0G_1". Figure 5 shows how models C and D fit the data of the IGROV-1 cell population in an extended 0–96 h interval, characterized by an increase of %G_0G_1, a decline of the growth rate and detection of apoptotic cells at 96 h.

In order to fit the whole 0–96 h time course, the steady state best fit G_1, S, G_2M parameters were applied in the 0–48 h interval (as shown in Fig. 2 for model C and similarly for model D), then two additional parameters were made variable in the 48–96 h interval only: θ, enabling cells to enter G_0, and μ_O, accounting for G_0 cell loss ($\theta = 1$ and $\mu_O = 0$ in the 0–48 h interval). The best fits shown in Fig. 5, obtained optimizing θ and μ_O, are satisfactory for both types of models, although somewhat underestimating %G_2M at 72 and 96 h. This suggests the presence of quiescent cells also in G_2, that could be accounted introducing an additional compartment for them. We conclude that time courses of the number of cells and cell cycle percentages are not sensitive to discriminate between the two types of models.

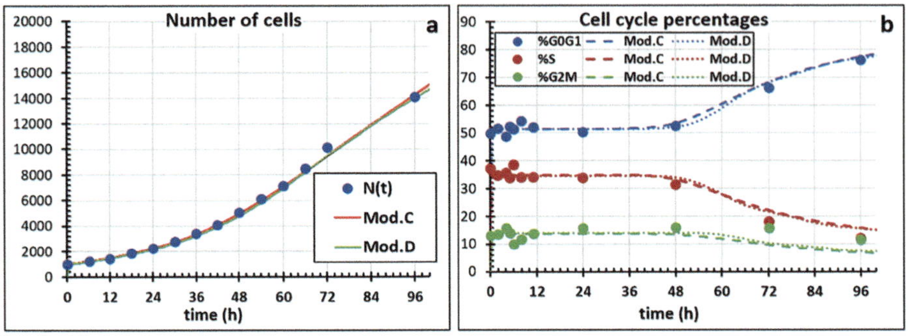

Fig. 5 Contemporary fit of the number of cells N(t) (panel **a**) and cell cycle percentages $\%G_0G_1(t)$, S(t) and $G_2M(t)$ (panel **b**) in the 0–96 h interval, with Models C (dashed lines) and D (dotted lines). Respective steady state G_1, S G_2M parameters were fixed (obtained from a previous steady state fit in 0–48 h) and the approach to confluence was fitted varying parameters θ and μ_0 in the 48–96 h interval

3.3 *More Flow Cytometric Experiments: BrdU Labeling*

In order to challenge/identify the models let's consider additional experiments with the FC platform. The BrdU labeling method provides valuable data for this purpose. The method can be applied both *in vitro* and *in vivo* and its use to evaluate the potential doubling time of tumours in patients was widespread in the last two decades of the last century [20]. It is based on the labeling of S phase cells with BrdU, which are then followed in time while exiting S, entering G_2M, divide and enter G_1 for a new cycle. Usually these BrdU+ cells remain distinguishable for at least one cell cycle time, providing data suitable for cell cycle modelling or testing consistency of proposed models. On the other hand, manipulation *in vitro* of the cells for BrdU labeling may cause a temporary cell cycle delay, which may be considered to refine models, and identification of the subpopulation of BrdU+ may be challenging in some instances.

Figure 6 shows the time courses of two kind of data obtained in a pulse labeling BrdU experiment: the percentage of labelled cells (%Brdu+(t)) and of residual undivided labelled cells (%ResU+(t)). The percentage of labelled cells (%BrdU+(t)) includes all BrdU+ cells at time "t", i.e. those still undivided and their descendant, while the percentage of residual undivided labelled cells (%ResU+(t)) refers to undivided cells among those that were in S phase at $t = 0$. Notice that such data can be readily obtained by simulation with the above models: given a set of parameter values and obtained the initial cell cycle distribution as before, the simulation of the time course will be made two times, one for BrdU+ cells (starting with S phase cells only, which have been labelled at $t = 0$), and one for BrdU- cells (starting with G_1 and G_2M phase cells only, which remain unlabelled). The time course of the whole population is trivially obtained by summing the number of BrdU+ and BrdU- in each phase at each time. Refinement of the modelling may

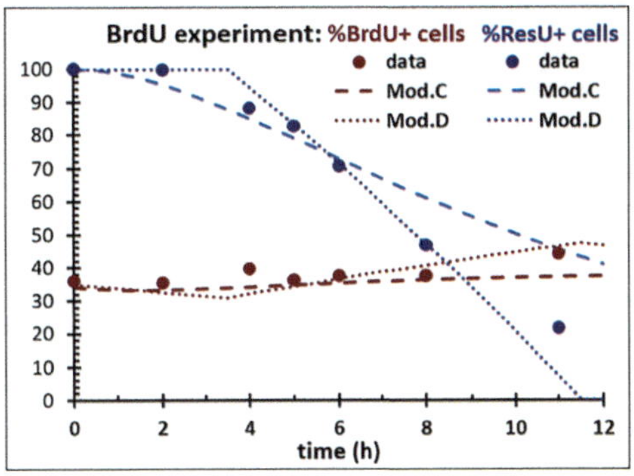

Fig. 6 BrdU labeling experiment in IGROV-1 cells: time course of the percentage of labelled cells (%BrdU+) and of residual undivided labelled cells (%ResU+). Model C (dashed lines) and D (dotted lines) predictions with the best fit parameters obtained by contemporary fit of N(t) and %G1(t), %S(t) and %G_2(M(t) in the exponential phase of growth

include consideration of the finite duration of the "pulse" (exposure to BrdU lasts 15–30 min, i.e. 1–2% of a typical cell cycle time).

Figure 6 evidences the difficulties of the model C to reproduce the time course of %ResU+, while model D shows the expected two periods of %ResU+(t): the first one when no division was observed (%ResU+ =100%) corresponding to the time to traverse G_2M phase before dividing, and the second period characterised by a progressive decrease of %ResU+, as the cells gradually arrive in mitosis and divide, leaving the first cycle. The bending of the data respect to the sharp division of the two periods predicted by model D, demonstrates that the duration of S and G_2M is not the same for all cells, as assumed by model D.

3.4 More Experiments: The Distribution of Intermitotic Times and Generation Wise Data

Of course, variability of phase durations is expected, even in a situation of uniform exposure to nutrients and unrestricted space, as in our experiment with IGROV-1 cells *in vitro*. The cellular content of any protein at a given time in a given cell cycle point should be variable because it is the result of millions of intertwined molecular reactions occurring from the birth in each cell, which also starts with a variable mass from the mitosis of the mother cell. Experimentally, the variability of cell cycle time can be directly measured by time-lapse microscopy (or "microcinematography"). In a pioneering work in 1963 from our laboratory [21], researchers applied microcinematography to measure the intermitotic times of individual cells by recording the time between two successive divisions. This experiment and several similar ones demonstrated that a minimum time for completing the cell cycle is necessary and variability was high with a frequency distribution skewed on the right. To account for that, Smith and Martin and others [22, 23] proposed a model (model SM) with two phases: the first (named "A-state") with first order exit kinetics acting like the G_1 phase of model B (but actually covering only a part of G_1) followed by the second phase with fixed duration, covering the rest of the cell cycle. Model SM is

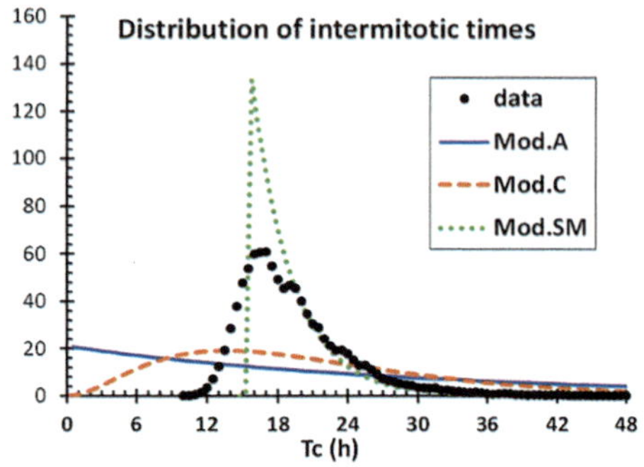

Fig. 7 Frequency distribution of cell cycle times (F(Tc)) measured with TLM in IGROV-1 cells with 0.5 h intervals, normalized to 1000 cells and fitted with models A, C and SM

equivalent to model D with $\theta = 0$, $k_O > 0$ and $\mu_O = 0$. In the Smith-Martin's view the stochastic exit from the A-state was introduced to explain the observed inter-cell variability of the cell cycle times and desynchronization. However also model SM, as the simpler models A, B, C described above, fails to reproduce a typical frequency distribution of Tc (F(Tc)) (Fig. 7) as that observed in the IGROV-1 cell line, obtained with a modern equipment for time lapse microscopy (TLM).

In the effort to reproduce this observation within the first order transition approach, model C can be modified splitting each phase in sub-compartments, characterised by the same exit rate [24]. In this way, the duration of each phase comes to follow an Erlang distribution and model parameters become the exit rates from sub-compartments (k_1 for G_1 sub-compartments, k_2 for S, k_3 for G_2M) and the number of sub-compartments (n_1, n_2, n_3) in each phase (model C_E). Instead the age structure of model D makes it suitable for the implementation of any distribution of phase durations. We implemented different two-parameter distributions [25], keeping the mean and coefficient of variation of phase durations as model parameters: T_{G1}, T_S, T_{G2M}, CV_{TG1}, CV_S, CV_{TG2M}. Already Sisken and Morasca observed also that 1/Tc was normally distributed [21] and a thorough examination of several similar datasets by Castor demonstrated that best fits were obtained assuming reciprocal-normal distributions [26]. We will consider an extension of model D with reciprocal-normal distributions of phase durations (model D_R). Figure 8a shows that models C_E and especially D_R provide a good fit of F(Tc).

Using TLM it is also possible to measure directly the expansion of the cell population in a generation-wise perspective [15]. This is achieved by tracking each cell detected at the start of the experiment in several fields of view and their descendants. In this way, it is possible to measure at each time the percentage of cells belonging to gen0, gen1 and successive generations.

Cell generation percentages for IGROV-1 cells in the exponential state are shown in Fig. 8b. In order to reproduce TLM data in the modelling framework, it is enough to replicate the cell cycle, so that the daughters of dividing cells instead of re-entering in the same cell cycle enter in a new cell cycle, with the same parameters. The state variables become $N(ph,a,gen,t)$, giving the number of cells in phase "*ph*" (G_1, S or G_2M), generation "*gen*", sub-compartment (model C-E) or age interval

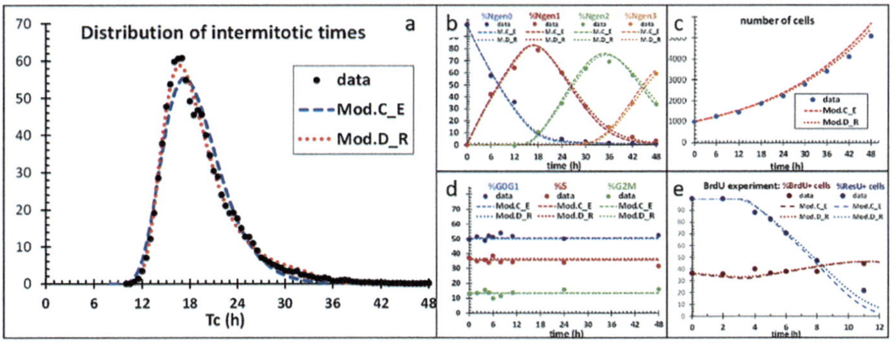

Fig. 8 Model C_E and D_R predictions with the best fit parameters obtained by contemporary fit of all data of IGROV-1 cells in the exponential phase of growth: (**a**) frequency distribution of cell cycle times (TLM); (**b**) percent of cells in subsequent generations (TLM); (**c**) absolute cell count; (**d**) %G_1(t), %S(t), %G_2M(t) (FC); (**e**) %BrdU+, %ResU (FC). Model C_R parameters: k_1, k_2, k_3, n_1, n_2, n_3; model D_R parameters: T_{G1}, T_S, T_{G2M}, CV_{TG1}, CV_S, CV_{TG2M}

(model D_R) "*a*", at time "*t*", and $N(G_0,gen,t)$ for the number of G_0 cells. For each cell ending G_2M of generation "i", two newborn cells enter at $a_{G1} = 0$ in generation "i + 1". For what concerns F(Tc), a parallel simulation of G_1, S and G_2M transit starting with all cells at $a_{G1} = 0$ gives the time course of the number of undivided cells, from which the cumulative of F(Tc) and F(Tc) itself are readily derived.

Figure 8 shows the best fit achieved fitting all data together with models C_E and D_R. Both models provided satisfactory fit of the data, thus unifying the views of the measures in the five panels with a single reconstitution of the underlying cell cycling of the cell population.

The two modelling approaches differs in F(Tc), where the reciprocal-normal functions adopted in model D_R allow an accurate fit of the shape of the experimental F(Tc), not achievable in model C_E. In general, the flexibility for the choice of the type of the distribution of the age-structured model gives a clear advantage for modelling different situations, particularly when a wide intercell variability is present as in the case of most tumour cell populations. Model C_E is in fact equivalent to an age-structured model with Erlang distributions of phase durations but links this variability to the number of compartments within each phase, which is not practical for fitting purposes.

On the other hand, model C_E fits the other data in panels 8b to 8e no less well than model D_R, with similar values of mean and variance of G_1, S and G_2M duration. This demonstrates that these types of data are poorly sensitive to the highest moments of the distribution, a result that was confirmed by evaluating other distributions in the age-structured model (not shown).

3.5 Modelling Treatment

The models discussed in the previous section can be readily adapted for studying cell proliferation in different contexts. For instance, the generation structure can be adapted for simulation of differentiation chains, as in studies of the lineages of the hematopoietic system, by reinterpreting generations as sequential differentiation levels and introducing a parameter of differentiation probability for the passage from a level to the next. In our laboratory, model D_R was adopted as basal representation of tumour proliferation in studies of the effects of various anticancer treatments. Treatment induces two kinds of perturbations of the basal proliferation process: cell cycle delay/arrest and cell kill. These are the functional end points of the cellular molecular reactions induced by the chemical/physical/biological agent (drug, radiation, antibody or others) used to treat the disease. Both delay/arrest and cell kill have time and dose/schedule dependence. Moreover, cells are diversely sensitive to treatment in the cell cycle phases and put in action different mechanisms of response in each phase, typically arresting cell cycle progression at checkpoints in G_1 and G_2M and reducing the rate of DNA replication in S. Then, arrested cells may repair the damage and resume cell cycle progression, but not rarely leaving some damage that lead to novel arrest and possibly death also in the subsequent generations. These complex phenomena can be modelled in the framework of the above models adding two minimal "checkpoint modules" (CpM) describing the dynamics of checkpoint arrest/resumption in G_1 and G_2M and a parameter rendering the reduction of the DNA synthesis rate in phase S [10, 13, 27]. Introducing such modules in model D_R we reproduced and explained the effects of X-ray treatment at increasing doses in the IGROV-1 cell line (see [15] for more details). We will show here how those data for a sublethal (0.5Gy) and a high (5Gy) dose are modelled with first order transition models, without (Fig. 9) and with (Fig. 10) introducing CpMs. The best fit of unperturbed growth considered above for each model provided the fixed parameters values of the basal proliferation and the time zero asynchronous cell cycle distribution.

In models A, B and C without CpM cell cycle delay/arrest and cell kill were rendered by lowering the previously determined steady state values of the exit rates k (model A) or k_1, k_2, k_3 (model C) and introducing non-zero values for the death rates μ (model A) or μ_1, μ_2, μ_3 (model C) in the presence of cell kill. These variations may be different for gen0 cells (directly exposed to treatment) and their descendants or even be time-dependent. Let's call k_{gen0} the k value in gen0, k_{gen1} the k value in gen1 and so on for all model parameters. Thus, a preliminary study is required to identify the relevant variables and the need to introduce a time-dependence for some parameter.

0.5Gy was a sublethal dose with a nearly exponential N(t), with a lower growth rate respect to control. The time course of the number of cells was well caught also by the simple model A with a single variable $k_{gen1,2}$ (with $k_{gen2} = k_{gen1}$) and leaving unperturbed k in gen0, gen3 and following (Fig. 9 upper left panel). However, the insight into the generation percentages revealed the inconsistence of

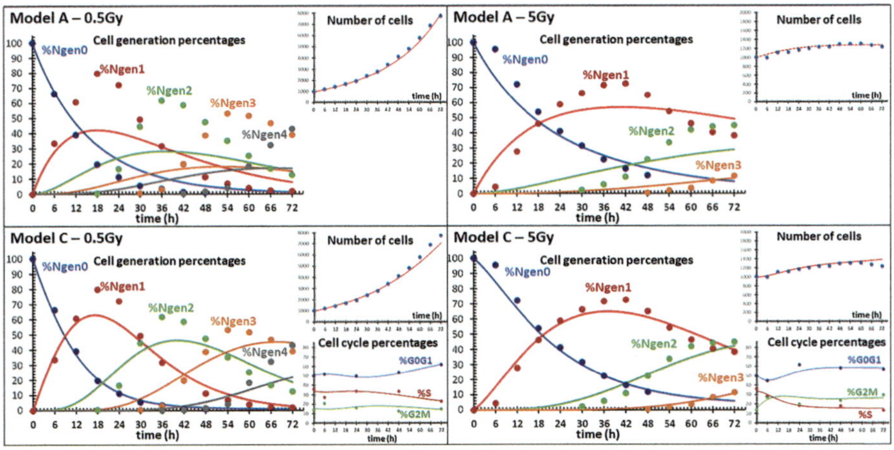

Fig. 9 Treatment challenge. Data (circles) and best fit (lines) with models A (upper panels) and C (lower panels) of the time course of cell number within generations (time-lapse measures) and cell cycle phases (flow cytometry measures) after exposure to 0.5Gy (left panels) and 5Gy (right panels) of IGROV-1 cells *in vitro*. Treatment variables **Model A 0.5Gy:**$k_{gen1,2}$ (i.e. with $k_{gen2} = k_{gen1}$), **5Gy:**k_{gen0}, $k_{gen1,2}$, $\mu_{gen0,1,2}$; **Model C 0.5Gy:**$k_{1gen1,2,3}$, $k_{2gen1,2,3}$, $k_{3gen1,2,3}$, $\theta_{gen3,4,5}$, **5Gy:**k_{1gen0}, $k_{2gen0,1,2,3}$, k_{3gen0},$k_{3gen1,2,3}$, $\mu_{1gen0,1,2,3}$. The other parameters were kept fixed as previously determined in untreated conditions

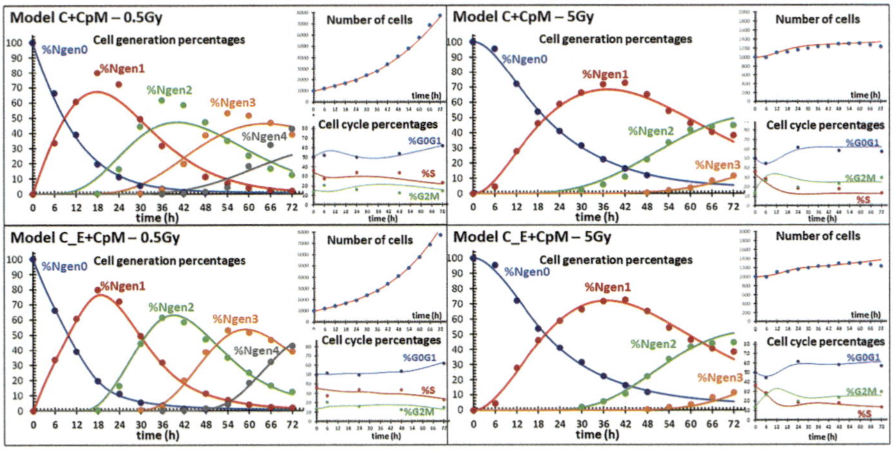

Fig. 10 Treatment challenge. Data (circles) and best fit (lines) with models C + CpM (upper panels) and C_E + CpM (lower panels) of the time course of cell number within generations (time-lapse measures) and cell cycle phases (flow cytometry measures) after exposure to 0.5Gy (left panels) and 5Gy (right panels) of IGROV-1 cells *in vitro*. Treatment variables **Model C + CpM 0.5Gy:**$pBl_{G1,gen1,2}$, $pBl_{G2,gen0,1,2}$, $Rec_{G1,gen1,2}$, $Rec_{G2,gen0,1,2}$, $k_{Sgen0,1}$, $\theta_{3,4,5}$, **5Gy:**$pBl_{G1,gen0}$, $pBl_{G1,gen1,2,3}$, $pBl_{G1,gen0,1,2,3}$, $Rec_{G1,gen0,1,2,3}$, $Rec_{G2,gen0,1,2,3}$, $k_{Sgen0,1}$, $\mu_{G1gen0,1,2,3}$, **Model C_E + CpM 0.5Gy:**$pBl_{G1,gen0,1,2}$ ($=pBl_{G2,gen0,1,2}$), $Rec_{G1,gen1,2}$ ($=Rec_{G2,gen0,1,2}$), $k_{Sgen0,1}$, $\theta_{3,4,5}$, **5Gy:**$pBl_{G1,gen0,1,2,3}$, $pBl_{G2,gen0,1,2,3}$, $Rec_{G1,gen0,1,2,3}$, $Rec_{G2,gen0,1,2,3}$, $k_{Sgen0,1}$, $\mu_{G1gen0,1,2,3}$. The other parameters were kept fixed as previously determined in untreated conditions

that model, which predicted a too high number of cells remaining in gen0 after 6 h and anticipated by hours the onset of cells in gen2 and subsequent generations. With the cytotoxic 5Gy dose, a good fitting of N(t) was obtained with model A, using a death parameter $\mu_{0,1,2,3}$ and two distinct k for gen0 and gen1,2,3, but the model again failed to predict correctly the passage through the subsequent generations (Fig. 9 upper right panel).

Models B and C reduced but did not eliminate the inconsistences of model A. Fitting 0.5Gy data with model C required to lower the exit rates in all cell cycle phases in gen1, gen2 and gen3, demonstrating the presence of delays in all phases, stronger in G_2M than in G_1 and S (Fig. 9 lower left panel). Again, no effects were suggested for gen0 cells. Parameter θ, unrelated to treatment, was included to render the approach to confluence, observed in 0.5Gy as in control. Best fit of 5Gy data required adding a cell death parameter in G_1 and a gen0 delay in all phases as well as reinforcing delays in gen1 and subsequent generations (Fig. 9 lower right panel).

Figure 10 shows the fits of models C and C_E with CpM modules in G_1 and G_2M, while S-phase delay, when required, was again modelled by reduction of the exit rate k_S. G_1 and G_2M CpMs are single compartments collecting arrested cells in the respective phase. The dynamics of arrest, repair and kill is reproduced with three parameters: the probability of enter the block (*pBl*), the recycling (*Rec*) and death (μ) rates. *Rec* and μ orient the fate of arrested cells, which may continue cycling in the next phase or die.

C + CpM model (Fig. 10 upper panels) provided a closer fit to %gen data respect to model C shown in Fig. 9, but cell cycle percentages were fitted less well, particularly at 0.5Gy. Best C + CpM model of 0.5Gy data required G_1 and G_2M blocks from gen1 and no S-phase effects, while 5Gy data required to extend blocks to gen0. Recycling rates from blocks were higher in G_2M respect to G_1 blocked cells and lower in 5Gy respect to 0.5Gy. Death was limited to 5Gy G_1 blocked cells with a rate similar to the G_1 recycling rate. Overall, the results suggest that C + CpM fitted better and offered a more consistent interpretation of the data than model C at cytotoxic doses, while model C performed somewhat better at low doses.

C_E + CpM model overcame the limitations of both C and C + CpM models and enabled the best fits (Fig. 10 lower panels). All 0.5Gy data were well fitted with only three variables: two parameters for G_1 and G_2M CpMs (a weak block probability and a recycling rate in gen0,1,2, the same for G_1 and G_2M) and one for S phase delay (Fig. 10 lower left panel). G_1 and G_2M CpMs adopted for fitting 5Gy data were characterized by strong block probabilities but different recycling rates (lower in G_1 than in G_2M) and death rates (present only in G_1 and higher than recycling rate) with a total of six variables. Similar results were obtained with D_R + CpM model (not shown). Actually, a correct prediction of the passage from one generation to the next requires models using a proper distribution of cell cycle duration (Fig. 8) and could be obtained with C_E (or D_R) models also without CpM (not shown). However, the models without CpM did not fit properly cell cycle percentages (not shown) and thus only C_E + CpM (or D_R + CpM) enabled contemporary satisfactory fit of both %gen and $\%G_1/\%S/\%G_2M$ at all doses. The fact that this can be achieved without increasing the number of variables add support to the preference for these models.

4 Conclusions

Our analysis shows that too simple models, like models A, B, C and D, give an inadequate representation of the process, especially due to the absence of consideration or unrealistic description of the variability of the cell cycle times. These flaws lead to erroneous previsions of the effect of a treatment in cancer research, possibly mitigated only at very high concentrations/doses, in the presence of a dominating cell kill process and reduced/absent proliferation.

The evolution of age-structured and first order transition models to adequately account for of the intercellular variability converge to similar representation of the proliferation process in models D_R or C_E, somewhat unifying at this advanced level the two points of view. Using discrete-time equations, both models split G_1, S and G_2M into a number of sub-compartments, but these have a different meaning. In model D_R sub-compartments are directly associated to the changing probability of ending a phase at any age, depending on the probability distribution of the duration (with common right-skewed two-parameter distributions found suitable, with preference to the reciprocal-normal distribution). This gives an "agnostic" description the progression within each cell cycle phase, without any assumption on the mechanisms below. Differently, model C_E assumes a random passage through equal sub-compartments in each phase, which does not match with the current biological knowledge of the phenomenon. Then, model C_E requires a previous identification of the appropriate number of sub-compartments that cells must go through in each phase.

In conclusion both C_E and D_R can be used to describe the effects of perturbing agents adding appropriate checkpoints modules in the same way and with similar results but D_R has practical and theoretical advantages. It is a good candidate as a general framework for modelling proliferation in different contexts, specifically in studies of the antiproliferative effects of treatment in anticancer research.

References

1. Bertuzzi, A., Gandolfi, A., Giovenco, M.A.: Mathematical models of the cell cycle with a view to tumor studies. Math. Biosci. **53**, 159–188 (1981)
2. Eisen, M.: Mathematical Models in Cell Biology and Cancer Chemotherapy, vol. 30. Springer Science & Business Media (1979)
3. Steel, G.G.: Growth Kinetics of Tumors: Cell Population Kinetics in Relation to the Growth and Treatment of Cancer. Oxford University Press (1977)
4. Trucco, E.: Mathematical models for cellular systems the von Foerster equation. Part I. Bull. Math. Biophys. **27**, 285–304 (1965)

5. Von Foerster, J.: Dome remarks on changing populations. In: Stohlman, F. (ed.) Kinetics of Cellular Proliferation. Greene and Stratton, New York (1959)
6. Lebowitz, J.L., Rubinow, S.I.: Grain count distributions in labeled cell populations. J. Theor. Biol. **23**, 99–123 (1969)
7. Diekmann, O., Lauwerier, H.A., Aldenberg, T., Metz, J.A.J.: Growth, fission and the stable size distribution. J. Math. Biol. **18**, 135–148 (1983)
8. Basse, B., Ubezio, P.: A generalised age- and phase-structured model of human tumour cell populations both unperturbed and exposed to a range of cancer therapies. Bull. Math. Biol. **69**, 1673–1690 (2007)
9. Bertuzzi, A., Gandolfi, A., Germani, A., Vitelli, R.: A general expression for sequential DNA-fluorescence histograms. J. Theor. Biol. **102**, 55–67 (1983)
10. Lupi, M., Matera, G., Branduardi, D., D'Incalci, M., Ubezio, P.: Cytostatic and cytotoxic effects of topotecan decoded by a novel mathematical simulation approach. Cancer Res. **64**, 2825–2832 (2004)
11. Lupi, M., Cappella, P., Matera, G., Natoli, C., Ubezio, P.: Interpreting cell cycle effects of drugs: the case of melphalan. Cancer Chem. Pharmacol. **57**, 443–457 (2006)
12. Montalenti, F., Sena, G., Cappella, P., Ubezio, P.: Simulating cancer-cell kinetics after drug treatment: application to cisplatin on ovarian carcinoma. Phys. Rev. E. **57**, 5877–5887 (1998)
13. Sena, G., Onado, C., Cappella, P., Montalenti, F., Ubezio, P.: Measuring the complexity of cell cycle arrest and killing of drugs: kinetics of phase-specific effects induced by taxol. Cytometry. **37**, 113–124 (1999)
14. Ubezio, P., Lupi, M., Branduardi, D., Cappella, P., Cavallini, E., Colombo, V., Matera, G., Natoli, C., Tomasoni, D., D'Incalci, M.: Quantitative assessment of the complex dynamics of G1, S, and G2-M checkpoint activities. Cancer Res. **69**, 5234–5240 (2009)
15. Falcetta, F., Lupi, M., Colombo, V., Ubezio, P.: Dynamic rendering of the heterogeneous cell response to anticancer treatments. PLoS Comput. Biol. **9**, e1003293 (2013)
16. Chiorino, G., Metz, J.A., Tomasoni, D., Ubezio, P.: Desynchronization rate in cell populations: mathematical modeling and experimental data. J. Theor. Biol. **208**(2), 185–199 (2001)
17. Bertuzzi, A., Gandolfi, A., Sinisgalli, C., Iacoviello, D.: Steel's potential doubling time and its estimation in cell populations affected by nonuniform cell loss. Math. Biosci. **143**, 61–89 (1997)
18. Arino, O.: A survey of structured cell population dynamics. Acta Biotheor. **43**, 3–25 (1995)
19. Webb, G.: Theory of Nonlinear Age-Dependent Population Dynamics. Marcel Dekker, New York (1985)
20. Begg, A.C., McNally, N.J., Shrieve, D.C., Kärche, H.: A method to measure the duration of DNA syntheses and the potential doubling time from a single sample. Cytometry. **6**, 620–626 (1985)
21. Sisken, J.E., Morasca, L.: Intrapopulation kinetics of the mitotic cycle. J. Cell Biol. **25**, 179–189 (1965)
22. Burns, F.J., Tannock, I.F.: On the existence of a Go-phase in the cell cycle. Cell Prolif. **3**, 321–334 (1970)
23. Smith, J.A., Martin, L.: Do cells cycle? Proc. Natl. Acad. Sci. **70**, 1263–1267 (1973)
24. Takahashi, M.: Theoretical basis for cell cycle analysis: II. Further studies on labelled mitosis wave method. J. Theor. Biol. **18**, 195–209 (1968)
25. Ubezio, P., Rossotti, A.: Sensitivity of flow cytometric data to variations in cell cycle parameters. Cell Tissue Kinet. **20**, 507–517 (1987)
26. Castor, L.N.: A G1 rate model accounts for cell-cycle kinetics attributed to 'transition probability'. Nature. **287**, 857–859 (1980)
27. Celora, G.L., Bader, S.B., Hammond, E.M., Maini, P.K., Pitt-Francis, J.M., Byrne, H.M.: A DNA-structured mathematical model of cell-cycle progression in cyclic hypoxia. J. Theor. Biol. **545**, 111104 (2022)

GPSR Compliance

The European Union's (EU) General Product Safety Regulation (GPSR) is a set of rules that requires consumer products to be safe and our obligations to ensure this.

If you have any concerns about our products, you can contact us on ProductSafety@springernature.com

In case Publisher is established outside the EU, the EU authorized representative is:

Springer Nature Customer Service Center GmbH
Europaplatz 3
69115 Heidelberg, Germany

Batch number: 10732090

Printed by Printforce, the Netherlands